Once Upon A GEMS Guide

Connecting
Young People's
Literature
to
Great Explorations
in Math and Science

C
O
N
T
E
N
T
S

What is GEMS?

*W*atching GEMS activities inside a classroom, you'll find students exploring a strange green substance said to come from a distant planet ... buzzing like bees in a hive ... blowing bubbles to test predictions ... solving a crime with chemistry, or tackling acid rain pollution problems in model lakes.

*T*he basis for the GEMS approach is that students learn best by doing—an approach backed by overwhelming educational evidence. Activities first engage students in direct experience and experimentation, before introducing explanations of principles and concepts. Utilizing easily obtained and inexpensive materials, GEMS activities allow teachers without special background in science or mathematics to successfully present hands-on experiences.

*G*EMS activities captivate the imagination while illuminating essential scientific themes, concepts, and methods. The enthusiasm is catching as teachers and students share in the joy of discovery. GEMS is a growing resource for activity-based science and mathematics. Developed at the University of California at Berkeley's Lawrence Hall of Science, and tested in thousands of classrooms nationwide, more than 30 GEMS teacher's guides offer a wide spectrum of learning opportunities from preschool through tenth grade.

*T*he experiential nature of activity-based science and mathematics makes it especially effective for reaching ALL students. Doing GEMS activities gives youngsters a positive experience with science, and a sense of their own ability to succeed, thus building student confidence. This makes GEMS an outstanding and proven resource for use with students considered at risk of failure.

*E*mphasis on teamwork and cooperative learning, the use of a wide variety of learning formats, and reliance on direct experience rather than textbooks makes GEMS highly appropriate for use with populations that have been historically underrepresented in science and mathematics pursuits and careers. In GEMS activities, students are encouraged to work together to discover more, explore a problem, or solve a mystery, rather than fixating on the so-called right answer, or engaging in negatively competitive behavior. Cooperative (or collaborative) learning is one of the most effective strategies for bridging and appreciating differences and diversities of background and culture. It is also one of the most effective ways to help prepare students for the workplaces of the future.

*T*he GEMS series interweaves a number of educational ideas and goals. GEMS guides encompass important learning objectives, summarized on the front page of each guide under the headings of skills, concepts, and themes. These objectives can be directly and flexibly related to science and mathematics curricula, state frameworks, and district guidelines.

*A*lthough GEMS activities are designed, written, and tested extensively in schools for specific grade levels, **they can be extended to lower or higher levels.** Guides often contain suggestions for ways to modify and adapt the activities for younger or older students. Many GEMS guides suggest ways to extend the activities across the curriculum, into language arts, social studies, art, and other areas. More recent editions include a Literature Connections section.

*T*he GEMS handbook *To Build A House: GEMS and the Thematic Approach to Teaching Science* provides concrete examples from the GEMS series to illustrate the latest approaches in science education. Other handbooks include a *Teacher's Handbook*, a *Leader's Handbook*, *A Parent's Guide to GEMS*, and this literature handbook.

*S*ince classroom testing began in 1984, more than 300,000 teachers and at least five million students have enjoyed GEMS activities. A national network of teachers and educators take part in GEMS Leadership workshops and receive a regular newsletter, the *GEMS Network News*. GEMS is a growing series. New guides and handbooks are being developed constantly and current guides are revised frequently. Your comments and suggestions are always welcome.

*G*reat Explorations in Math and Science—share the joy of discovery!

The GEMS Staff

Lawrence Hall of Science
University of California
Berkeley, CA 94720

Chairman: Glenn T. Seaborg
Director: Marian C. Diamond

Initial support for the origination and publication of the GEMS series was provided by the A.W. Mellon Foundation and the Carnegie Corporation of New York. GEMS has also received support from the McDonnell-Douglas Foundation and the McDonnell-Douglas Employees Community Fund, the Hewlett Packard Company Foundation, and the people at Chevron USA. GEMS also gratefully acknowledges the contribution of word processing equipment from Apple Computer, Inc. This support does not imply responsibility for statements or views expressed in publications of the GEMS program.

Under a grant from the National Science Foundation, GEMS Leader's Workshops have been held across the country. For further information on GEMS leadership opportunities, or to receive a publication brochure and the *GEMS Network News*, please contact GEMS at the address and phone number below.

Printed in the United States of America.
International Standard Book Number:
0-912511-78-8

COMMENTS WELCOME

Great Explorations in Math and Science (GEMS) is an ongoing curriculum development project. GEMS guides are revised periodically, to incorporate teacher comments and new approaches. We welcome your criticisms, suggestions, helpful hints, and any anecdotes about your experience presenting GEMS activities. Your suggestions will be reviewed each time a GEMS guide is revised. Please send your comments to:

GEMS Revisions
Lawrence Hall of Science
University of California
Berkeley, CA 94720-5200.

Our phone number is (510) 642-7771.

Acknowledgments

The primary authors of this handbook are **Jacqueline Barber, Lincoln Bergman, Kimi Hosoume, Jaine Kopp, Cary Sneider,** and **Carolyn Willard**.

Any errors or oversights are the responsibility of the GEMS staff. Opinions and matters of interpretation are not necessarily endorsed by the consultants acknowledged below or by any organizations or agencies that have provided funding for GEMS.

This GEMS handbook owes much of its existence to the careful and dedicated work of **Valerie Wheat**, a writer, editor, researcher, and librarian who spent many months helping GEMS staff members find appropriate books, research publication information, write annotations, gather resources, and compile listings. Valerie energetically took on all the myriad tasks, both global and detailed, that go into a compendium such as this. In honor of her efforts, the GEMS Principal Editor wrote her this limerick:

> Here's to Valerie, last name of Wheat
> Who accomplished a fantastic feat
> She had what it took
> To bring forth this handbook
> In your hands at last it's complete.

In addition we would like to thank the following individuals who reviewed our first full-length draft, made important and constructive comments, and suggested many excellent books we had missed:

Marian Drabkin, the former Lawrence Hall of Science Librarian, provided important comments and suggestions, especially relating to the folkloric and related literary resources for the *Investigating Artifacts* listings, and in many other categories as well. We also very much appreciated the thoughtful and helpful review and, in many cases, detailed and specific suggestions of: **Jocelyn Berger** of Lane's Books in Oakland; **Marilyn C. Carpenter** of Arcadia, California, Education Consultant; **Ann Jensen**, Reference Librarian/Assistant Head, Engineering Library, UC Berkeley; **Elizabeth C. Overmyer**, Children's Specialist, Bay Area Library and Information System (BALIS); **Kathleen M. Rose**, science teacher, Lakeshore School, San Francisco; **Chiyo Masuda**, teacher at Albany Middle School; **Patsy**

Sherman, an early childhood educator at Los Medanos Community College in Pittsburg, California, as well as director of the Clayton Valley Parent Preschool in Concord, California; and **Barbara Bannister** of Portland, Oregon, who shared her own classification of children's literature by various science themes with us. All of these reviewers were kind enough to share their experience and expertise, and we did our best to integrate their comments in appropriate ways, granted the inevitable differences of emphasis and opinion.

Deborah Lee Rose, author of children's books (who has assisted the GEMS project and Lawrence Hall of Science in the past through her U.C. Berkeley public affairs expertise) provided her sage advice on ways the final version of this handbook might be most useful and accessible. We would also like to thank **Starr LaTronica** and the staff of the Children's Room at the Berkeley Public Library for their frequent assistance. Very special thanks go to former GEMS staff member **Nancy Kadzierski** for her proofreading in preparation of this 1994 edition.

Carl Babcock's editorial, design, and desktop publishing skills helped make this lengthy handbook attractive and readable, as do the illustrations of **Rose Craig**, and the art from GEMS guides designed by **Carol Bevilacqua** and **Lisa Klofkorn**.

In addition, we want to recognize the contributions of **Cynthia Ashley**, former Administrative Coordinator of the GEMS project, whose enthusiasm for young people's literature helped launch this project.

Finally, we trust that the input and counsel of all the individuals noted here, as well as our own cooperative group process, rather than contributing to what might be termed the "too many cooks" phenomena, has in fact spiced and seasoned this literary "Stone Soup" to a "T," and we hereby serve it, piping hot, to you!

LHS GEMS

Once upon a time there was a GEMS guide, and then there were 10, then 20, then 30 GEMS guides … each enhancing a growing curriculum series of clear, step-by-step instructions for teachers to present exciting science and mathematics discovery activities in the classroom.

With the growth of the GEMS series new needs emerged. Teachers wanted more background information on the GEMS philosophy and method of guided-discovery learning. There was a demand for more information on how the GEMS series fits into new ideas on education and assessment. Parents wanted to know how GEMS activities could help them foster a family love of learning and inquiry. Many other topics, including the great advantages of interweaving literature with science and mathematics, came to the fore.

This handbook connecting young people's literature to the GEMS series is the result of a growing understanding of the profound educational benefits that can emerge from the many possible combinations of science and mathematics with literature and the language arts.

The "chemical reactions" resulting from such connections can be exciting, memorable, and of great benefit. "Making connections across the curriculum" is far more than just a popular slogan, it is an enormously effective, practical, and valuable educational principle. Teachers know that science and mathematics classes and concepts can be greatly reinforced, enriched, and highlighted when connections to literature (as well as the other arts and the social sciences) are intertwined with hands-on activities. Insights that might otherwise have gone unnoticed spring out; new light is shed on underlying characteristics and concepts within nature and society.

Reading, writing, and the dramatic arts can help make science and mathematics more accessible and relevant. At the same time, the scientific and mathematical knowledge we gain about the workings of the natural world and the skills and processes we refine through investigation, such as thinking logically and evaluating evidence, can sharpen our understanding of and appreciation for literature.

Literature connections can also help change the misconception of science as non-feeling and narrowly technological, so students come to appreciate the human side of science and mathematics. The important ways in which both literature and the sciences are **creative** can emerge, thus helping give students an exciting sense of the real nature of science.

Ethical issues are increasingly raised in science, and literature has much to teach in this connection. As young people face a technological future influenced by extraordinary developments in so many fields— biotechnology and genetic engineering, for example—the positive sense of life and human worth that resounds through all literature can shed its humanitarian light on science and mathematics as well.

Fundamental issues relating to cultural diversity, gender inclusivity, social equality, and ecological urgency arch across all learning. Their exploration in literature can give us a lens through which to better comprehend the way these issues affect science and mathematics education, the educational system, and the worldwide environmental crisis.

Making Real Connections

There are many different and effective ways to make books and literature come alive in connection with science and mathematics. Teachers tell us of ways they imaginatively blend literature with science and mathematics. Their thoughts and some of their letters have been included in this handbook. In addition to book lists and annotations, we have included a number of review articles, special features, anecdotes, and resource suggestions throughout the handbook.

While we include books in various categories, there are obviously innumerable excellent books for young people whose content does not intersect with these categories. **This handbook is by no means meant to be exhaustive, dogmatic, or in any other way rigid or restrictive. Just the opposite!** The books listed here are keyed to a specific GEMS guide, mathematics strand, or science theme. Even within that limitation, we have selected books we consider to be **particularly apt** for making connections, rather than all possible books.

For example, almost any story that concerns bees could presumably be listed as a literature connection for the GEMS unit *Buzzing A Hive*. Such a listing is not the purpose of this handbook. Instead we listed books that, when presented in conjunction with a GEMS unit, will make a meaningful connection to an important component of that GEMS guide. In some cases, such books may not be about bees at all—they could be about the life of a meadow, nonverbal ways of communication, or cooperation, because these subjects (among many others) are represented in that GEMS guide.

We have endeavored to de-emphasize superficial connections, and instead sought to bring out the underlying ideas and processes in the GEMS activities through their links with literature. While many of the books naturally reflect the main subject of the

GEMS guide, the ways we suggest to make the connections endeavor to go deeper. The special features and articles contain detailed examples, but in many of the annotations we suggest ways the GEMS guide and a theme in literature intersect, and how you might make this connection come alive with your students.

Throughout, we tried to bring our own teaching experience and collective creative imaginations to bear on our selections, aiming to find intriguing, resonating ideas embodied in both the science/mathematics and the literary selections—ideas that will capture students' imaginations and encourage them to further explore, read, write, think, experiment, and create. The listings, annotations, and articles are what we consider to be examples or models of **meaningful** connections. We hope teachers everywhere will be encouraged and empowered to weave their own curricula with science/mathematics activities and literature selections to reflect their planned emphases and favorite books.

In some cases, we may have reviewed a seemingly logical and well-known book choice and found it makes only a surface connection, or perhaps is questionable in relation to other criteria we deem important, such as stereotypical depiction of girls/women; issues of racial equality and cultural diversity; or presenting science or mathematics in ways that intimidate, obfuscate, or discourage. Sometimes we have included a particularly suitable and otherwise excellent book with an appropriate criticism relating to what we consider its drawbacks; in other cases, we have chosen not to list it.

We tried to select those books that have the strongest, richest, most natural connections, and even then, especially in some categories, have not included them all. Of course, we may have not known about a book that makes a particularly excellent connection to one of the GEMS guides or to the science and mathematics categories. **We welcome your suggestions and will review all of your recommendations before the next edition of this handbook is published.**

How To Use This Book

There are three major sections of this handbook. In each section, literary selections are listed **alphabetically by title**, and include an estimated age range and an annotation of the book. The three major sections are divided so you can make literature connections if you are teaching a particular GEMS guide, or following a particular math strand or science theme. The three major sections are:

1) Books connected to **GEMS Guides**
2) Books connected to **Math Strands**
3) Books connected to **Science Themes**

Why Three Sections?

We adopted this somewhat unusual form of organization to address differing curricular needs among the diverse educational community; to make sure the handbook reflected modern, progressive trends in science and mathematics education; and to ensure that the literature selections addressed underlying ideas that intersect throughout science and mathematics. These underlying ideas often have corresponding or complementary constructs in language arts, social studies, and many other areas of study.

GEMS Guides

The GEMS guides are listed alphabetically by title.

Before the annotated book listings for each GEMS guide is a description of the student activities in the GEMS guide, along with the themes, concepts, and skills highlighted in that guide. A brief essay suggests some general ideas to keep in mind in choosing literature connections for that GEMS guide, followed by cross references to other titles, math strands, or science themes that relate to that guide. Then, we list our literature connections.

The same format is followed for each of the 45 GEMS guides. (See the inside back cover for a listing of all guides, and the "What is GEMS?" section at the front for more general information on GEMS.)

Math Strands

This section is organized according to major strands in mathematics, similar to those defined in numerous educational documents, such as the Mathematics Framework for California Public Schools, and are one way to classify the major categories involved in a study of mathematics at the pre-college level.

Eight key areas, or strands, of mathematics are defined in this GEMS handbook, but it is important to emphasize that in building a mathematics curriculum, appropriate concepts and skills from previous lessons and strands are always interwoven. There are no clearly defined boundaries between the strands—every mathematics lesson is likely to contain a combination of several strands.

Each Mathematics Strand is defined as well as an explanation of the mathematical content and skills that are involved in that strand. A brief essay with general ideas on ways the strand can be connected to young people's literature follows along with cross references to other mathematics strands, science themes, and/or specific GEMS guides. Specific book listings and annotations conclude the section on each strand.

Science Themes

This section is organized according to ten major Science Themes. The themes we selected are major, recurring concepts that occur throughout the science curriculum and provide a framework for improved student comprehension and retention of over-arching ideas common to all scientific disciplines.

The thematic approach to teaching science has been emphasized by many leading science educators in recent years, and is on the agenda in many states. The GEMS handbook, *To Build A House*, describes the thematic approach.

As with the mathematics strands, the Science Themes section begins with a definition of what each theme means, then a brief essay on general ways the theme can be explored through young people's literature, cross references, followed by a listing of books with annotations that relate to that theme.

Where to Start

The place you start depends upon your curricular needs, interests, and preferences.

If, for example, you were planning to present the GEMS "Discovering Density" activities to your class, and wanted to include several literature connections, you would find the guide in the GEMS Guide section of this handbook, read the description of what the guide contains (if you are not familiar with it already), then the "Making Literature Connections" ideas (which can also help trigger your own ideas), then peruse the book listings in the age categories relevant to your class. A cross-references box may tell you that books listed under the theme "Patterns of Change," for example, could also make some fruitful connections, as might the mathematics strand of "Logic." You could then check those categories too, in the other main sections of the handbook.

In many school districts, the science curriculum is increasingly being organized by major themes. For example, if you are preparing a series of lessons with a variety of subjects, all focused around theme "Systems and Interactions," you would go to the section on Science Themes and look up "Systems and Interactions." You could read a brief overview of this theme and discover ways to make literature connections to it. Then, you could look through the book listings to see if any fit into your plans. This might also help you see ways to connect the theme to your own favorite books or resources. You would also find cross references, some of them to GEMS guides, or to a mathematics strand, or to another science theme. Ideally, several of the GEMS guides mentioned could be part of the overall theme you plan to present, and, if so, you would definitely want to look under the specific guide to find more literature connections. Even if you are not presenting these GEMS units, checking the listings under the cross references makes sense, because it will provide you with more possible books with which to construct the literature component of the "Systems and Interactions" theme.

There are numerous interconnections and imaginative interweavings that we hope this handbook will help foster. For example, let's say you are working with your class on the mathematics of measurement. Just as in the examples above you will find books listed in the Math Strands section under the measurement strand, and you will also discover connections to the Science Theme of "Scale" and to many GEMS guides, including *Group Solutions* which has a section of cooperative "Coin Count" games that would make a great addition to a class investigation of money.

Also, don't forget to use the index to good advantage. We have endeavored to make it as helpful as possible. If you have favorite books (or authors) we list that you like to use with your classes, you can see which GEMS activities, Science Themes, or Mathematics Strands the book relates to, and combine the book with a science or math activity in a new way.

Interspersed among the annotated listings of books are short articles, reviews, features, helpful hints, teacher anecdotes, and other special items. A number of these connect science or math activities to a particular literary selection in somewhat greater detail than in the annotations. We hope these articles enliven this necessarily extensive reference handbook, and urge that you read them. Sometimes the recounting of actual experiences, or a focus on the way an excellent book reflects many themes, can spark new ideas and convey a sense of the educational effectiveness and enthusiastic energy that can be derived from "making literature connections."

Special Considerations

Although GEMS guides are designed and tested for specific age ranges, all can be modified for presentation at lower or higher levels. Many guides contain age modification suggestions. This is one of the reasons why a literary selection that usually would be considered outside of the age range listed for the particular GEMS guide might nevertheless be listed with it. There are quite a number of cases like this.

Age Range Flexibility

It is important to recognize that even though a particular literary selection might be classified in, for example, a younger age range, it could be listed

under a GEMS guide for older students, if there are some special connections that outweigh the age factor. A good example is the GEMS guide *Oobleck: What Do Scientists Do?* Designed for grades 4–8, it can be modified successfully for use from kindergarten through adult! The same is true for the Dr. Seuss book, for somewhat younger students, *Bartholomew and the Oobleck*. On page 208, a teacher describes how she successfully used the Dr. Seuss book with her quite sophisticated junior high class. Or, for another example related to the GEMS guide *Earth, Moon, and Stars*, an article on page 90 suggests how older students can analyze drawings of the moon and its phases in a wide selection of younger children's books to evaluate accuracy and their own new-found knowledge of the moon. In cases where a teacher is taking a GEMS activity designed for fifth grade and adapting it for second graders, the books listed for younger students would be useful.

In cases where we stretched an age range, we mentioned it in the annotation and described ways the book can be used. Occasionally, there may be a book usually considered for older students that is listed under a GEMS guide for younger students. We noted this as well and suggested, for example, that a teacher may want to read a relevant portion of the book to the students. Of course, if that GEMS activity were adapted upward in grade level by a teacher, then the literary connection for older students would be helpful.

Finally, a group sense of creative flexibility guided us, along with full confidence in a teacher's ability to gauge the abilities of the students. The assignment of age ranges to young people's literature may have certain rules and precedents, but it is still highly subjective. We chose to err on the side of flexibility and inclusion, rather than impose narrow limits. We would rather list a fine and involving book whose age range may be somewhat different from the GEMS guide than miss an opportunity to make a strong connection. Such flexibility seems especially warranted, given the wide variation between schools and grade levels across the country, the startling and rising rates of illiteracy, and the deep inroads television and the modern industrial pace have made into the joys of reading.

Out of Print?

Despite the frustration that *can* be involved in trying to find out-of-print books, we chose to include them to a limited extent for the following reasons. If the book in question makes a particularly strong or unique connection to a GEMS guide, science theme, or mathematics strand, we included it, with a clear notation that it is out of print. In some cases, these may be books that, though temporarily out of print, tend to be reprinted on a regular basis. In many other cases, a book may be out of print but was in such widespread use at one time that it is very likely it is available in many school or public libraries.

Fiction, Nonfiction, and Points In-Between

For the most part, the literary selections in this handbook are fiction. From a story with full-page pictures to a novel, from talking animal communities to science fiction fantasies, these selections connect the creative and imaginative attraction of literary works to science, mathematics, and the fascinating natural world around us.

Nevertheless, we have included some works traditionally classified as nonfiction. We did so for several reasons. In many cases, the visual presentation or narrative method of a particularly apt nonfiction book puts it on the borderline between fiction and nonfiction. Such books are in some ways especially useful because they contain a great deal of scientific information, but present it within an imaginative or semi-fictional format. Their images or unique points-of-view involve the reader in a different, but somewhat comparable way, to "pure" fiction.

In looking for matches to particular GEMS guides, sometimes there were not a great many fiction literary connections we could find, but there were several excellent works of nonfiction that expanded or deepened the content of the guide in an important way. These we included. For example, under the

GEMS *Acid Rain* guide we included *Love Canal: My Story* by Lois M. Gibbs, the woman who organized protests against the toxic waste at Love Canal, Pennsylvania. This book fits in well with the sense of student environmental responsibility and empowerment that is fostered in *Acid Rain* and several other GEMS guides.

Gibbs' story highlights another reason for the inclusion of some nonfiction—the great value of biography. The depiction of positive role models who overcome great obstacles can be inspiring for all students. While other reference materials contain extensive lists of biographies, our listing is very limited. We listed a book when it related to a significant content area in a GEMS guide, and also when its inclusion strengthened our representation of the participation of women and people of diverse cultures in science and mathematics.

Multicultural Diversity and Equity

Many of the books and several feature articles consider ways that issues related to multicultural diversity, equality, and gender are involved in science and mathematics education. Progress in these areas is an important priority of the GEMS project and of the Lawrence Hall of Science (LHS).

Many LHS programs have devoted special attention to, and are fully supportive of, efforts to ensure the full participation in science and mathematics of peoples of diverse cultures and backgrounds, women, and other historically underrepresented groups in science and mathematics. The EQUALS and Family Math projects pioneered the creation of programs and mechanisms that reach out to communities with accessible mathematics activities and a message of equal opportunity. These programs have established a strong national network and an international reach and reputation. The MESA (Mathematics, Engineering, and Science Achievement) program (also based at LHS) has built an extensive network to encourage the scientific and technical career aspirations of students from historically underrepresented groups in science and mathematics. In several projects, GEMS has joined forces with MESA, providing GEMS workshops to MESA mentors/teachers.

Yet such efforts need to be multiplied much more than a thousandfold. One of the most effective ways all of us can learn more about the diverse cultures in this country and its schools is through direct contact and exchange, by sharing the unique aspects of each culture and recognizing the great value of its contribution, along with an underlying sense of the essential humanity that unites us all.

In addition to the natural growth and social interaction that goes on in the multicultural and international classroom, books that put forth positive messages of equality, respect, dignity, and peace can play a very important role as children develop their own values and ideas. In recent years, numerous excellent books celebrating many different cultures have been published. Our selections in no way represent the multitude of these fine books; we chose those among them that also had a strong connection to science and mathematics concepts. We did, however, endeavor to include a diverse spectrum, as one way to represent our commitment to quality *and* equality in education.

In Your Hands

We know that many of you already have your own favorite books and unique ways to integrate them into the science and mathematics curriculum—we look forward to hearing about them! Using the books in this handbook and your own favorites in your own ways will demonstrate how powerful these connections can be. The educational benefits will extend throughout the curriculum and over the years, helping nurture all the abilities, talents, and infinite potential of the growing child.

We hope you find *Once Upon A GEMS Guide: Connecting Young People's Literature to Great Explorations in Math and Science* an interesting and useful handbook. We very much welcome your ideas and suggestions for ways we could improve this handbook (please see page 412), or any other aspects of the GEMS program. Thanks for your interest in GEMS!

EMS guides are listed here in alphabetical order by title, for your convenience in locating a specific guide.

We are of course aware that many teachers are guided by grade level considerations. For your quick reference, the grade levels for all the GEMS guides are highlighted in the full-page chart on page 2. GEMS guide grade levels are also noted at the top of the descriptions for each guide that precede the literature connections for that guide.

Please note that many of the GEMS guides can be adapted for lower or higher grade levels. For the browsers among you, we might note that even though books listed in the next GEMS guide over may be for another grade level, many of our literature connection listings include a considerable age span, and you might find something just right for your students.

Note: A number of the books cited are currently out of print. However, many of them should be readily available at your local school or public library.

COMING SOON!

OUTCOMES & INSIGHTS
The GEMS Assessment Handbook

The new GEMS assessment handbook, scheduled for publication in 1995, will spell out expected student learning outcomes for all GEMS guides, along with descriptions of the "built-in" assessment opportunities already contained in many GEMS activities. In addition, it will feature more than 15 extensive examples of active assessment of student learning, in a variety of assessment modes. These examples will include diverse student work, and critical analysis of it, to assist teachers in their own assessment efforts as they evaluate the effectiveness of GEMS and other activity-based lessons they present.

If we were to choose a young people's literature connection for assessment, it would be *First Grade Takes A Test* by Miriam Cohen, Dell Publishing, New York, 1980. It points out, in deceptively simple fashion, some of the main limitations of standardized testing:

George looked at the test. It said: Rabbits eat
☐ lettuce   ☐ dog food   ☐ sandwiches
George raised his hand. "Rabbits have to eat carrots, or their teeth will get too long and stick into them," he said. The teacher nodded and smiled, but she put her finger to her lips. George carefully drew in a carrot so the test people would know.

In the "classic" testing sense, George would be "wrong" because he did not pick one of the answers. But George is correct nonetheless. His real-world knowledge clearly exceeds the limitations of the test. *First Grade Takes A Test* makes a great segue into more hands-on, flexible and active assessment methods such as those explored in the new GEMS assessment handbook.

Acid·Rain

Grades 6–10   **Eight Sessions**

Students gain scientific inquiry skills as they learn about acids and the pH scale, make "fake lakes," determine how the pH changes after an acid rainstorm, present a play about the effects of acid rain on aquatic life, and hold a town meeting to discuss solutions to the problem. Students play a "startling statements" game and conduct a plant-growth experiment. Students are encouraged to analyze complex environmental issues for themselves.

Session 1

Students write down everything they have heard about acid rain and any questions they have about it. These lists are added to, and modified, throughout the unit, as knowledge changes and as new questions arise. Students set up an experiment in which they observe the effect of a common acid (vinegar) of various concentrations on seed germination. The seed experiment continues throughout much of the unit. As homework, students search their homes for things they can identify as acids.

Session 2

Students learn more about what acids are, and how to measure their strength, by conducting tests on common substances. The class adds their data to a large pH chart, which is used as a reference throughout the unit.

Session 3

Students play a "Startling Statements" game. The class compares its answers to those of scientists. Students read an article about acid rain as homework.

Session 4

Students experiment with model or "fake lakes," to investigate the effects of different soil types on the pH of lakes and what happens to lake pH after an acid rainstorm. Students add a buffer tablet to some of the model lakes to replicate a common treatment for acidic lakes.

Session 5

Students observe what effects the buffer had on acidity in each model lake. They collect results and draw conclusions from the seed germination experiment. The students prepare for an emergency town meeting to address the problem of acid rain by choosing which interest group in the imaginary town they'd like to represent.

Session 6

The class brainstorms possible solutions to the problem of acid rain, followed by a dramatic reading of a play in which plants and animals threatened by acid rain meet to discuss their future. The play also addresses human responses to environmental issues, which provides the basis for discussing the role played by members of a community. This leads into a town meeting planning session in which the interest groups prepare for the town meeting.

Session 7

The town meeting takes place. As Mayor, the teacher facilitates an interchange between interest groups. The focus is on finding mutually acceptable solutions and outlining more clearly the areas of conflict and disagreement.

Session 8

Students put aside their town meeting roles, and participate in a straw poll to determine which possible solutions they **personally** think are best. They review the ongoing lists of questions and statements about acid rain, then generate ideas for new scientific experiments or technologies that could help solve the problems posed by acid rain.

The Morning Wind   From The West

Is Bringing The Smoke
Of The   Paper   Mills
               And
               The
                       Smoke
Of Burning
     Plastic   And   Coal

Some Days   The   Air Is So Good
It Is The Best Cure For Bad Dreams
But This Morning
     This Air   Is
          Bad
               For All Living Things
          The
          Breeze   Brings   Ills
          The Wind Seems Sad
          To Be Blowing Our Way

We Stay In For The Rest Of The Day

— Arnold Adoff
In for Winter, Out for Spring

Skills

Observing, Measuring and Recording Data, Experimenting, Classifying, Drawing Conclusions, Synthesizing Information, Role Playing, Problem Solving, Critical Thinking, Decision-Making, Brainstorming Solutions

Concepts

Acid, Base, Neutral, pH, Neutralize, Buffering Capacity, Ecosystem, Environmental Issues, Controversy in Science, Relationship Between Scientific and Social Issues, and Effect of Acid Rain on Plants, Animals and Lakes

Themes

Systems & Interactions, Models & Simulations, Stability, Patterns of Change, Evolution, Scale, Energy, Matter

MAKING CONNECTIONS

In these books there exists a situation (fictional or real) in which the **impact of humans** causes some **imbalance in the ecosystem**. Some books are specifically about **acid rain** while others are about **other environmental problems** such as chemical spills, the vanishing rain forest, or diminishing habitats for animals. Running through many of the books is a strong message that people acting in a timely fashion can turn these problems around. We have included several nonfiction books: two of them about acid rain and surface water pollution and two biographies of **influential women ecological activists**.

CROSS REFERENCES

LITERATURE CONNECTIONS

The Berenstain Bears and the Coughing Catfish

by Stan and Jan Berenstain
Random House, New York. 1987
Out of print
Grades: K–6

> A respected bear scientist helps a wise catfish and the Berenstains to clean up a lake where pollution is life-threatening. A sunken treasure chest, some imaginative scientific devices, and the creation of a lake-life museum are part of the story. Although this book has the juvenile humor common to the series, the language level is fairly high and might even stretch into the 7th and 8th grades with a suitable introduction.

Danny Dunn and the Universal Glue

by Jay Williams and Raymond Abrashkin; illustrated by Paul Sagsoorian
McGraw-Hill, New York. 1977
Grades: 4–9

> Danny and his friends bring evidence to a town meeting that waste from a local factory is polluting the stream. Discussion of societal issues such as tax revenue and jobs that the factory contributes to the town make a good connection to the town meeting in Session 7.

The Day They Parachuted Cats on Borneo: A Drama of Ecology

by Charlotte Pomerantz; illustrated by Jose Aruego
Young Scott Books/Addison-Wesley, Reading, Massachusetts. 1971
Out of print
Grades: 4–7

> This cautionary verse, based on a true story, explores how spraying for mosquitoes in Borneo eventually affected the entire ecological system, from cockroaches, rats, cats, and geckoes to the river and the farmer. The strong, humorous text makes the book successful as a read-aloud or as a play to be performed. A good example of the interacting elements of an ecosystem.

The Earth is Sore: Native Americans on Nature

adapted and illustrated by Aline Amon
Atheneum, New York. 1981
Out of print
Grades: 4–Adult

> This collection of poems and songs by Native Americans celebrates the relationship between the Earth and all creatures and mourns abuse of the environment. Illustrated with black and white collage prints made from natural materials.

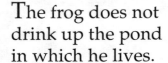

The frog does not
drink up the pond
in which he lives.

— *Teton Sioux proverb*

Just A Dream

by Chris Van Allsburg

Houghton Mifflin, New York. 1990

Grades: 1–6

When he has a dream about a future Earth devastated by pollution, Walter begins to understand the importance of taking care of the environment. Session 7 encourages students to take responsibility for environmental concerns, and to empower themselves.

Kid Heroes of the Environment:
Simple Things Real Kids Are Doing To Save the Earth

edited by Catherine Dee; illustrated by Michele Montez

Earth Works Press, Berkeley, California. 1991

Grades: 3–12

Twenty-nine profiles of individuals and organizations working at home, at school, locally, and nationally, to help the environment through "kid power." Strong connections include "Oliver the Otter," an oil spill awareness and publishing project; "Tree's Company," tree planting and research on the greenhouse effect; "Toxic Avengers" which shut down a waste storage facility dumping hazardous waste; and "Testing the Water," a Boy Scout project to monitor streams for acid rain.

The Last Free Bird

by A. Harris Stone; illustrated by Sheila Heins

Prentice-Hall, Englewood Cliffs, New Jersey. 1967

Out of print

Grades: K–4

Here's a moving plea for protecting the natural beauty and habitats of the land and "the last free bird." The easy reading level and the interplay between the watercolor illustrations and the text make this book effective for younger children. An interesting activity would be to read this book about vanishing bird habitats and supplement it with *Urban Roosts* (in the "Structure" section) which shows birds adapting to urbanized and seemingly inhospitable environments.

Love Canal: My Story

by Lois M. Gibbs

State University of New York at Albany Press, Albany, New York. 1982

Out of print

Grades: 6–12

Autobiography of the housewife who organized a neighborhood association that eventually resulted in a clean up of the Love Canal toxic waste site and relocation of the families living there. She went on to form the Citizen's Clearinghouse for Hazardous Waste based in Arlington, Virginia.

The Missing 'Gator of Gumbo Limbo: An Ecological Mystery

by Jean C. George
HarperCollins, New York. 1992
Grades: 4–7

Sixth-grader Liza K and her mother live in a tent in the Florida Everglades. She becomes a nature detective while searching for Dajun, a giant alligator who plays a part in a waterhole's oxygen-algae cycle, and is marked for extinction by local officials. The book is full of detail about the local habitats and species and the forces that impact on them, as well as environmental information that relates to this GEMS unit. Lisa and her amateur naturalist neighbor discuss, for example, pesticide and phosphate pollution and the differences between green and blue-green algae: "Blue-green algae, on the other hand, is an announcer of doom. It says the water is polluted with phosphorus and nitrogen from septic tanks, lawns, and road runoff. That day it also said to me there was no Dajun. If he was there, he would have bulldozed that clump of blue-green algae to shore. It suffocates the fish and turtles he lives on. Somehow he knows this and weeds it out." Reading this extraordinary book would make a wonderful extension to the "lake animals play" in Session 6 of this GEMS guide.

One Day in the Tropical Rain Forest

by Jean C. George; illustrated by Gary Allen
HarperCollins, New York. 1990
Grades: 4–7

When a section of rain forest in Venezuela is scheduled to be bulldozed, a young boy and a scientist seek a new species of butterfly for a wealthy industrialist who might preserve the forest. As they travel through the ecosystem rich with plant, insect, and animal life, everything they see on this one day is logged beginning with sunrise at 6:29 a.m. They finally arrive at the top of the largest tree in the forest and fortuitously capture a specimen of an unknown butterfly.

Our Endangered Planet: Rivers and Lakes

by Mary Hoff and Mary M. Rogers
Lerner Publications, Minneapolis. 1991
Grades: 4–8

An attractive and user-friendly reference book covering the dangers of surface water pollution with many illustrations and photographs. Other relevant titles in this series (all published in 1991) include: *Groundwater, Population Growth,* and *Tropical Rain Forests.*

SO_X and NO_X

There once was an oxide called SO_X
Who, along with another named NO_X,
Unleashed acid rain,
As this guide will explain,
Causing eco-illogical shocks.

From smokestacks and auto exhaust
Exacting a terrible cost
Acid rain's killing lakes
Do we have what it takes—
To make certain that no more get lost?

The tough problems posed by pollution
Are crying out loud for solution
To bring NO_X and SO_X down
Let us meet in our town
For we each have a key contribution.

— *Lincoln Bergman*

Rachel Carson

by Leslie A. Wheeler

Silver Burdett Press/Simon & Schuster, New York. 1991

Grades: 5–12

> The life and work of the biologist and conservationist are examined in light of the important role her writing played in initiating the modern environmental movement. The second half of the book provides fascinating background on influences and obstacles that contributed to her writing *Under the Sea Wind, The Sea Around Us, The Edge of the Sea*, and the landmark work, *The Silent Spring*, which caused an uproar with its descriptions of the pollution of earth and sea by chemical pesticides and the potential effects on humans and wildlife.

Rain of Troubles:
The Science and Politics of Acid Rain

by Lawrence Pringle

Macmillan, New York. 1988

Grades: 5–12

> Acid rain's discovery, formation, transportation, its effects on plant and animal life, and how economic and political forces have delayed action are discussed. This book provides good background for the town meeting activities in the GEMS guide.

Restoring the Earth: How Americans Are Working to Renew Our Damaged Environment

by John J. Berger

Alfred A. Knopf, New York. 1985

Doubleday, New York. 1987

Out of print

Grades: 7–12

> This book profiles individuals and groups active in conservation. It is written at the adult level and does not include any photographs or illustrations. Especially relevant to the GEMS activities is the first section, which deals with restoration of a polluted lake, a trout stream, and a dead marsh. "Mother Nashua" is the true story of the clean-up effort, also portrayed in Lynne Cherry's picture book *A River Ran Wild*. Other sections relate to land use and waste disposal, human settlements and their environmental impact, and wildlife preservation.

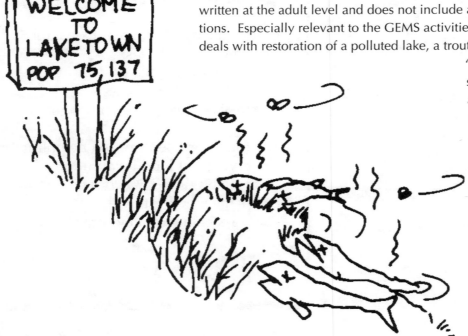

WELCOME TO LAKETOWN POP 75,137

The River
by David Bellamy; illustrated by Jill Dow
Clarkson Potter/Crown, New York. 1988
Grades: *3–5*

Plant and animal life co-exist in a river—then they must struggle for survival when a human-made catastrophe strikes. Details about stream ecology include a description of the effects of waste water discharged from a factory pipe and how the bacteria, algae, and oxygen interact in the dam area and beyond where the waste has been diluted. The ending seems overly optimistic with the river "back to normal" a month after the waste is released. "Everyone hopes the factory owners will be more careful in the future."

A River Ran Wild: An Environmental History

by Lynne Cherry
Harcourt, Brace, Jovanovich, San Diego. 1992
Grades: 1–5

This is the true story of the Nashua River Valley in North-Central Massachusetts from the time that the Native Americans settled there, naming it
River With the Pebbled Bottom. The book traces the impact on the river of
the industrial revolution and the eventual clean-up campaign mounted by
a local watershed association. The graphic borders are packed with
historical information, showing the tools and artifacts that represent
cultural changes that affected the water quality: the proliferation of textile
and paper mills and attendant pollution, the increased presence of plastics
and chemicals in manufacturing, and waste disposal.

And Still the Turtle Watched

by Sheila MacGill-Callahan; illustrated by Barry Moser
Dial Books, New York. 1991
Grades: K–5

A turtle carved by Native Americans on a rock watches, with sadness, the
changes humans bring over the years. After the rock is cleaned of spray
paint and installed indoors at a botanical garden, the turtle's vision is
restored and he communicates his wisdom to the many children visiting.
Moser's watercolor paintings are dramatic.

The Talking Earth

by Jean C. George
HarperCollins, New York. 1983
Grades: 6–12

Billie Wind, a Seminole, is known for her curiosity and criticized for
doubting the traditional wisdom of her people. Her sister says, "You are
too scientific. You are realistic like the white men." Poling through the
Florida Everglades sloughs and then a river in a dugout canoe, she fends
for herself with an otter, a panther cub, and a turtle as companions and
guides. Viewing the destruction after a hurricane, she hears the message
of the animal spirits, "we must love the earth or it will look like this … life
can be destroyed unless we work at saving it."

Who Really Killed Cock Robin?

by Jean C. George
HarperCollins, New York. 1991
Grades: 3–7

A compelling ecological mystery examines the importance of keeping
nature in balance, and provides an inspiring account of a young environmental hero who becomes a scientific detective.

ANIMAL DEFENSES

Grades: Preschool–K

Two Sessions

*I*n this activity, children add defensive structures to an imaginary defenseless animal. Then, in a classroom drama, the animal encounters a *Tyrannosaurus rex*. In the second session, the children learn about the defenses of modern-day animals.

Session 1

The class discusses dinosaur defenses, then the teacher enacts drama of a defenseless imaginary animal facing a tyrannosaurus. Students add their own cut-out defenses to the imaginary paper animal. Silhouettes on overhead projector can be used to enhance drama.

Session 2

Defenses of modern animals are discussed, with emphasis on the physical defenses they *have* (structures) and the things they *do* (behaviors). Can include bringing in a toy animal, making up scenarios of protection. Final discussion of ways children can defend themselves.

Skills

Observing, Identifying, Creative Thinking, Using Scissors (optional)

Concepts

Animal Protection, Predator-Prey, Distinction between Defensive Structures and Defensive Behaviors

Themes

Systems & Interactions, Models & Simulations, Evolution, Scale, Structure, Diversity & Unity

Notes: The guide includes modifications for Grades 1 and 2, with mention of the food chain; classifying the kinds of defenses animals have; making a diorama; and taking a field trip to the zoo.

MAKING CONNECTIONS

All of the books listed below look at ways animals **defend** themselves by using specific **behaviors**, **structures**, and **surroundings**. Animals featured in the books range from small fish to dinosaurs. Stories include the relationships between a clownfish and an anemone with protective tentacles, a mouse and a discarded pumpkin shelter, and a day in the life of a camouflaged lizard. In a direct connection with the students' animals "meeting" a *Tyrannosaurus rex* in the GEMS activities, two books address youthful enthusiasm for dinosaurs by describing what these prehistoric creatures were like and how they used their distinctive features, such as spines, horns, and teeth.

CROSS REFERENCES

LITERATURE CONNECTIONS

Curious Clownfish

by Eric Maddern; illustrated by Adrienne Kennaway

Little, Brown & Co., Boston. 1990

Grades: K–3

> A baby clownfish wants to leave the protection of the anemone whose stinging tentacles he keeps clean. During his adventure exploring a coral reef he encounters a sea slug, porcupine fish, dragon fish, crab, cuttlefish, and a terrifying eel. They all demonstrate their defense mechanisms and he is grateful to return to the anemone. The bold neon illustrations depict a beautiful undersea world as well as clearly showing defensive behaviors.

Dinosaurs are Different

by Aliki

Harper & Row, New York. 1985

Grades: K–3

> Explains how the various orders and suborders of dinosaurs were similar and different in structure and appearance. A good catalogue of dinosaur defenses.

Dinosaurs, Dinosaurs

by Byron Barton

Thomas Y. Crowell, New York. 1989

Grades: K–2

> In prehistoric days there were many different kinds of dinosaurs, big and small, those with spikes and those with long sharp teeth. Perfect to read either before or after Session 1 of the GEMS guide.

Seal Lullaby

Oh! hush thee, my baby, the night is behind us.

And black are the waters that sparkled so green.

The moon, o'er the combers, looks downward to find us

At rest in the hollows that rustle between.

Where billow meets billow, there soft be thy pillow;

Ah, weary wee flipperling, curl at thy ease!

The storm shall not wake thee, no shark overtake thee.

Asleep in the arms of the slow-swinging seas.

— *Rudyard Kipling*

Eric Carle's Animals Animals

compiled by Laura Whipple; illustrated by Eric Carle

Philomel/Putnam & Grosset, New York. 1989

Grades: K–5

Anthology of over 50 poems from many cultures on both wild and domestic animals illustrated with Carle's joyous color collages. The poems cover a wide range of topics, and some, such as those on the barracuda, porcupine, and narwhal, focus particularly on animal defenses.

A House for Hermit Crab

written and illustrated by Eric Carle

Picture Book Studio, Saxonville, Massachusetts. 1987

Grades: Preschool–2

One day Hermit Crab moves out of the house he has outgrown and finds a bigger shell "house" that is perfect but plain. He collects sea anemones, starfish, coral, snails, sea urchins, lantern fish, and pebbles to adorn it. When he outgrows that home, he finds a bigger one with new possibilities for decorating—barnacles, clown fish, sand dollars, electric eels, and so on. An additional page of text in front and back gives more information about the crab's habit and defenses used by other creatures such as poisonous spines or camouflage.

Lizard in the Sun

by Joanne Ryder; illustrated by Michael Rothman

William Morrow, New York. 1990

Grades: Preschool–2

The friendly narration guides you through your day as a lizard; you are camouflaged from hungry birds and hidden from insects that become your next meal. Children enjoy seeing the natural world from a lizard's viewpoint, and learn firsthand facts about a lizard's lifestyle.

The Mixed-Up Chameleon

by Eric Carle

Harper & Row, New York. 1975

Grades: Preschool–2

A bored chameleon wishes it could be more like all the other animals it sees, but soon decides it would rather just be itself. Protective coloration (the chameleon changes color according to the surface on which it rests) and energy (when the chameleon is warm and full, it turns one color, when cold and hungry, it turns another) are woven into the story, as are a discussion of attributes of various other animals.

Mousekin's Golden House

written and illustrated by Edna Miller
Prentice-Hall, Englewood Cliffs, New Jersey. 1964
Simon & Schuster, New York. 1987
Grades: K–3

> Mousekin acquires an unusual tool for self defense in the forest—a discarded jack-o-lantern in which he hides from a hungry young owl, a cat, and a box turtle. As the cut-out spaces in the pumpkin slowly melt together, he has an even cozier golden house.

Pretend You're A Cat

by Jean Marzollo; illustrations by Jerry Pinkney
Dial Books, New York. 1990
Grades: Preschool–1

> Wonderful illustrations and friendly verse ask children, "Can you chatter and flee? Disappear in a tree? ... What else can you do like a squirrel?" Young readers love to read the verse and then act out the animal's behavior. Animals include a cat, pig, snake, bear, horse, and seal. Great springboard to a discussion of similarities and differences among animal behaviors.

Swimmy

by Leo Lionni
Alfred A. Knopf, New York. 1987
Grades: K–4

> A clever little black fish discovers a way for her school of little red fish to swim together and be protected from larger predators. With Swimmy as the "eye," the fish swim in formation masquerading as a big fish.

ANIMALS IN ACTION

Grades 5–9   **Five Sessions**

While observing animals in a large classroom corral, the class experiments with the corral environment and adds different stimuli. Teams of students generate hypotheses, conduct experiments, and hold a scientific convention to discuss their findings. Simple investigations that students can do with classroom animals include: How do animals move? What do they prefer to eat? How do they respond to light and sound? These questions are addressed through behavior experiments students conduct with rats, crickets, guinea pigs, cardboard boxes, and common classroom objects. Students learn to distinguish between direct observations and assumptions, as they differentiate **evidence** (what they observe) and **inference** (conclusions they draw from their observations.)

Session 1

An animal corral is made from seven large boxes. Students arrange chairs around the corral, discuss the meaning of the word "behavior" and observe an animal placed in the corral. Discussion of student observations clarifies the meanings of assumption and anthropomorphism. An observation chart is made.

Session 2

Students explore the concepts of stimulus and response by adding foods, shelters, and other objects to the corral and observing animal reactions. Students generate hypotheses to explain the behavior of wild animals.

Session 3

Students observe a small organism (guppies, crayfish, isopods, snails, etc.) in an environment containing a variety of stimuli. The teacher guides students through the process of planning an experiment, introducing the concept of a fair test.

Session 4

Students conduct and evaluate their own animal behavior experiments. Students work in teams of four: a Recorder, a Materials Manager, an Animal Manager, and an Assistant Animal Manager.

Session 5

Students hold a scientific convention to discuss their experimental observations and results.

Skills

Observing, Comparing, Communicating, Relating

Concepts

Biology, Animals, Objective Observation, Animal Behavior, Humane Treatment of Animals, Stimulus and Response

Themes

Systems & Interactions, Patterns of Change, Structure, Diversity & Unity

Notes: Includes a Small Animal Resource Guide (which includes guppies and gambusia, crayfish, milkweed bugs, isopods, crickets, grain beetles and meal worms, ladybugs, butterfly and moth larvae, and garden snails), and a Code of Practice on Use of Animals in Schools. Going Further suggestions include making observations at home, an additional session to improve or trade experiments, making a list of humane animal treatment guidelines, and seeing the film "Never Cry Wolf."

MAKING CONNECTIONS

The literature chosen for this unit focuses on **observation of animals, animal behavior**, and **animal management** and **conservation**. Some stories take real-life behaviors of manta rays, wolves, and carrier pigeons, and weave captivating tales and mysteries from a boy's or girl's perspective.

The role of animals in different environments is brought up through a book on endangered species written by a sixth grader. Important societal issues are addressed in two familiar fantasy books about intelligent rats, and a farmyard pig and spider. Two books of animal poems will encourage the reading of verse, especially as a duet.

CROSS REFERENCES

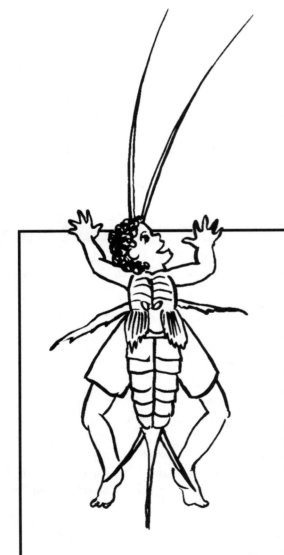

Cricket

A cricket's ear is in its leg.
A cricket's chirp is in its wing.
A cricket's wing can sing a song.
A cricket's leg can hear it sing.

Imagine if your leg could hear.
Imagine if your ear could walk.
Imagine if your mouth could swing.
Imagine if your arm could talk.

Would everything feel upside down
And inside out and wrongside through?
Imagine how the world would seem
If you became a cricket, too.

— *Mary Anne Hoberman*
Bugs

LITERATURE CONNECTIONS

A Caribou Alphabet
by Mary Beth Owens
Dog Ear Press, Brunswick, Maine. 1988
Farrar, Straus & Giroux, New York. 1990
Grades: K–5

> An alphabet book depicting the characteristics and ways of caribou. While at first glance, this book may seem a primary-level "A, B, C" book, it includes a compendium of information about caribou, including intricacies of their behavior, habitat requirements, and physical features that is directed to an older audience.

Charlotte's Web
by E.B. White; illustrated by Garth Williams
Harper & Row, New York. 1952
Grades: 4–7

> This classic story tells of the friendship between a wise gray spider named Charlotte and a pig named Wilbur. Charlotte saves Wilbur from being slaughtered, sometimes with the help of ravenous Templeton the rat. Although the story centers around the anthropomorphised animals, Charlotte offers many cogent observations on web spinning and egg sacs, the natural cycle of life-death-reproduction, and the lasting value of friendship.

Chipmunk Song
by Joanne Ryder; illustrated by Lynne Cherry
E.P. Dutton, New York. 1987
Grades: Preschool–5

> A lyrical description of a chipmunk as it goes about its activities in late summer, prepares for winter, and settles in until spring. You are put in the place of a chipmunk and participate in food gathering, hiding from predators, hibernating, and more. Roots, tunnels, stashes of acorns and other facets of the imagined environment loom large and lifelike.

Eric Carle's Animals, Animals
compiled by Laura Whipple; illustrated by Eric Carle
Philomel/Putnam and Grosset, New York. 1989
Grades: K–5

> Anthology of over 50 poems from many cultures on both wild and domestic animals. Illustrated with Carle's joyous color collages.

Fireflies in the Night
by Judy Hawes; illustrated by Ellen Alexander
HarperCollins, New York. 1963, 1991
Grades: K–4

> A young girl visits her grandfather and tells of their investigations of fireflies on summer nights. Describes how and why fireflies make their light, how to catch and handle them, and several uses for firefly light.

The Frog Alphabet Book

by Jerry Pallotta; illustrated by Ralph Masiello

Charlesbridge Publishing, Watertown, Massachusetts. 1990

Grades: K–3

> A beautifully illustrated book that shows the diversity of frogs and other "awesome amphibians" from around the world.

Frogs, Toads, Lizards, and Salamanders

by Nancy W. Parker and Joan R. Wright; illustrated by Nancy W. Parker

Greenwillow/William Morrow, New York. 1990

Grades: 3–6

> Physical characteristics, habits, and environment of 16 creatures are encapsulated in rhyming couplets, text, and anatomical drawings, plus glossaries, range maps, and a scientific classification chart. A great deal of information is presented, the rhymes are engaging and humorous, and the visual presentation terrific. "A slimy Two-toed Amphiuma/terrified Grant's aunt from Yuma" (she was picking flowers from a drainage ditch).

Gay-Neck: The Story of a Pigeon

by Dhan Gopal Mukerji; illustrated by Boris Artzybasheff

E.P. Dutton, New York. 1927

Grades: 6–12

> A carrier pigeon is sent from India to serve as a messenger in World War I. There is a great deal of detail about bird behavior as a boy observes an eagle's eyrie and describes his bird's mating and offspring. Gay-Neck tells part of the story describing the predator-prey relationship between swifts, owls, and other species; and gives a unique account of war and the "machine-eagles" (airplanes) he saw. The healing by a lama of "the fear and hate caught on the battlefields" makes a nice ending to the book. Newbery Medal winner.

The Girl Who Loved Caterpillars

adapted by Jean Merrill; illustrated by Floyd Cooper

Philomel Books/Putnam & Grosset, New York. 1992

Grades: 2–6

> Based on a twelfth century Japanese story, this book is a wonderful and early portrait of a highly independent and free-spirited girl, Izumi, who loves caterpillars. Although a famous and highly refined noblewoman who loves butterflies lives next door, Izumi says, "Why do people make such a fuss about butterflies and pay no attention to the creatures from which butterflies come? It is caterpillars that are really interesting!" (She also likes toads, worms, insects, and many other creatures that she keeps in her room, spending hours observing their movements and watching them grow.) Izumi is interested in the "original nature of things," and in doing things naturally. Clever poetry is interspersed as part of the plot. Great connection to the observation activities in *Animals in Action*. An excellent and relevant portrayal of an independent-thinking female role model.

Joyful Noise: Poems for Two Voices

by Paul Fleischman; illustrated by Eric Beddows
Harper & Row, New York. 1988
Grades: K–Adult

This series of poems celebrating insects are meant to be read aloud by two readers at once, sometimes merging into a duet. It includes grasshoppers, water striders, mayflies, fireflies, book lice, moths, water boatmen, digger wasps, cicadas, honeybees, beetles, crickets, and metamorphosis. The combination of rich and fascinating scientific detail with poetry, humor, and a sense of the ironic contrasts and divisions of labor in the lives and life changes of insects is powerful and involving. Kids in upper grades might love performing these for the class. (Two of his poems are included in this handbook: "Honeybees," is on page 56, and "Book Lice" is on page 240.) Newbery medal winner.

> People love to wonder, and that is the seed of science.
>
> — *Ralph Waldo Emerson*

Living With Dinosaurs

by Patricia Lauber; illustrated by Douglas Henderson
Bradbury Press/Macmillan, New York. 1991
Grades: 3–6

The time is 75 million years ago. The place is prehistoric Montana. Still alive are the giant reptiles and fishes of the sea; the birds and pterosaurs in the sky; the dinosaurs, tiny mammals, crocodiles and plants of the lowlands; and the predators of dinosaur nesting grounds in the dry uplands. There is a great deal of detail on their habits, measurements, reproduction, feeding requirements, and a clear and elementary description of how a fossil forms and evolves. The colorful paintings are dynamic and involving.

Mark Twain's Short Stories

by Mark Twain
Penguin Books USA, New York. 1985
Grades: 5–Adult

The famous short story "The Notorious Jumping Frog of Calaveras County" is the tale of a frog jumping contest that gets rigged. A bet that one frog can jump farther than another leads to the belly of one frog being weighed down with buckshot so the frog cannot leap at all. (The buckshot is later poured out of the frog, so the frog is not harmed in the story.) The frogs are prodded to make them jump, which ties in with the stimulus and response activities in this guide. Of course, Mark Twain's sense of humor is always an added bonus. This story is a wonderful introduction to a great American writer.

Mrs. Frisby and the Rats of NIMH

by Robert C. O'Brien; illustrated by Zena Bernstein
Atheneum, New York. 1971
Grades: 4–12

A mother mouse learns that the rat colony near her home is actually a group of escapees from a NIMH research institute. These rats, injected with DNA and other substances, have acquired great intelligence, learned to read and write, and are planning to develop their own civilization. In addition to offering a great plot, the book helps us to visualize nature from the scale of a small animal, to imagine communications between birds and rodents, to consider the impact of animal experimentation, and to comment on the technological top-heaviness of modern day human society. Newbery award winner.

My Father Doesn't Know About the Woods and Me

by Dennis Haseley; illustrated by Michael Hays
Atheneum/Macmillan, New York. 1988
Grades: 2–5

A young boy walks in the woods with his father and secretly fantasizes about being a fish, a wolf, and a hawk. At the end, he sees that his father may share some of the qualities of a deer. Peaceful, sun-dappled illustrations and a mystical quality of wondering what it would be like to be an animal flying, swimming, running, etc., balance the intense observations. Could use with older students to stimulate creative writing about such experiences.

My Side of the Mountain

by Jean C. George
E.P. Dutton, New York. 1959
Penguin Books, New York. 1991
Grades: 5–12

Classic story of a boy who runs away and spends a year living alone in the Catskill mountains, recording his experiences in a diary. He struggles for survival and is supported by animal friends. Ultimately he realizes he needs human companionship. In his diary, he makes many notes about specific animal behaviors, such as his weasel friend, the Baron, who "chews with his back molars, and chews with a ferocity I have not seen in him before. His eyes gleam, his lips curl back from his white pointed teeth..." The highly detailed observations and his growing understanding of the intricacies and interconnected web of nature make this book an outstanding literature connection to any of the GEMS animals activities. Winner: Newbery Honor Book, ALA Notable Book, Hans Christian Andersen International Award.

Nessa's Fish

by Nancy Luenn; illustrated by Neil Waldman
Atheneum, New York. 1990
Grades: 3–5

Nessa and her grandmother walk half a day from their Arctic village to go fishing in a lake. When the grandmother becomes ill, Nessa has to use all her ingenuity and bravery to protect their catch from animal poachers—a fox, wolves, and a bear.

Never Cry Wolf

by Farley Mowat
Atlantic Monthly/Little, Brown & Co., New York. 1963
Bantam Books, New York. 1984
Grades: 6–Adult

Wolves are killing too many of the Arctic Caribou, so the Wildlife Service assigns a naturalist to investigate. Farley Mowat is dropped alone onto the frozen tundra of Canada's Keewatin Barrens to live among the wolf packs to study their ways. His interactions with the packs, and his growing respect and understanding for the wild wolf will captivate all readers.

Nicky The Nature Detective

by Ulf Svedberg; illustrated by Lena Anderson
R&S Books/Farrar, Straus & Giroux, New York. 1983
Grades: 3–8

Nicky loves to explore the changes in nature. She watches a red maple tree and all the creatures and plants that live on or near the tree through the seasons of the year. Her discoveries lead her to look carefully at the structure of a nesting place, why birds migrate, who left tracks in the snow, where butterflies go in the winter, and many many more things. This book is packed with information.

On the Frontier with Mr. Audubon

by Barbara Brenner
Coward, McCann & Geoghegan, New York. 1977
Grades: 6–9

Based on an unedited diary of 1820-26, this fictionalized journal tells of Joseph Mason, a 13-year-old assistant who really traveled with John J. Audubon for 18 months. The work gives a detailed account of their daily life, hunting, drawing birds and their habitats, a stay at a plantation, and travel by flatboat, keelboat, and steamer. Black and white illustrations include reproductions of paintings and drawings by Audubon and other artists of his day.

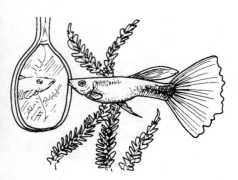

Once There Was a Tree

by Natalia Romanova; illustrations by Gennady Spirin
Dial Books/Penguin, New York. 1983
Grades: Preschool–5

Rich and detailed color illustrations trace the evolution of a tree which was struck by lightning, cut down, and reduced to a stump. The stump is visited, inhabited, or used by a succession of beetles, birds, ants, bears, frogs, earwigs, and humans who consider it theirs. As a new tree grows from the old stump, the question remains: "Whose tree is it?"

One Day in the Prairie

by Jean C. George; illustrated by Bob Marstall
Thomas Y. Crowell, New York. 1986
Grades: 4–7

An approaching tornado threatens a prairie wildlife refuge. The prairie dogs and other animals in the community sense and react to the danger. Delicate black and white drawings help capture the tension.

One Day in the Tropical Rain Forest

by Jean C. George; illustrated by Gary Allen
HarperCollins, New York. 1990
Grades: 4–7

When a section of rain forest in Venezuela is scheduled to be bulldozed, a young boy and a scientist seek a new species of butterfly for a wealthy industrialist who might preserve the forest. As they travel through the ecosystem rich with plant, insect, and animal life, everything they see on this one day is logged, beginning with sunrise at 6:29 a.m. They finally arrive at the top of the largest tree in the forest and fortuitously capture a specimen of an unknown butterfly.

Out in the Night

by Karen Liptak; illustrated by Sandy F. Fuller
Harbinger House, Tucson. 1989
Grades: 4–6

Nocturnal animals are shown in their habitats in 14 locations all over the world. Their habits are portrayed, including the sounds they make. The illustrations are especially good for botanical detail on the habitats ranging from a suburb of London to a desert in the Sahara.

Owl Moon

by Jane Yolen; illustrated by John Schoenherr
Philomel/Putnam, New York. 1987
Grades: Preschool–5

On a moonlit winter night, a father and daughter go on a search to see the elusive Great Horned Owl. The suspense of the hunt, along with the lyrical language and stunning illustrations of a rural scene at night make one feel a part of the journey. They seek the owl in its habitat, observe its behavior, and imitate its call.

The Roadside

by David Bellamy; illustrated by Jill Dow
Clarkson N. Potter, New York. 1988
Grades: 3–5

> Construction of a six-lane highway in a wilderness area disrupts the balance of nature and forces animals there to struggle for existence.

The Snail's Spell

by Joanne Ryder; illustrated by Lynne Cherry
Penguin, New York. 1988
Grades: Preschool–5

> You become a snail and in the process learn about the anatomy and locomotion of a snail. Though the picture-book format is excellent for younger students, older students will also enjoy imagining their life as a snail. Winner of the New York Academy of Science's Outstanding Science Book for Young Children Award.

The Song in the Walnut Grove

by David Kherdian; illustrated by Paul O. Zelinsky
Alfred A. Knopf, New York. 1982
Grades: 4–6

> A curious cricket meets a grasshopper. Together they learn of each other's daytime and nighttime habits while living in an herb garden. The friendship between them grows when the cricket rescues grasshopper from being buried in a pail of grain and they learn to appreciate each other's differences. This story weaves together very accurate accounts of insect behavior and their contributions to the ecology of Walnut Grove.

Wild Mouse

by Irene Brady
Charles Scribner's Sons, New York. 1976
Out of print
Grades: K–6

> This precisely written diary describes the behavior of a white-footed mouse who has babies. The entry for the day of the birthing begins: "He is a she! I pulled out the drawer of the coffee mill because I heard scratching inside and I'm watching a small miracle." The closely detailed observation, and lovely precise drawings, make this an excellent connection to this guide.

Will We Miss Them?

by Alexandra Wright; illustrated by Marshall Peck
Charlesbridge Books, Watertown, Massachusetts. 1992
Grades: 2–5

> A sixth grader writes about "some amazing animals that are disappearing from the earth." Each double-paged spread in the book asks the question: "Will we miss...?" and gives basic information on 13 animal species and how their habitats may be threatened. The illustrations are strong, a simple map shows approximate locations of threatened species, and the book presents the hopeful message that we don't have to miss these animals.

BUBBLE·FESTIVAL

Presenting Bubble Activities in a Learning Station Format

Grades K–6

*T*his GEMS Festival guide includes 12 class-room table-top activities with set-up instructions. Learning stations encourage independent thinking and cooperative learn-ing—bringing fun and excitement to the classroom. From "Bubble Shapes" and "Bubble Measurement" to "Bubble Skeletons" and "Body Bubbles," the most intriguing and classroom-appropriate bubble activi-ties are featured. The guide includes suggestions to assist teachers in flexibly presenting the challenges, tips on classroom logistics, ways to further explore scientific content, and writing and literature exten-sions. A section on setting up an all-school Bubble Festival is included.

Skills

Observing, Measuring, Recording, Experimenting, Cooperating, Classifying, Collecting and Analyz-ing Data, Noticing and Articulating Patterns, Predicting, Inferring, Drawing Conclusions

Concepts

Substances, Properties, Chemical Composition, Evaporation, Air Current Pressure, Surface Tension, Patterns, Polygons, Polyhedrons, Angles, Light, Color, Minimal Surface Area, Diameter, Radius, Volume

Themes

Scale, Structure, Matter, Energy, Stability, Patterns of Change, Systems & Interactions, Diversity & Unity

Mathematics Strands

Measurement, Geometry, Pattern, Functions

Notes: *Bubble Festival* differs from the GEMS teacher's guide *Bubble-ology* in several ways. *Bubble Festival* includes different activities and incorporates a learning station approach while *Bubble-ology* is a sequenced series of classroom sessions. Also, the grade range of *Bubble-ology* is 5–9.

MAKING CONNECTIONS

Included in the books listed below are several fantasy books involving bubbles appropriate for younger students. There is a question-answer type book focused on questions about **bubbles, soap, water,** and **other bath-related topics.** Several other books focus on the mathematical topics of **measure-ment** (using both standard and non-standard measur-ing tools), **volume,** and **topology,** and on the topic of **color,** all of which relate to the activities in *Bubble*

Festival. Don't miss *The Wise Woman and Her Secret,* a great book for preparing students for the activities in *Bubble Festival.* This book shows the **power of curiosity, asking questions,** and **making careful observations.**

CROSS REFERENCES

LITERATURE CONNECTIONS

Anno's Math Games III

by Mitsumasa Anno
Philomel Books/Putnam & Grosset, New York. 1991
Grades: 2–5

Picture puzzles, games, and simple activities introduce mathematical concepts and invite active participation. Measurement activities relate to the "Bubble Measurement" activity.

Bubble Bubble

by Mercer Mayer
Parents' Magazine Press, New York. 1973
Rainbird Books/Publishers Group West, Emeryville, California. 1991
Grades: Preschool–2

A little boy buys a magic bubble maker and creates and controls bubbles in the shapes of animals. The animals are progressively larger and more scary, so they appear to chase each other away. Watch for the surprise ending!

Hailstones and Halibut Bones: Adventures in Color

by Mary O'Neill; illustrated by John Wallner
Doubleday, New York. 1961, 1989
Grades: All

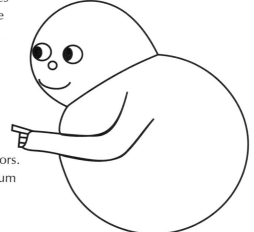

In 12 two-page poems the author presents her impression of various colors. Her perceptions go far beyond visual descriptions, painting a full spectrum of images. Good extension to the "Bubble Colors" activity.

How Big Is A Foot?

by Rolf Myller
Dell/Bantam, New York. 1962, 1990
Grades: K–5

When the king asks the apprentice carpenter to build the queen a bed for her birthday, he readily agrees and asks for the measurements. The king obliges and measures her bed using his feet. Somehow the bed that gets made is much smaller. This delightful story clearly shows the need for a standard unit of measurement. Good extension to the "Bubble Measurement" activity.

The Magic Bubble Trip
by Ingrid & Dieter Schubert
Kane/Miller Book Publishers, New York. 1981
Grades: K–3

> James blows a giant bubble that carries him away to a land of large hairy frogs where he has a fun, fanciful adventure.

Monster Bubbles
by Dennis Nolan
Prentice-Hall, Englewood Cliffs, New Jersey. 1976
Out of print
Grades: Preschool–1

> Various monsters take turns blowing bubbles in amounts from one to twenty. This simple, wordless book is a fun counting book for the youngest students.

Mr. Archimedes' Bath
by Pamela Allen
HarperCollins, New York. 1980
Grades: K–2

> Upset by his bath overflowing and puzzled by the changing water level, Mr. Archimedes first tries to blame one of his three bath companions (a kangaroo, a goat, and a wombat). He then resorts to scientific testing and measuring to find out about his bath. Good connections to the measurement activities in this guide, and to questions about volume and density.

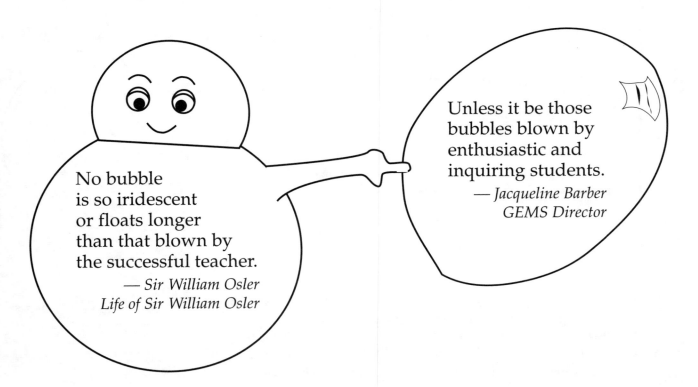

No bubble
is so iridescent
or floats longer
than that blown by
the successful teacher.
— *Sir William Osler*
Life of Sir William Osler

Unless it be those
bubbles blown by
enthusiastic and
inquiring students.
— *Jacqueline Barber*
GEMS Director

Rubber Bands, Baseballs and Doughnuts: A Book about Topology

by Robert Froman; illustrated by Harvey Weiss
Thomas Y. Crowell, New York. 1972
Out of print
Grades: 4–10

An introduction to the world of topology through active reader participation. Topics can be related to the behavior of soap film in the bubble skeletons, walls, and windows activities.

Splash! All About Baths

by Susan K. Buxbaum and Rita G. Gelman.; illustrated by Maryann Cocca-Leffler
Little, Brown, and Co., Boston. 1987
Grades: K–6

Before he bathes, Penguin answers his animal friends' questions about baths such as, "What shape is water?" "Why do soap and water make you clean?" "What is a bubble?" "Why does the water go up when you get in?" "Why do some things float and others sink?" and many other questions. Answers to questions are both clear and simple. Received the American Institute of Physics Science Writing Award.

The Wise Woman and Her Secret

by Eve Merriam; illustrated by Linda Graves
Simon & Schuster, New York. 1991
Grades: K–3

A wise woman who lives in the hills past the hollow is sought out by many people for her wisdom. They look for her secret in her barn and her house, but only little Jenny who lags and lingers and loiters and wanders finds it. As the wise woman tells her, "The secret of wisdom is to be curious—to take the time to look closely, to use all your senses to see and touch and taste and smell and hear. To keep on wandering and wondering." Great way to prepare your students for the experience of *Bubble Festival* by emphasizing and valuing the role of curiosity, asking questions, and using all the senses when gathering data. For the teacher, this book also provides a nice sense of the "discovery" element essential to hands-on science.

THROUGH A BUBBLE BRIGHTLY

Through a bubble brightly
Kaleidoscopic swirl
Prismatic soapy rainbows
Colorize the world
Translucent fragile bubbles
Bouncing on the breeze
Through a bubble brightly
No greater fun than these
Shining gemstone brilliant
Whirling wetly wild
Through a bubble brightly
Clear as eyes of child.

— *Lincoln Bergman*

Grades 5–9

Ten Sessions (Six activities)

Students combine fun with an exploration of important concepts in chemistry and physics through imaginative experiments with soap bubbles. Students devise an ideal bubble-blowing instrument; test dish-washing brands to see which makes the biggest bubbles; determine the optimum amount of glycerin needed for the biggest bubbles; employ the Bernoulli principle to keep bubbles aloft; use color patterns to predict when a bubble will pop; and create bubbles that last for days.

Activity 1

Students experiment to determine what objects/materials can be used to blow bubbles, and which make little and which make big bubbles. They design and draw bubble-makers for specialized uses.

Activity 2

Students experiment to determine which brand of dish-washing liquid makes the biggest bubbles. Introduces concept of a "fair test," and provides students with a way to quantify how well a soap solution makes bubbles. Students analyze results and make a bar chart on a chalkboard.

Activity 3

Students are introduced to some of the properties of bubble-making substances. Students observe how soap affects water's surface tension and investigate the role of evaporation in bubble formation, as they test what effect a differing amount of glycerin has on bubble size.

Activity 4

Students are introduced to a basic principle in aerodynamics, learning how Bernoulli's principle helps keep bubbles aloft (and helps enable planes to fly). Going Further includes a bubble obstacle course.

Activity 5

Using color changes as an important clue, students discover how to "Predict-a-Pop" by observing color sequences as they learn about the phenomena of constructive and destructive interference. Going Further includes research on Thomas Young and the combined wave/particle theory of light.

Activity 6

Students conduct open-ended experiments to determine how to make bubbles that last a long time. They keep data records and report results.

Skills

Observing, Measuring and Recording Data, Experimenting, Classifying, Drawing Conclusions, Controlling Variables, Calculating Averages, Graphing Results

Concepts

Technology, Engineering, Chemical Composition, Substances, Properties, Surface Tension, Hygroscopicity, Optimum Amount, Bernoulli's Principle, Aerodynamics, Pressure, Patterns, Light and Color, Interference, Air Currents, Evaporation, Environments

Themes

Systems & Interactions, Stability, Patterns of Change, Scale, Structure, Matter

MAKING CONNECTIONS

Several books pertain to **technology** and **inventions**, relating nicely to Activity 1. There are also two books which relate to **aerodynamics** and **dynamic lift**, a perfect complement to Activity 4. There's a book of poems about **color** providing a good extension to Activity 5. A question and answer type book focuses on bubbles, soap, water, and baths in general. An excellent "ecological mystery" explores many subjects, including phosphates in detergents. We expected to find many more connections to the wide-ranging bubble- and physical science-related activities in this guide, and look forward to your suggestions.

CROSS REFERENCES

LITERATURE CONNECTIONS

Better Mousetraps:
Product Improvements That Led to Success
by Nathan Aaseng
Lerner Publications, Minneapolis. 1990
Grades: 5–10

The book's focus is on "improvers, refiners, and polishers" and not on pioneers or trailblazers. To dramatize the results of safety testing, Elisha Otis set up an elevator at a big exposition in New York and had an assistant intentionally cut the cable with Otis aboard! The safety device brought the elevator to a halt in midfall. Getting heavy machinery to travel over muddy ground was the challenge faced by Caterpillar Tractor Company—what was learned in product development was applied to tank technology in World War I. The chapter on Eastman Kodak introduces the concept of a brand name, showing how Eastman promoted the names "Kodak" and "Brownie."

Chitty Chitty Bang Bang: The Magical Car
by Ian Fleming; illustrated by John Burningham
Alfred A. Knopf, New York. 1964
Grades: 6–Adult

> Wonderful series of adventures featuring a magical transforming car, an eccentric explorer and inventor, and his 8-year-old twins. It's a nice combination of technical and scientific information, much of it accurate, with a magical sense of how some machines seem to have a mind of their own. This one definitely does; it flies when it encounters traffic jams, becomes a boat when the tide comes in, senses a trap, and helps catch some gangsters. The inventing challenges in the book relate well to Activity 1.

Danny Dunn and the Heat Ray
by Jay Williams and Raymond Abrashkin; illustrated by Owen Kampen
McGraw-Hill, New York. 1962
Out of print
Grades: 4–6

> Danny and his friend explore various science fair project possibilities. They choose one that demonstrates how airplanes fly. The story includes an explanation of dynamic lift that relates perfectly to Activity 4.

Glorious Flight
by Alice and Martin Provensen
Viking Press, New York. 1983
Grades: 2–4

> This is the true story of Louis Bleriot, a pioneer of aviation, who developed and flew a plane over the English Channel in 1909. The evolution of the various prototypes of flying machines from Bleriot I which "flaps like a chicken," Bleriot II, a glider without a motor, Bleriot VII, a "real aeroplane that will fly," to Bleriot XI which makes the 36-minute flight is shown. The charming watercolor illustrations of the French village, the Bleriot family in period costume, and the very rudimentary aircraft emphasize the audacity of the attempt. Relates to Activity 4 of the GEMS guide, which explores principles of aerodynamics. Caldecott award winner.

Who sees with equal eye, as God of all,
A hero perish or a sparrow fall,
Atoms or systems into ruin hurl'd,
And now a bubble burst, and now a world.

— Alexander Pope
Essay on Man

Hailstones and Halibut Bones: Adventures in Color
by Mary O'Neill; illustrated by John Wallner
Doubleday, New York. 1961, 1989
Grades: All

> Twelve two-page poems of impressions of various colors. The perceptions go far beyond visual descriptions, painting a full spectrum of images. Connects nicely to Activity 5 of the GEMS guide, in which students very carefully observe bubble color changes.

The Missing 'Gator of Gumbo Limbo:
An Ecological Mystery
by Jean C. George
HarperCollins, New York. 1992
Grades: 4–7

> Sixth-grader Liza K and her mother live in a tent in the Florida Everglades. She becomes a nature detective while searching for Dajun, a giant alligator who plays a part in maintaining a waterhole's quality, and is marked for extinction by local officials. The ecology of the area is changing along with new populations, condominiums, the draining of the swamps, etc. One of the dangers noted by Liza and her neighbor James James is the presence of PCP (from herbicides) and phosphates (from detergents) in the water. "I checked the detergents in the supermarket to see if the companies really had gotten rid of the phosphates. It's almost true. Some laundry detergents have less than 0.5 percent, but some still have 5 and 6 percent." Ties in with Activities 2 and 3 of the GEMS guide, which deal with chemical composition of bubble solutions.

The Paper Airplane Book
by Seymour Simon; illustrated by Byron Barton
Viking Press, New York. 1971
Grades: 4–8

> Experiments in making paper airplanes with explanations of the aerodynamic principles involved, which connect well to Activity 4 of this GEMS guide.

The Rejects:
People and Products that Outsmarted the Experts
by Nathan Aaseng
Lerner Publications, Minneapolis. 1989
Grades: 5–10

> Part of a quirky series about inventors and innovators. Some of the rejects by "experts" include Graham Crackers, Birdseye, Xerox, and Orville Redenbacher popcorn. Redenbacher had a background in plant breeding and hired a genetics expert to improve upon popcorn.

Overheard
in a Bubble Chamber

You are
a magnetic monopole
with an intrinsic spin
of your own.

I am a quark
completely charmed.

May we occupy
the same quantum state
forever.

— Lillian Morrison

Ruby Mae Has Something To Say

by David Small
Crown Publishers, New York. 1992
Grades: 2–6

This zany saga traces Ruby Mae Foote's path from Nada, Texas, to the United Nations. Her message of world peace cannot be given until Billy Bob, Ruby Mae's nephew, transforms her tongue-tied and sometimes incomprehensible speech into earthshaking eloquence with a Rube Goldberg-type invention—a hat she wears called the Bobatron. The invention connects quite nicely with the bubble-maker activities. (The photo of the author on the book jacket wearing a Bobatron adds another laugh.)

Splash! All About Baths

by Susan K. Buxbaum and Rita G. Gelman; illustrated by Maryann Cocca-Leffler
Little, Brown and Co., Boston. 1987
Grades: K–6

Penguin answers his animal friends' questions about baths such as, "What shape is water?" "Why do soap and water make you clean?" "What is a bubble?" "Why does the water go up when you get in?" "Why do some things float and others sink?" and other questions. Answers to questions are both clear and simple. American Institute of Physics Science Writing Award winner.

The Toothpaste Millionaire

by Jean Merrill; illustrated by Jan Palmer
Houghton Mifflin, Boston. 1972
Grades: 5–8

Incensed by the price of a tube of toothpaste, twelve-year-old Rufus tries making his own from bicarbonate of soda with peppermint or vanilla flavoring. Assisted by his friend Kate and his math class (which becomes known as Toothpaste I), his company grows from a laundry room operation to a corporation with stocks and bank loans. Beginning on page 47, Rufus designs a machine for filling toothpaste tubes, which relates well to Activity 1 of the GEMS guide.

The Unsung Heroes:
Unheralded People Who Invented Famous Products

by Nathan Aaseng
Lerner Publications, Minneapolis. 1989
Grades: 5–10

This off-beat book tells the story of various products that changed our world and their originators. Includes Coca-Cola ("Dr. Pemberton's Back-yard Brew"), Bingo, Hoover Vacuum Cleaners ("The Sickly Janitor"), pneumatic tires (Dunlop Corporation), McDonald's, and others. The archival photographs are great fun and the positive message is that anyone's crazy idea might be a valid invention. Accompanies Activity 1 nicely, where students become inventors.

BUILD IT! FESTIVAL

Grades K–6

Eight Activities

*T*his wide assortment of classroom learning station activities focuses on mathematics, especially relating to construction, geometric challenges, and spatial visualization. Many activities connect strongly to the real world and potential careers. Introductory activities involve students in free exploration of the materials and lay the foundation for such mathematical challenges as Create-A-Shape, Dowel Designs, Polyhedra, Symmetry, Tangrams, and What Comes Next? Depending on the stations chosen, materials required include pattern blocks, polyhedra, and wooden cubes. Template patterns for folding and constructing shapes and creating tangrams are included. A background section on geometry is provided.

This festival guide can be adapted for Grades 7, 8, and above.

Introductory Activity

Pairs of students take turns being architects and builders as they design and build structures using pattern blocks. Students develop spatial concepts as the "architect" builds a structure and then gives directions to the "builder." The builder visualizes the structure and uses the spatial relationship descriptions to build the structure.

Activity 1

Can you visualize the flat shape of a cereal box before it was folded into a box and filled? This activity invites students to investigate dimension and structure. Using two-dimensional paper shapes, students explore the construction of three-dimensional shapes.

Activity 2

The builder in all students emerges as they create two- and three-dimensional structures out of newspaper dowels. These creations simulate the beams inside buildings. Using a recyclable material (newspaper), students roll dowels with the aid of tape and a piece of a plastic straw. After a number of dowels are made, students use masking tape to attach dowels to dowels to create shapes and structures.

Activity 3

Students construct three-dimensional creations at this learning station with high-interest building materials: snap-together triangles, squares, and pentagons. The figures that the students make serve as models of polyhedra. Polyhedra are three-dimensional, closed figures that are formed by polygonal surfaces.

Activity 4

Students fill shapes in a variety of ways with colorful pattern blocks. Using small wooden pattern blocks, students fill star, triangle, hexagon, and dodecagon shapes that are outlined on pieces of card stock. As students fill the same shape more than one way, they discover relationships among the pattern block shapes.

Activity 5

Students explore bilateral symmetry as they make designs with pattern blocks. Using a card that has only half of a design on it, students first fill in the existing half with pattern blocks. Using additional pattern blocks they complete the other half of the design.

Activity 6

Students figure out or "decode" patterns constructed out of pattern blocks and then continue building the pattern. The patterns vary from simple ones, such as an "ABAB" pattern (green triangle, orange square, green triangle, orange square, etc.) to more advanced patterns, such as the pattern of squares (one square, two-by-two square, three-by-three square, etc.). At each learning station, there can be a variety of patterns. After building pre-made patterns, students can have fun creating their own!

Activity 7

Students are given cards with tessellations on them. The students use pattern blocks to cover the tiling on the card. They then attempt to continue building the tessellation so it fills the entire board. Students can also use the pattern blocks to create their own tessellations.

Activity 8

Students make their own tangrams. They create sets of the seven "tans" either by cutting out a set of shapes from a master or by following directions on how to fold and cut a square sheet of paper. Before attempting to recreate the square, students have opportunities to create animals, letters, vehicles, buildings, designs, and their own unique images with the "tans."

Skills

Observing, Building Models, Constructing, Communicating, Finding and Articulating Patterns, Cooperating, Predicting, Finding Multiple Solutions, Drawing Conclusions, Learning Geometric Definitions, Problem Solving

Concepts

Shape Recognition, Polygons, Polyhedra (edges, faces, vertices), Spatial Visualization, Surface Area, Pattern, Symmetry, Congruence, Similarity, Architectural Design, Tessellation

Science Themes

Structure, Scale, Models & Simulations, Diversity & Unity

Mathematics Strands

Geometry, Logic and Language, Number, Discrete Mathematics, Functions

Nature of Science and Mathematics

Cooperative Efforts, Interdisciplinary Connections, Real-Life Applications

MAKING CONNECTIONS

Books that have to do with building or construction make excellent connections to *Build It! Festival*, including books at the appropriate age level about famous works of architecture and/or architects. The books by David Macaulay listed here are highly recommended. Several books that focus on decoding patterns are listed and connect nicely to Activity 6 and Activity 8. Any books in which geometric shapes, designs, symmetry, or spatial visualization play a part would fit in nicely. We welcome your additional suggestions, particularly for stories that describe a **process** of construction similar to what your students experience in the activities in this guide. You may find the article about tessellations on page 306 of interest.

CROSS REFERENCES

LITERATURE CONNECTIONS

Anno's Math Games III

by Mitsumasa Anno

Philomel Books/Putnam & Grosset, New York. 1991

Grades: 4–10

> Picture puzzles, games, and simple activities introduce the mathematical concepts of abstract thinking, circuitry, geometry, and topology. The book invites active participation. An exploration of triangles includes origami shapes, while a section on ever-popular mazes encourages logical thinking.

Block City

by Robert Louis Stevenson; illustrated by Ashley Wolff

Andersen Press, London. 1988

Grades: K–2

> Robert Louis Stevenson's poem "Block City" comes alive for young children through vibrant illustrations. Encourages imaginative building with blocks. Nice classic poetic connection to pattern block and other construction activities.

Bridges

by Ken Robbins

Dial Books, New York. 1991

Grades: K–5

> From delicate webs of steel spanning a vast river to stone arches reaching over a highway, bridges expand our world by joining one place with another. This book of hand-tinted photographs of bridges includes many types with descriptions of their design and use.

Castle

by David Macaulay

Houghton Mifflin, Boston. 1977

Grades: K–6

> One of a number of outstanding and beautifully illustrated books that focus on human structures, the work of building, architecture, shapes, and related content. Any and all of these books make a great connection to *Build It! Festival*. Other David Macaulay books in this category, with the same publisher, in the K–6 grade range are: *Cathedral* (1973); *City* (1974); *Pyramid* (1975); and *Unbuilding* (1980).

Chicken Soup With Rice

by Maurice Sendak

Scholastic, New York. 1986

Grades: Preschool–2

> This classic of rhyming verse about eating chicken soup throughout the year connects well to a discussion of repetitive patterns and the cycle of months in a year.

It was built
against the will
of the immortal gods,
and so it did not last
for long.

— *Homer*
The Iliad

Too low they build,
who build beneath
the stars.
— *Edward Young*
Night

Eight Hands Round: A Patchwork Alphabet

by Ann Whitford Paul; illustrated by Jeanette Winter
HarperCollins, New York. 1991
Grades: 2–6

Introduces the letters of the alphabet with the names of early American patchwork quilt patterns and explains the origins of the designs by describing the activity or occupations from which they came. The designs are rich in geometric patterns.

A Grain of Rice

by Helena C. Pittman
Bantam Books, New York. 1992
Grades: 2–5

A hard-working farmer's son wins the hand of the Emperor's daughter through clever use of mathematical knowledge of what results when a grain of rice is doubled every day for 100 days. Connects to Activity 6 and is an excellent literary introduction to the concept of exponential growth.

Grandfather Tang's Story

by Ann Tompert; illustrated by Robert A. Parker
Crown, New York. 1990
Grades: K–5

Grandfather tells Little Soo a story about shape-changing fox fairies who try to outdo each other until a hunter brings danger to both of them. The seven shapes that grandfather uses to tell the story are the pieces of an ancient Chinese puzzle, a tangram. Students can make their own tangrams, replicating the animals in the story or creating their own. This book is a wonderful and powerful way to connect mathematics to literature because in itself it embodies the connection, and because creating and solving tangrams is an involving activity for all ages.

If You Look Around You

by Fulvio Testa
Dial Books for Young Readers/E.P. Dutton, New York. 1983
Grades: K–3

Geometric shapes, two-dimensional and three-dimensional, points and lines are depicted in scenes of children in- and out-of-doors. Nice real-world connections to geometry.

Jim Jimmy James

by Jack Kent
Greenwillow Books/William Morrow, New York. 1984
Out of print
Grades: K–2

> One boring rainy day, Jim Jimmy James makes friends and plays with his shadow. A very elementary look at the concept of reflection. As a follow-up, children can partner with a friend and play shadow games with each other. Shadows and reflections are among the earliest phenomena related to shape and geometry that children experience. (Note: Some of the illustrations in the book are not accurate reflections.)

The Keeping Quilt

by Patricia Polacco
Simon & Schuster, New York. 1988
Grades: K–5

> A homemade quilt ties together the lives of four generations of an immigrant Jewish family, remaining a symbol of their enduring love and faith. Strongly moving text and pictures. A resource to begin a quilt project. Quilts are creative real-life examples of fitting shapes into a defined space, including tessellations—the intriguing mathematical and creative art of *exactly* fitting similar shapes into a defined space. Sidney Taylor award winner.

Tessellation Jingle

Build tessellations, bit by bit,
Repeating patterns, perfect fit,
Like checkerboards or bathroom tiles—
Make patterns stretch for miles and miles.
Remember, there can be no gap;
All shapes must fit, not overlap.

—Lincoln Bergman

The King's Chessboard

by David Birch; illustrated by Devis Grebu
Dial Books, New York. 1988
Grades: K–6

> A proud king learns a valuable (and exponential) lesson when he grants his wise man a request for rice that doubles with each day and square on the chessboard. Connects to Activity 6 and, more generally to the mathematics strands of number, pattern, and function.

My Cat Likes to Hide in Boxes

by Eve Sutton; illustrated by Lynley Dodd
Scholastic, New York. 1973
Grades: K–2

> This delightful book has rhymes about cats all over the world and "my cat" who likes to hide in boxes! The predictable pattern encourages reading participation. The idea of boxes and using shapes as homes is an early connection to structure and geometry.

Rome was not built
in one day.
— *John Heywood*
Proverbs

Opt: An Illusionary Tale
by Arline and Joseph Baum
Viking Penguin, New York. 1987
Grades: 2–6

A magical tale of optical illusions in which objects seem to shift color and size while images appear and disappear. You are an active participant in this book as you are guided through the land of Opt. Explanations of the illusions and information on how to make your own illusions are included.

The Paper Airplane Book
by Seymour Simon; illustrated by Byron Barton
Viking Press, New York. 1971
Grades: 3–6

A user-friendly book on the aerodynamics of airplanes, complete with instructions on how to construct paper airplanes. Emphasis on the structure of airplanes and how changes in structure/shape impact the forces in flight. Additional experiments are included.

The Patchwork Quilt
by Valerie Flournoy; illustrated by Jerry Pinkney
Dial Books, New York. 1985
Grades: K–5

Using scraps cut from the family's old clothing, Tanya helps her grandmother piece together a quilt of memories. When Grandma becomes ill, Tanya's whole family also gets involved in the project and they all work together to complete the quilt. Quilts, like geometry, are fascinating explorations of shapes and how they fit together.

The Phantom Tollbooth
by Norton Juster; illustrated by Jules Feiffer
Random House, New York. 1989
Grades: 2–8

Milo has mysterious and magical adventures when he drives his car past The Phantom Tollbooth and discovers The Lands Beyond. On his journey, Milo encounters amusing situations that involve numbers, geometry, measurement, and problem solving. The continuous play on words is delightful.

Round Trip
by Ann Jonas
Greenwillow Books/William Morrow, New York. 1983
Grades: K–3

Illustrated solely in black and white, this story of a trip between the city and the country is read at first in the standard way, then, on reaching the end, the book is flipped over as the story continues. Lo and behold, the illustrations turned upside down are transformed to depict the new scenes of the story. The strong black/white contrast helps provide a startling demonstration of the ways shapes and images fit into each other and can change, depending on one's perspective.

Rubber Bands, Baseballs and Doughnuts: A Book about Topology

by Robert Froman; illustrated by Harvey Weiss

Thomas Y. Crowell, New York. 1972

Out of print

Grades: 4–8

> This book introduces the world of topology through active reader participation. The activities provide concrete examples and insights into abstract concepts.

Sadako and the Thousand Paper Cranes

by Eleanor Coerr; illustrated by Ronald Himler

Dell Books, New York. 1977

Grades: 3–6

> In this true story, a young Japanese girl is dying of leukemia as a result of radiation from the bombing of Hiroshima. According to Japanese tradition, if she can fold 1,000 paper cranes, the gods will grant her wish and make her well, but she was able to fold only 644 paper cranes before she died. In her honor, a Folded Crane Club was organized and each year on August 6, members place thousands of cranes beneath her statue to celebrate Peace Day. This moving story can introduce a class origami project to make 1,000 cranes or other origami figures, and of course connects strongly to social studies and current events issues.

The Secret Birthday Message

by Eric Carle

Harper & Row, New York. 1986

Grades: Preschool–2

> Instead of a birthday package, Tim gets a mysterious letter written in code. Full-color pages, designed with cut-out shapes, allow children to fully participate in this enticing adventure. This book could serve as an exciting way to launch a series of lessons on shapes, which could also include a project where the students make "shape books."

Shadowgraphs Anyone Can Make

by Phila H. Webb and Jane Corby

Running Press, Philadelphia. 1991

Grades: K–6

> Illustrates how to make shadowgraphs of various animals and humans. A simple verse accompanies each shadowgraph. Students can create shadows and experiment with the size of the shapes by holding hands nearer or farther from the light.

The Shapes Game

by Paul Rogers; illustrated by Stan Tucker

Henry Holt & Co., New York. 1989

Grades: Preschool–2

> Fun-to-say riddles and pictures that are kaleidoscopes of brilliant colors take young children from simple squares and circles through triangles, ovals, crescents, rectangles, diamonds, spirals, and stars.

> **H**eaven is not reached
> at a single bound;
> But we build the ladder
> by which we rise
> From the lowly earth
> to the vaulted skies,
> And we mount to its summit
> round by round.
>
> — *Josiah Gilbert Holland*
> *Gradatim*

Shapes, Shapes, Shapes
by Tana Hoban
Greenwillow Books/William Morrow, New York. 1986
Grades: Preschool–5

Color photographs of familiar objects, such as a chair, barrettes, and manhole cover, are a way to study round and angular shapes.

Spaces, Shapes and Sizes
by Jane J. Srivastava; illustrated by Loretta Lustig
Thomas Y. Crowell, New York. 1980
Grades: 1–6

This inviting and well-presented nonfiction book about volume shows the changing forms and shapes a constant amount of sand can take. The book includes estimation activities, an investigation of volume of boxes using popcorn, and a displacement activity. The reader will want to try the activities.

The Tipi: A Center of Native American Life
by David and Charlotte Yue
Alfred A. Knopf, New York. 1984
Grades: 5–8

This excellent book describes not only the structure and uses of tipis, but Plains Indian social and cultural life as well. Some of the cultural language and oversimplification are less vital than they might be, but it is written in an accessible style. There are good charts, exact measurements, and information on the advantages of the cone shape. The central role played by women in constructing the tipi and in owning it are discussed. While this book includes some mention of the negative consequences of European conquest, noting that in some places tipis were outlawed, it is weak in this important area, and should be supplemented with other books.

There was an Old Man with a beard,
Who said: "It is just as I feared!
    Two Owls and a Hen,
    Four Larks and a Wren
Have all built nests in my beard."

— *Edward Lear*
Limerick

The Village of Round and Square Houses
by Ann Grifalconi
Little, Brown & Co., Boston. 1986
Grades: K–4

A grandmother explains to her listeners why the men live in square houses and the women live in round ones in their African village on the side of a volcano. The village of Tos really exists in the remote hills of the Cameroons. This book can begin an exploration of shape and structure. Caldecott Honor book.

Wings & Things
by Stephen Weiss; illustrated by Paul Jackson
St. Martin's Press, New York. 1984
Adult reference

Contains more than 30 paper origami models that fly. The great variety of shapes and flight patterns is especially appealing.

BUZZING A HIVE

Grades: K–3

Six Sessions

In this extensive and fascinating unit, students learn about the complex social behavior, communication, and hive environment of the honeybee by making paper bees, a bee hive, flowers with pollen, and bee predators. They also role-play bees in a bee hive drama and perform bee dances. (Live bees are not a part of this unit.)

Lesson 1

Students explore structure of the honeybee body, learn bee body parts, make paper honeybees, and put them on a paper sky.

Lesson 2

Students pretend cotton balls are bees, roll the balls in flowers to collect pollen, make paper flowers, place flowers on mural for bees to visit, and learn about pollen and nectar. The children then pretend they are bees (sucking fruit juice through straws) using their proboscis to obtain nectar to make honey.

Lesson 3

Students discover that a honeycomb is made of wax, learn how bees hang in chains while making honeycomb cells, give wax scales to their paper bees, and work together as a class to build a paper hive.

Lesson 4

Students learn about bee metamorphosis, role-play nurse bees feeding larvae, and find out what is hidden in a hive as they taste honeycomb and glue paper pollen, eggs, larvae, and pupae to the cells of their paper hive. Then children use their paper bees to dramatize young bees at work in the hive and older bees gathering food outside.

Lesson 5

Students learn about bee predators and honey robbers, hear a Bee Enemies story, and make a paper skunk. They role-play guard bees and learn how bees work together to protect the hive.

Lesson 6

Students learn more about pollen gathering, the work of older bees, role-play bee school (to learn how bees use landmarks to find their way back to hive), flying in a beeline, and communicating messages with dances.

Skills

Observing, Comparing, Matching, Communicating, Role Playing

Concepts

Biology, Entomology, Honeybee (Structure, Pollen, Nectar, Hive, Metamorphosis, Life Cycle, Enemies, Protection, Social Organization, Cooperation, Communication, Flight Pattern)

Themes

Systems & Interactions, Models & Simulations, Stability, Patterns of Change, Evolution, Structure, Energy, Matter, Diversity & Unity

Notes: Has modifications for preschool and kindergarten, numerous art activities, large posters, drama (such as children playing nurse bees and baby bees), as well as information on the parts of a flower, the skunk as a predator, honeycomb, bee dances, and "beeline."

MAKING CONNECTIONS

This GEMS unit guides children through the fascinating world of a **honeybee community**, their **food gathering, life cycle, defenses,** and **interdependence with flowers**. We thought there would be tons of literature on bees as the focus, but found most of the connections to involve the **ecology** of the honeybee or how a bee's life is intertwined with other living things. *In the Tall, Tall Grass* and *The Rose in My Garden* use rhyming verse to illustrate the interconnectedness of meadow life with bees as important players. *Where Butterflies Grow* describes the butterfly life cycle; a nice comparison to bee **metamorphosis**. And in a classic book, **honey** is central to a story about a familiar bear and his encounter with bees.

CROSS REFERENCES

HONEYBEES

Being a bee   Being a bee
           is a joy.
is a pain.
           I'm a queen
I'm a worker
I'll gladly explain.   I'll gladly explain.
           Upon rising, I'm fed
           by my royal attendants,
I'm up at dawn, guarding
the hive's narrow entrance
           I'm bathed
then I take out
the hive's morning trash
           then I'm groomed.
then I put in an hour
making wax
without two minute's time
to sit still and relax.
           The rest of my day
           is quite simply set forth:
Then I might collect nectar
from the field
three miles north
           I lay eggs,
or perhaps I'm on
larva detail
           by the hundred.

feeding the grubs
in their cells,
wishing that I were still
helpless and pale.
           I'm loved and I'm lauded,
           I'm outranked by none.
Then I pack combs with
pollen—not my idea of fun.
           When I've done
           enough laying
Then, weary, I strive
           I retire
to patch up any cracks
in the hive.
           for the rest of the day.
Then I build some new cells.
slaving away at
enlarging this Hell,
dreading the sight
of another sunrise,
wondering why we don't
all unionize.

Truly, a bee's   Truly, a bee's
is the   is the
worst   best
of all lives.   of all lives.

— *Paul Fleischman*
Joyful Noise: Poems for Two Voices

LITERATURE CONNECTIONS

The Flower Alphabet Book

by Jerry Pallotta; illustrated by Leslie Evans
Charlesbridge Press, Watertown, Massachusetts. 1988
Grades: Preschool–2

> Beautiful alphabet picture book showing
> many varieties of flowers and plants in
> accurate detail. Would make a good early
> primary accompaniment to the meadows
> ideas in *Hide a Butterfly* and the activities
> relating to nectar, pollen, and flowers in
> *Buzzing a Hive.*

A House is a House for Me

by Mary Ann Hoberman;
illustrated by Betty Fraser
Viking Penguin, New York. 1978
Grades: K–3

> Lists in rhyme the dwellings of various
> animals, peoples, and things such as a
> shell for a lobster or a glove for a hand.
> Nice extension to "Lesson 3: Building A
> Bee Hive."

In the Tall, Tall Grass

by Denise Fleming
Henry Holt and Co., New York. 1991
Grades: Preschool–3

> Rhyming text and vibrant collage
> illustrations look at the world of
> creatures you might see in the
> long, tall grass. Their behaviors
> are captured in catchy rhymes as
> caterpillars lunch, hummingbirds sip and
> dip, bees strum, ants lug, and moles ritch,
> ratch and scratch.

Michael Bird-Boy

by Tomie dePaola
Simon & Schuster, New York. 1975
Grades: K–3

> A young boy who loves the countryside determines to
> find the source of the black cloud that hovers above it.
> When he discovers the source of this pollution, a factory
> making "genuine" artificial honey syrup, he helps the
> "boss-lady" set up beehives so she can make natural
> honey without creating pollution.

Pretend You're A Cat

by Jean Marzollo; illustrated by Jerry Pinkney
Dial Books, New York. 1990
Grades: Preschool–1

Wonderful illustrations and friendly verses ask children to pretend they are different animals and to act out the animal's behavior. "Can you buzz? Are you covered with fuzz?" Great springboard to a discussion of similarity and differences among animal behaviors. Animals include a cat, pig, snake, bear, horse, seal, and bee.

The Rose in My Garden

by Arnold Lobel; illustrated by Anita Lobel
Greenwillow Books, New York. 1984
Grades: Preschool–2

Each page adds a new rhyming line to a poem as a beautiful garden of flowers, insects, and animals grows. A surprise interaction among the garden residents takes place at the end of the book. Young readers will enjoy the repeated patterns in the story.

When I'm Sleepy

by Jane R. Howard; illustrated by Lynne Cherry
Dutton, New York. 1985
Grades: Preschool–2

A young girl speculates about sleeping in places other than her bed and is shown sleeping with twelve different animals in a nest, a swamp, standing up, hanging upside down, etc. The witty, glowing illustrations and active verbs capture a wide range of habitats and possibilities.

Where Butterflies Grow

by Joanne Ryder; illustrated by Lynne Cherry
Lodestar Books, New York. 1989
Grades: K–5

Here's an imaginative description of what it might feel like to grow from a tiny egg into a black swallowtail butterfly. Structure, metamorphosis, locomotion, camouflage, and feeding behaviors are all described from the point of view of the butterfly. There's also a page of gardening tips on how to attract butterflies. Outstanding illustrations include detailed drawings of metamorphosis.

Winnie-The-Pooh

by A.A. Milne; illustrations by Ernest H. Shepard
E.P. Dutton, New York. 1926
Dell Publishing, New York. 1954
Grades: K–5

This well-loved classic contains chapter after delightful chapter of the adventures of Christopher Robin, Pooh, and their animal friends. Relevant chapters include one in which bees are encountered; several in which logic is applied to solve a problem; and an adventure in navigation.

CHEMICAL·REACTIONS

Grades 6–10   **Two Sessions**

An ordinary ziplock bag becomes a safe and spectacular laboratory, as students mix chemicals that bubble, change color, get hot, and produce gas, heat, and an odor. They experiment to determine what causes the heat in this chemical reaction. This exciting activity explores chemical change, endothermic and exothermic reactions, and is a great introduction to chemistry.

Session 1

Discussion of what might happen when chemicals are mixed together with emphasis on safety. Students observe each chemical to be mixed, then put baking soda and calcium chloride into the ziplock bag, then add phenol red solution. Students record at least five observation of what happens. The concept of a chemical reaction is introduced and discussed.

Session 2

Students are challenged to conduct further experiments to determine what causes the heat in the chemical reaction. Results are discussed, with students drawing their own conclusions.

Skills

Observing, Recording Data, Experimenting, Making Inferences

Concepts

Chemistry, Evidence of Chemical Reaction, Endothermic/Exothermic Reactions, Chemical Safety

Themes

Systems & Interactions, Stability, Patterns of Change, Energy, Matter

Notes: Section on safety considerations, and background on the chemical nature of the reaction.

MAKING CONNECTIONS

Most of the books involve the use of **chemistry and chemical reactions** as a tool to solving a mystery or problem in a fictional situation. Occasionally the chemistry gets people in trouble, but, just as in real life, it's usually the injudicious use of chemistry that creates the problem! For the most part, these books are appropriate for students at the younger end of the age range for this guide.

Though chemistry is not the focus of any of these books, certain key concepts such as combining **reactants** to form new **products**, the need to **record data**, the idea that **more is not always better**, and that **some things are irreversible** appear throughout the stories. Also, fun kid-use of **chemical names** and **formulas** can make this sometimes intimidating aspect of chemistry friendlier.

We are on the lookout for good examples of books for older students that use chemistry to solve real problems or mysteries, but so far we haven't found them. Again, we welcome your help!

CROSS REFERENCES

LITERATURE CONNECTIONS

Everything Happens to Stuey
by Lilian Moore; illustrated by Mary Stevens
Random House, New York. 1960
Out of print
Grades: 4–7

> After smelling up the refrigerator with his secret formula, turning his sister's doll green with a magic cleaner, and having his invisible ink homework go awry, budding chemist Stuey is in trouble. In the end, he uses his knowledge to rescue his sister by fabricating a homemade flashlight. The illustrations, depiction of family life, and sex roles are dated, but the spirit of adventure is timeless.

Gorky Rises
by William Steig
Farrar, Straus & Giroux, New York. 1980
Grades: 2–5

> When Gorky's parents leave the house, he sets up a laboratory at the kitchen sink and mixes up a liquid mixture with a few secret ingredients— his mother's perfume and his father's cognac! The liquid proves to have magical properties which allow him to fly over the world. Students in Session 1 mix chemicals to make something new just as Gorky does. Although it is a picture-book format, the content makes it usable for older students.

June 29, 1999

by David Wiesner

Clarion Books, Houghton Mifflin, New York. 1992

Grades: 3–6

The science project of Holly Evans takes an extraordinary turn—or does it? This highly imaginative and humorous book has a central experimental component, and conveys the sense of unexpected results—one of the lessons of the reaction in Session 1.

The Lady Who Put Salt in Her Coffee

by Lucretia Hale

Harcourt, Brace, Jovanovich, San Diego. 1989

Grades: K–6

When Mrs. Peterkin accidentally puts salt in her coffee, the entire family embarks on an elaborate quest to find someone to make it drinkable again. Visits to a chemist, an herbalist, and a wise woman result in a solution, but not without having tried some wild experiments first.

The Monster Garden

by Vivien Alcock

Delacorte Press, New York. 1988

Grades: 5–8

Frankie Stein creates her own special monster from a "bit of goo" her brother steals from the lab. Scientific information is sprinkled throughout the book and Chapter II includes Frankie's experiment log. The book is a combination of fantasy, science fiction, and young adult novel with a strong main character, an arrogant older brother, and a "friend" who spills the secret. It stimulates thinking about the complex issues surrounding biotechnology and genetic engineering. Students who read this book after doing the GEMS *Chemical Reactions* activities could be asked to write or diagram a short and imaginative explanation for the chemical reactions that might have taken place as Frankie's monster was formed.

Susannah and the Poison Green Halloween

by Patricia Elmore; illustrated by Joel Schick

E.P. Dutton, New York. 1982

Grades: 5–7

Susannah and her friends try to figure out who put the poison in their Halloween candy. Tricky clues, changing main suspects, and some medical chemistry make this an excellent choice, with lots of inference and mystery.

COLOR ANALYZERS

Grades 5–9 **Four Sessions**

Students investigate light and color while experimenting with diffraction gratings and color filters. They use color filters to decipher secret messages, then create their own secret messages. A class set of red and green filters and diffraction gratings is included.

Session 1

Color analyzers are made for the class—by teacher or teacher and a few students—from filters, diffraction grating, and index cards. Students look at a secret message with green and red filters, then color in and decode other secret messages.

Session 2

Based on their experience in Session 1, students make a rule for how to make secret messages detectable with the red filter, then originate their own messages. They discuss how scientists use colored light filters.

Session 3

Students use a diffraction grating to observe that white light is actually made up of the color spectrum, then discuss where they think the colors come from. They discuss how scientists use diffraction gratings.

Session 4

Color filters and diffraction gratings are reviewed, then a demonstration/discussion follows to figure out the answer to the question "Why does an apple look red?"

Skills

Observing, Comparing, Describing, Classifying, Inferring, Predicting, Recording Data, Drawing Conclusions

Concepts

Physical Science: Properties of Light and Color, Color Filters, Diffraction Gratings

Themes

Systems & Interactions, Energy, Matter

Notes: Background information on light, how our eyes see color, diffraction/ interference, how a diffraction grating works, and how it differs from a prism. The book cover is part of the class activity, revealing a secret word when viewed through the proper filter. The back cover photograph of a galaxy demonstrates how astronomers sometimes view stars through filters to highlight parts of the image. Going Further activities include making a Mystery Box, Colored Wheels, and making a rainbow in the classroom.

MAKING CONNECTIONS

Many of the books are folktales and are connections since they are about colors. They introduce opportunities to extend the scientific study of light and color to discussions about color in social and cultural contexts. Even though some of these books were written for younger children, your 5–8 grade students may enjoy them too.

Be on the lookout for books about sunglasses, color-blindness, color vision in humans and animals, how certain colors absorb and others reflect light, rainbows, prisms, and different kinds of secret codes comparable to the secret messages students create in this unit. Send us your nominations for inclusion in the next edition of this handbook!

LITERATURE CONNECTIONS

The Adventures of Connie and Diego

by Maria Garcia; illustrated by Malaquias Montoya
Children's Book Press, San Francisco. 1987
Grades: K–5

The twins Connie and Diego are born different than all the other children because they have "colors all over their little bodies." The other children laugh and laugh so one day the twins decide to run away in search of "a place where no one will make fun of us." They encounter a bear, a whale, a bird (who loves their colors), and a tiger. They learn that although their surface appearance is different from that of other children, they are human beings, no matter what color they may be. This strong anti-racist message, while designed for younger students, becomes a lesson for people of all ages (and colors). This book could also be part of a class discussion about the colors we see, pigments, and the role that skin color plays in society. The English and Spanish verse appear together on each page.

Colors

by Gallimard Jeunesse and Pascale de Bourgoing;
illustrated by P.M. Valet and Sylvaine Perols
Cartwheel/Scholastic, New York. 1991
Grades: Preschool–2

You mix the colors of the rainbow by using the transparent color overlays and vibrant illustrations of animals, flowers, food, etc. Although designed for early grades, the hands-on pages of this book create several "discovery" experiments that could be of interest to older students as well.

CROSS REFERENCES

All the colors of the race

All the colors of the race
are
       in my face, and just behind my face:
             behind my eyes:
    inside my head.

And inside my head, I give my self a place
    at the end of a long
          line   forming
              it self into a
                circle.

And I am holding out my hands.

— Arnold Adoff

Hailstones and Halibut Bones: Adventures in Color

by Mary O'Neill; illustrated by John Wallner

Doubleday, New York. 1961, 1989 (new edition)

Grades: All

> Twelve two-page poems of impressions of various colors. The perceptions go far beyond visual descriptions, painting a full spectrum of images. In the GEMS *Color Analyzers* activities, students learn how white light is made up of light of different colors. They use red and green filters and diffraction grating to learn more about light and color. Many teachers have highly recommended these poems as an excellent literary accompaniment to the GEMS unit.

How the Birds Changed Their Feathers

by Joanna Troughton

Blackie & Son, London. 1976

Peter Bedrick Books, New York. 1991

Grades: K–4

> This South American tale tells how the birds used to be all white and then came to have different colors. When the cormorant kills the huge Rainbow Snake, the people skin it and jokingly challenge the bird to carry the skin if he wants to keep it. All the birds join together to carry it through the air and each bird keeps the part of the skin it has carried then its feathers become that color. The cormorant carries the head and becomes mostly black.

Dear GEMS

I am a fourth grade teacher at Windrush School in El Cerrito, California, and a BIG GEMS fan. I was just reading your newsletter, specifically about integrating literature with GEMS units. I love to do this and have a couple of suggestions.

With *Color Analyzers* I use the poetry book *Hailstones and Halibut Bones* by Mary O'Neill (Doubleday, 1961). We read the poems together, which cleverly relate colors to each other ("Pink is the niece of red") or some such idea, while beautifully describing things of that color. Great illustrations too! Then the kids write their own poems about a color. In art we're making our own color wheels by mixing patches of watercolors on a large sheet. We begin with the pure primary, then secondary, then add colors to those, so they get a huge variety. When these are dry, the kids cut the patches out in a shape they choose and arrange them on a sheet of black. The colors flow from one to the next as on a color wheel.

The Legend of the Indian Paintbrush

by Tomie dePaola
G.P. Putnam's Sons, New York. 1988
Grades: K–4

> After a Dream Vision, the Plains Indian boy Little Gopher is inspired to paint pictures as pure as the colors in the evening sky. He gathers flowers and berries to make paints but can't capture the colors of the sunset. After another vision, he goes to a hilltop where he finds brushes filled with paint, which he uses and leaves on the hill. The next day, and now every spring, the hills and meadows are ablaze with the bright color of the Indian Paintbrush flower.

The River That Gave Gifts, an Afro-American Story

by Margo Humphrey
Children's Book Press, San Francisco. 1987
Grades: K–5

> Four children in an African village make gifts for wise old Neema while she still has partial vision. Yanava, who is not good at making things, does not know what to give, and seeks inspiration from the river. As she washes her hands in the river, rays of light fly off her fingers, changing into colors and forming a rainbow. After all the other gifts are presented, she rubs her hands in the jar of river water giving a rainbow and the gift of sight to Neema. In addition to the themes of respect for elders and the validity of different kinds of achievement, the color theme evolves into evoking symbolism of each color in the rainbow. The book could be used as the start of a comparison of the scientific view of colors with the ways color is viewed and used in different cultures and art. This book could also help prompt an open-ended discussion about the relationship of **light** to **color**.

With *Fingerprinting* and *Crime Lab Chemistry*, I read aloud a mystery, for example, *From the Mixed-Up Files of Mrs. Basil E. Frankweiler* by E.L. Konigsburg (MacMillan, 1967) or one of John Bellair's books. We talk about clues, fingerprints, etc. Also last fall when we did these GEMS units, I had an El Cerrito policeman/detective visit our classroom and tell how he used clues, prints, etc. The children wrote their own mysteries, first in groups using a bag of "clues" I'd given them and then on their own, incorporating secret codes/ciphers, fingerprints, etc.

I know I've gotten off the track of using literature with GEMS, but I do find these units integrate so well with many curriculum areas. I was one of the trial testers for *Bubble Festival* this spring—we wrote bubble poems and did bubble art. GEMS units are terrific in themselves, but also as a springboard into many other areas.

THANK YOU!

Martha Vlahos

CONVECTION

A CURRENT EVENT

Grades 6–9
Three Sessions

Students explore this important physical phenomenon by observing and charting the convection currents in liquid. They explore convection in air and generalize their findings to describe wind patterns. Convection is related to the ways heat moves and to the movement of magma inside the earth.

Session 1

Students conduct three experiments, observing how food coloring moves through water that is being heated. They record their observations on a data sheet and share results in a class discussion.

Session 2

Through discussion, the students are able to generalize from their observations to describe a complete convection current. They apply their knowledge to guide an imaginary submarine through currents generated by a volcanic vent. They test their hypotheses by using drops of color as submarines.

Session 3

Three teacher demonstrations show how convection occurs in gases as well as liquids. Carbon dioxide gas is poured down a cardboard ramp to extinguish a flame; a flavored extract is heated and diffuses the scent in the room. Students apply what they've learned to predict air flow in a room and wind patterns in the environment.

Skills

Observing, Recording, Making Inferences, Applying, Generalizing

Concepts

Heat, Heat Transfer, Convection, Diffusion, Fluids, Wind, Ocean Currents

Themes

Systems & Interactions, Models & Simulations, Stability, Patterns of Change, Evolution, Scale, Structure, Energy, Matter, Diversity & Unity

Notes: Going Further includes activities and information on the movement of magma inside the Earth, continental drift, and convection inside the Sun. Related topics for literature connections include volcanoes, earthquakes, and submarines.

MAKING CONNECTIONS

Many of the books are about volcanoes. Volcanoes are connected to convection in the following way: The earth's crust is fractured into large segments called tectonic plates that move slowly, driven in part by convection currents in the magma they float on. Most volcanoes occur in the places where these big plates collide, rub together or separate.

Books about earthquakes could be connected in the same way, but we haven't found any.

Look for stories in which convection currents in water or air play a part, including books about sailboats, gliders, or birds. We found two books, one about kites (wind is caused by convection), and another about heat transfer and air conditioning.

CROSS REFERENCES

LITERATURE CONNECTIONS

Catch the Wind: All About Kites
by Gail Gibbons
Little, Brown & Co., Boston. 1989
Grades: K–6

When two children visit Ike's Kite Shop they learn about kites and how to fly them. Instructions for building a kite are given. The book relates especially to "Going Further" activities in which students do a thought experiment to determine where the best hang-gliding spot would be.

A Chilling Story: How Things Cool Down
by Eve and Albert Stwertka; illustrated by Mena Dolobowsky
Julian Messner/Simon & Schuster, New York. 1991
Grades: 4–8

Refrigeration and air conditioning are simply explained, with sections on heat transfer, evaporation, and expansion. Humorous black and white drawings show a family and its cat testing out these principles in their home.

Hill of Fire
by Thomas P. Lewis; illustrated by Joan Sandin
Harper & Row, New York. 1971
Grades: 2–5

A Mexican villager plowing a field opened up a crack in the earth that erupted within days into a new volcano, Paricutin. While told in simple language, the story is still appropriate for older students who are studying volcanoes and the concept of convection. The power of the volcano and the impact of change are strongly conveyed. The actual 1943 event is described in a historical note. This was only the second time in recorded history that the birth of a volcano has been directly witnessed by humans.

Magic Dogs of the Volcanoes

by Manlio Argueta; illustrated by Elly Simmons
Children's Book Press, San Francisco. 1990
Grades: 1–4

When the traditional magic dogs who protect the people of El Salvador, and who live on top of ancient volcanoes, are pursued by soldiers, the volcanoes play a trick. The male volcano fans himself with his steam hat, making the earth hot. The female volcano shakes her dress made of water and makes the soldiers wet so they sizzle and melt. This is an imaginative, multicultural extension to GEMS activities which introduce convection as one of the forces that contribute to volcanic eruption. Text in English and Spanish.

The Magic School Bus Inside the Earth

by Joanna Cole; illustrated by Bruce Degen
Scholastic, New York. 1987
Grades: K–6

On a special field trip to the center of the earth, Ms. Frizzle's class learns firsthand about different kinds of rocks and the formation of the earth and its structure. Reading this book would be a good introductory way for interested students to continue exploring topics touched on by the GEMS convection activities, and to begin learning more about the different geological layers and forces at work inside the Earth.

Paul's Volcano

by Beatrice Gormley; illustrated by Cat B. Smith
Avon Books, New York. 1987
Grades: 5–8

When Adam and Robbie see Paul's science fair project, a model of a volcano (complete with smoke and eruption sound track), they decide that it must become the symbol of their new club. The "Vulcans" conduct rituals with the model volcano, chanting their password "Magma, Magma" as they prepare to march in the July 4th parade. But mysterious, dangerous forces seem to be at work. What begins as a playful imitation of legends about people sacrificed to volcanoes turns into a series of unexplained and bizarre events, fear, and a final conflagration. Qualities of leadership and the meaning of accomplishment are explored as the strange events surge like lava down a mountainside. There is some scientific information throughout, including a description of the eruption of Mount St. Helens. In the end, the spirit of friendship triumphs over the evil genie of the volcano.

Volcano:
The Eruption and Healing of Mount St. Helens

by Patricia Lauber
Bradbury Press, New York. 1986
Grades: 4–7

Photographic account of how and why Mount St. Helens erupted in May 1980 and the destruction it caused. Two chapters discuss the survivors and colonizers, and the plant and animal life that returned to the area. Dormant volcanoes and their mechanics are explained in Chapter 5 with a positive look at the creative effects of an eruption. Although convection is not referred to directly, Chapter 5 includes a good basic introduction to the key topic of plate tectonics.

Who Has Seen the Wind?

Who has seen the wind?
    Neither I nor you;
But when the leaves lie trembling,
    The wind is passing through.

Who has seen the wind?
    Neither you nor I;
But when the trees bow down their heads,
    The wind is passing by.

—*Christina G. Rossetti*

CRIME LAB CHEMISTRY

Grades 4–8

Two Sessions

Challenged to determine which of several black pens was used to write a ransom note, students learn and use paper chromatography. Several mystery scenarios are suggested, with intriguing characters.

Notes: Using "candy chromatography," students solve an additional mystery using Reese's Pieces, and both dark-brown and light-brown M&Ms. The challenge is to figure out which of three people used the candy machine.

Session 1

Teacher explains plot and suspects, saying a ransom note was written, suspects named, and their pens seized. Chromatography is explained and demonstrated, then students receive pieces of the ransom note and conduct tests. Pigments, substances, mixture, and chromatogram are discussed and defined. Students are challenged to use chromatography to find out which pen was used to write the ransom note.

Session 2

The chromatograms are analyzed, now that they are dry, and students discuss who they think the culprit was and other results. Students and teacher discuss why some colors travel farther than others.

Skills

Experimenting, Analyzing Data, Making Inferences

Concepts

Chemistry, Chromatography, Separating Mixtures, Pigments, Solubility

Themes

Systems & Interactions, Stability, Patterns of Change, Structure, Energy, Diversity & Unity

MAKING CONNECTIONS

All of the books below depict **mysteries**. While some are simple and others more complex, they all share in the collection of **evidence** or clues, and use analysis to make **inferences** or conclusions. The distinction between evidence and inference is made in the *Crime Lab Chemistry* unit and comes alive through application to a diverse collection of mysterious situations, as in these stories. Several of the books involve the use of **chromatography** or its related concepts, such as **solubility**.

Some teachers use newspaper articles with their students, which describe a crime (usually unsolved), the evidence, and some possible inferences. There are books containing **nonfiction accounts of mysteries or scientific discoveries**, in which detective-like behavior was required. Such a mystery would be particularly apt if it involved chromatographic analysis of evidence.

There are numerous other books that focus on solving mysteries. The process of science is, after all, parallel to that of detection and making inferences to solve a problem. These other books include several well-known collections such as the "Encyclopedia Brown" series and "One-minute mysteries." You and your students probably have your own favorites.

We would especially welcome hearing about those that include details of scientific tests and evaluation of evidence similar to those in the GEMS guides.

CROSS REFERENCES

LITERATURE CONNECTIONS

Cam Jensen and the Mystery of the Gold Coins
by David A. Adler; illustrated by Susanna Natti
Viking Press, New York. 1982
Dell Publishing, New York. 1984
Grades: 3–5

Cam Jensen uses her photographic memory to solve a theft of two gold coins. Cam and her friend Eric carry around their 5th grade science projects throughout the book and the final scenes take place at the school science fair. (Other titles in this series include *Cam Jensen and the Mystery at the Monkey House* and *Cam Jensen and the Mystery of the Dinosaur Bones* in which she notices that three bones are missing from a museum's mounted dinosaur.)

Chip Rogers: Computer Whiz

by Seymour Simon; illustrated by Steve Miller
William Morrow, New York. 1984
Out of print
Grades: 4–8

> Two youngsters, a boy and a girl, solve a gem theft from a science museum by using a computer to classify clues. A computer is also used to weigh variables in choosing a basketball team. Although some details about programming the computer may be a little dated, this is still a good book revolving directly around sorting out evidence, deciding whether or not a crime has been committed, solving it, and demonstrating the role computers can play in human endeavors. By the author of the Einstein Anderson series.

Einstein Anderson Science Sleuth

by Seymour Simon; illustrated by Fred Winkowski
Viking Press, New York. 1980
Grades: 4–7

> In the "Universal Solvent" chapter, Einstein Anderson's friend Stanley tries to convince him that the cherry soda-looking liquid he has invented will dissolve anything. In Session 2 of this GEMS guide, students analyze chromatograms looking at variables such as the type of solvent used and the test substance used.

From the Mixed-Up Files of Mrs. Basil E. Frankweiler

by E.L. Konigsburg
Atheneum, New York. 1967
Dell Publishing, New York. 1977
Grades: 5–8

> Twelve-year-old Claudia and her younger brother run away from home to live in the Metropolitan Museum of Art and stumble upon a mystery involving a statue attributed to Michelangelo. This book is a classic, and has been recommended to GEMS by many teachers.

The Great Adventures of Sherlock Holmes

by Arthur Conan Doyle
Viking Penguin, New York. 1990
Grades: 6-Adult

> These classic short stories are masterly examples of deduction. Many of the puzzling cases are solved by Holmes in his chemistry lab as he analyzes inks, tobaccos, mud, etc. to solve the crime and catch the criminal. Nearly every Sherlock Homes story is suitable for this GEMS guide. These stories are available from many different publishers and in many editions.

The Missing 'Gator of Gumbo Limbo: An Ecological Mystery

by Jean C. George
HarperCollins, New York. 1992
Grades: 4–7

Sixth-grader Liza K and her mother live in a tent in the Florida Everglades. She becomes a nature detective while searching for Dajun, a giant alligator who plays a part in a waterhole's oxygen-algae cycle, and is marked for extinction by local officials. She is motivated to study the delicate ecological balance by her desire to keep her outdoor environment beautiful.

Motel of the Mysteries

by David Macaulay
Houghton Mifflin, Boston. 1979
Grades: 6–Adult

Presupposing that all knowledge of our present culture has been lost, an amateur archeologist of the future discovers clues to the lost civilization of "Usa" from a supposed tomb, Room #26 at the Motel of the Mysteries, which is protected by a sacred seal ("Do Not Disturb" sign). *Motel of the Mysteries* is an elaborate and logically constructed train of inferences based on partial evidence, within a pseudo-archaeological context. Reading this book, whose conclusions they know to be askew, can encourage students to maintain a healthy and irreverent skepticism about their own and other's inferences and conclusions, while providing insight into the intricacies and pitfalls of the reasoning involved. An extended review of *Motel of the Mysteries* is on pages 160–161.

Big Daddy Morebucks

The Mystery of the Stranger in the Barn

by True Kelley
Dodd, Mead, & Co., New York. 1986
Grades: K–4

A discarded hat and disappearing objects seem to prove that a mysterious stranger is hiding out in the barn, but no one ever sees anyone. A good opportunity to contrast evidence and inference.

The important thing is not to stop questioning. Curiosity has its own reason for existing. One cannot help but be in awe when contemplating the mysteries of eternity, of life, of the marvelous structure of reality. It is enough if one tries merely to comprehend a little of this mystery every day. Never lose a holy curiosity.

— *Albert Einstein*

Greta Garbanzo

The One Hundredth Thing About Caroline

by Lois Lowry
Houghton Mifflin, Boston. 1983
Dell Publishing, New York. 1991
Grades: 5–9

Fast-moving and often humorous book about 11-year-old Caroline, an aspiring paleontologist, and her friend Stacy's attempts to conduct investigations. Caroline becomes convinced that a neighbor has ominous plans to "eliminate" the children and Stacy speculates about the private life of a famous neighbor. Due to hasty misinterpretations of real evidence, both prove to be wildly wrong in their inferences. Also included are lots of interesting facts about dinosaurs, scenes at the natural history museum, and a good portrait of a paleontologist at work. Gathering evidence, weighing it, and deciding what makes sense are good accompanying themes. A somewhat inaccurate portrayal of "color blindness" is a minor flaw.

The Real Thief

by William Steig
Farrar, Straus & Giroux, New York. 1973
Grades: 4–8

King Basil and Gawain, devoted Chief Guard, are the only two in the kingdom who have keys to the Royal Treasury. When rubies, gold ducats, and finally the world-famous Kalikak diamond disappear, Gawain is brought to trial for the thefts. But is he the real thief? As the mystery unfolds, it becomes clear that it is important to investigate fully before making judgments or drawing conclusions.

Betty Biograph

Susannah and the Blue House Mystery

by Patricia Elmore
E.P. Dutton, New York. 1980
Scholastic, New York. 1990
Grades: 5–7

Susannah (an amateur herpetologist) and Lucy have formed a detective agency. They check into the death of a kindly old antique dealer who lived in the mysterious "Blue House." They attempt to piece together clues in hopes of finding the treasure they think he has left to one of them. The detectives evaluate evidence, work together to solve problems, and prevent a camouflaged theft from taking place.

Susannah and the Poison Green Halloween

by Patricia Elmore
E.P. Dutton, New York. 1982
Scholastic, New York. 1990
Grades: 5–7

Susannah and her friends try to figure out who put the poison in their Halloween candy when they trick-or-treated at the Eucalyptus Arms apartments. Tricky clues, changing main suspects, and some medical chemistry make this an excellent choice, with lots of inference and mystery.

The Tattooed Potato and Other Clues

by Ellen Raskin
E.P. Dutton, New York. 1975
Penguin Books, New York. 1989
Grades: 6–9

Answering an advertisement for a portrait painter's assistant in New York City involves a 17-year-old in several mysteries and their ultimate solution, such as the "Case of the Face on the Five Dollar Bill" where the smudged thumbprint of the counterfeiter is a clue.

The Westing Game

by Ellen Raskin
E.P. Dutton, New York. 1978
Avon, New York. 1984
Grades: 6–10

The mysterious death of an eccentric millionaire brings together an unlikely assortment of 16 beneficiaries. According to instructions contained in his will, they are divided into eight pairs and given a set of clues to solve his murder and thus claim the inheritance. Newbery award winner.

Who Really Killed Cock Robin?

by Jean C. George
HarperCollins, New York. 1991
Grades: 3–7

A young hero in this compelling ecological mystery examines the importance of keeping nature in balance. This is an inspiring account of an environmental hero who becomes a scientific detective.

DISCOVERING DENSITY

Grades 6–10

Five Activities

Students attempt to layer various liquids in a straw, leading them to explore the concept of density. The teacher introduces the formula for determining density. Students have fun creating secret formula sheets, while reinforcing their practical understanding of this important concept. A final session includes "Puzzling Scenarios" as students explore real-life connections of density.

Activity 1

Students are challenged to layer four "mystery liquids" of different densities in a clear straw. Students record their attempts and gather data to solve the challenge, learning that liquids have a property that enables some to float on top of others.

Activity 2

Students again attempt to layer liquids, but this time the liquids are all made with the same ingredients (water, salt, and food coloring) with the only difference being the amount of salt. The layering is more difficult, as division between layers is less distinct. Students learn to distinguish density from weight.

Activity 3

Students prepare salt solutions based on "secret formulas," then calculate the densities of the liquids they prepared. The results are used to predict how the liquids will layer in Activity 4.

Activity 4

Teams of students exchange the trays of liquids they mixed according to "secret formulas" in Activity 3. Teams try to determine each others' secret formulas and test each others' predictions.

Activity 5

Students predict which will layer on top, hot water or cold water, then take part in a class demonstration to find out the effect of temperature on density. Class discusses and tries to solve "puzzling scenarios" drawn from everyday situations in which density is involved, such as warm and cold air in the house, refrigerators, floating in salt water, ice cream floating in root beer, as well as scientific uses, such as submarine navigation, aircraft construction, and radiation shielding.

Skills

Measuring, Observing, Predicting, Using Proportions, Calculating Density

Concepts

Densities of Liquids

Themes

Systems & Interactions, Models & Simulations, Stability, Patterns of Change, Scale, Structure, Matter, Diversity & Unity

Notes: Going Further ideas include linking density to ecology by discussing fresh and salt water in the San Francisco Bay, the Amazon, Mono Lake, and low coral islands. The GEMS guide *Convection: A Current Event* is recommended as a companion activity.

MAKING CONNECTIONS

A number of the books listed pertain to **relative density in gases**, with most of these touching on **hot air ballooning** in some way. These range from historical accounts of balloon excursions to exciting fictional adventures. The technical and conceptual considerations involved in hot air ballooning are often discussed in these books.

Two books are mysteries in which a knowledge of the density in gases allows the reader to solve the mystery. In an ecological mystery, a knowledge of the difference in density between fresh and salt water is important. One fascinating adventure book deals with gold panning and **relative density in solids**. Another book touches on **sink/float situations**. There is also a nonfiction picture book that does a wonderful job of conveying the **concept of volume**.

CROSS REFERENCES

LITERATURE CONNECTIONS

Archimedes and His Wonderful Discoveries
by Arthur Jonas; illustrated by Aliki
Prentice-Hall, Englewood Cliffs, New Jersey. 1963
Out of print
Grades: 3–6

The focus of the book, after a brief biographical chapter, is on the problems Archimedes solved and how his discoveries and theories are still used today. Chapter 2 tells how Archimedes tested the king's crown to see if it was really made of gold by comparing the amount of water displaced by the crown and by a lump of real gold. Chapter 3 describes Archimedes' research on the density of liquids. Other chapters illustrate his discovery of levers and some tools we use today which act as levers, his experiments relating to measurement in astronomy and mathematics, and how his science was used to formulate war machines.

Balloon Ride
by Evelyn C. Mott
Walker & Co., New York. 1991
Grades: K–4

Young Megan and Joy, the pilot, describe the preparation of a hot air balloon trip by the all-woman crew. The course of the journey is illustrated with color photographs. Words such as "gondola" and "altimeter" are introduced. Descriptions of shadows, weather, and how a balloon flies are included.

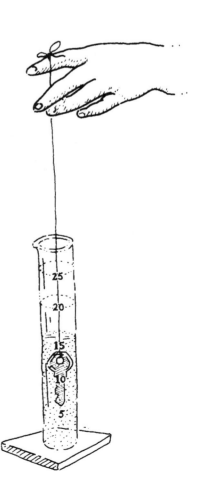

The Big Balloon Race

by Eleanor Coerr; illustrated by Carolyn Croll
Harper & Row, New York. 1981
Grades: 1–3

A young girl's mother enters a hot-air balloon race in 1882. The young girl falls asleep in the balloon basket, accidentally ending up on the flight. During the exciting flight, they evade a number of hazards and young Ariel saves the day. Basic facts about balloon dynamics are described: the use of hydrogen in some balloons, the effect of cold air outside on the temperature of the gas, and the altitude of the balloon.

By the Great Horn Spoon!

by Sid Fleischman; illustrated by Eric von Schmidt
Atlantic Monthly Press/Little, Brown and Co., Boston. 1963
Grades: 5–8

In this adventure novel about sailing around Cape Horn to the California gold rush, several passages exemplify the way density figures into the gold panning process. On page 97, Quartz Jackson teaches the boy Jack about panning for gold, explaining "Gold's heavy ... even the yellow dust sinks to the bottom if you keep workin' the pan." Pages 123–127 provide some good technical detail on the panning process, "grain for grain, gold is eight times heavier than sand." An ingenious way to use an umbrella as a filter for gold is also described.

Einstein Anderson Tells a Comet's Tale

by Seymour Simon; illustrated by Fred Winkowski
Viking Press, New York. 1981
Grades: 4–7

Chapter 3 has a mystery that can deepen your students' understanding of density. The mystery can be solved if they realize that adding helium to a balloon makes it heavier; and even though the balloon may rise quickly in the air, it won't go all the way to the moon!

Encyclopedia Brown and the Case of the Secret Pitch

by Donald J. Sobol; illustrated by Leonard Shortall
LodeStar Books/E.P. Dutton, New York. 1965
Bantam Books, New York. 1978
Grades: 3–5

In the "Case of the Balloon Man," the evidence in a kidnapping investigation hinges on how high a balloon filled with air (and not helium) could rise.

Hot-Air Henry

by Mary Calhoun; illustrated by Erick Ingraham
William Morrow, New York. 1981
Grades: K–3

Henry, a spunky Siamese cat, stows away on a hot air balloon and accidentally gets a solo flight. He learns there is more to ballooning than just watching as he deals with air currents, power lines, and manipulating the gas burner. Though the format and style of the book is aimed at the primary grades, the information on ballooning and the concept that hot air is less dense than cool air is presented. Good extension to the activity in which students learn what density is in liquids.

> Observation is a passive science, experimentation an active science.
>
> — *Claude Bernard*
> *Introduction a l'Etude*
> *de la Medecine Experimentale*

The Missing 'Gator of Gumbo Limbo: An Ecological Mystery

by Jean C. George
HarperCollins, New York. 1992
Grades: 4–7

Sixth-grader Liza K and her mother live in a tent in the Florida Everglades. She becomes a nature detective while searching for Dajun, a giant alligator who plays a part in maintaining a waterhole's quality, and is marked for extinction by local officials. She and James James, a neighbor and amateur naturalist, observe and note many changes in the local ecology including the presence of salt in the town and condominium wells. "James James wants to close the canals … The fresh water will back up, sink into the ground where it ought to be, and push down the salt water." By layering liquids in the GEMS *Discovering Density* activities, students discover that fresh water floats on salt water.

Spaces, Shapes and Sizes

by Jane J. Srivastava; illustrated by Loretta Lustig
Thomas Y. Crowell, New York. 1980
Grades: 1–6

An inviting and well-presented nonfiction book about volume. It shows the changing form a constant amount of sand can take, and includes estimation activities, an investigation of the volumes of boxes using popcorn, and a displacement activity. Your students will want to try the listed activities, and the book is likely to inspire further investigations.

Splash! All About Baths

by Susan K. Buxbaum and Rita G. Gelman; illustrated by Maryann Cocca-Leffler
Little, Brown and Co., Boston. 1987
Grades: K–6

Penguin answers his animal friends' questions about baths such as, "What shape is water?" "Why do soap and water make you clean?" "What is a bubble?" "Why does the water go up when you get in?" "Why do some things float and others sink?" and other questions. Answers to questions are both clear and simple. American Institute of Physics Science Writing Award winner.

Supersuits

by Vicki Cobb; illustrated by Peter Lippman

J.B. Lippincott, Philadelphia. 1975

Grades: 4–7

The severe environmental conditions that require special clothing for survival such as freezing cold, fire, underwater work, and thin or nonexistent air are described. In "Going Under the Sea," the mechanics of diving bells and suits to deal with air pressure are explained and one of the supersuits pictured is "really a submarine shaped like a person." The section on ballooning, page 70, gives an exciting account of an almost fatal incident of oxygen starvation in an early balloon venture in 1862.

The Twenty-One Balloons

by William Pene duBois

Viking Press, New York. 1947

Grades: 5–12

Professor Sherman leaves San Francisco in 1883 to cross the Pacific by balloon. Three weeks later he is picked up in the Atlantic clinging to the wreckage of a platform flown through the air by 2l balloons, after having passed through Krakatoa just before the historic volcanic eruption. In Chapter 2, balloon happenings include two brothers (weighing 75 and 58 pounds) who tie themselves to a balloon with a lifting pull of 60 pounds, and the flight of a 400-pound cupola which had been tied to balloons with a total of 900 pounds lifting pull. This section does include a few pages of negatively stereotypical Native American dialogue, although the conclusion drawn is that the Indians would not have done such a stupid thing as the "dumb white man." Chapter 3 is all about the design and outfitting of the balloon and the mechanics of ballast. In Krakatoa, the professor describes a Balloon Merry-Go-Round and a Balloon Life Raft designed to carry 80 people. The Krakatoa section is the most dated. However, the nuggets of balloon dynamics and detailed quasi-technical illustrations remain fascinating. Newbery award winner.

Grades 5–9

Fifteen Sessions (Six activities)

Students learn about astronomy and answer questions such as: If the Earth is a ball, why does it look flat? Why does the moon change its shape? How can I find constellations and tell time by the stars? Activities include observing and recording changes in the sky and creating models to explain observations.

Activity 1

The term and meaning of "model" are introduced and discussed. Students learn about some ancient models of the world, then design their own.

Activity 2

Students respond to the "What Are Your Ideas About the Earth?" questionnaire on earth's shape and gravity, then discuss the ideas with the teacher.

Activity 3

Students learn how to estimate the distance between the Sun and Moon in the sky. In six 15-minute sessions they observe the day-visible moon, and its phases every other day, then summarize their observations.

Activity 4

Students model moon phases and eclipses of the moon using a light and polystyrene "moonballs."

Activity 5

Students make star clocks, learn about constellations, practice using the clocks in class, then use them at home with the night sky. They discuss the motion of the stars.

Activity 6

Students make star maps and learn how to use them to find constellations in the night sky.

Skills

Creating/Using Models, Synthesizing, Visualizing, Observing, Explaining, Measuring Angles, Recording, Estimating, Averaging, Using Instruments, Drawing Conclusions, Using a Map

Concepts

Astronomy, History of Astronomy, Spherical Earth, Gravity, Moon Phases, Eclipses, Measuring Time, The North Star, Earth's Daily Motion, Constellations, Horizon, Zenith

Themes

Systems & Interactions, Models & Simulations, Stability, Patterns of Change, Evolution, Scale, Structure

Notes: Cross-cultural examples, such as those shown in "Ancient Models of the World," could be widely extended to include other creation and astronomical myths of ancient peoples.

MAKING CONNECTIONS

Books that explain the physical world through myths and legends are a perfect accompaniment to Activity 1. Of course, certain myths and legends can also be matched with other sessions of the guide. For instance, legends about the moon or stars can be connected to the class sessions on moon phases or constellations.

Many of the books here are not myths or legends. These can be related to Activities 2 through 6, depending on their focus. *How to Dig a Hole to the Other Side of the World* works well with Activity 2. Stories that include eclipses of the moon or the sun would fit well with Activity 4. Any book about the Big Dipper enhances Activity 5, in which students learn how to use the Big Dipper to tell time and to find the North Star. Books about other constellations relate best to Activity 6.

Several books, such as those about comets or planets, are connected more generally to the subject of astronomy. Such books can be used to lead students from their investigations of the Earth, Moon and stars to other elements of our Solar System.

Watch for any books that picture the Moon in the sky, even if they are primary level books. In this connection, see the feature article on page 90 about challenging older students to evalute the astronomical accuracy of these books.

CROSS REFERENCES

Digging for China

"Far enough down is China," somebody said.
"Dig deep enough and you might see the sky
As clear as at the bottom of a well.
Except it would be real—a different sky.
Then you could burrow down until you came
To China! Oh, it's nothing like New Jersey.
There's people, trees, and houses, and all that,
But much, much different. Nothing looks the same."

I went and got the trowel out of the shed
And sweated like a coolie all that morning,
Digging a hole beside the lilac-bush,
Down on my hands and knees. It was a sort
Of praying, I suspect. I watched my hand
Dig deep and darker, and I tried and tried
To dream a place where nothing was the same.
The trowel never did break through to blue.

Before the dream could weary of itself
My eyes were tired of looking into darkness,
My sunbaked head of hanging down a hole.
I stood up in a place I had forgotten,
Blinking and staggering while the earth went round
And showed me silver barns, the fields dozing
In palls of brightness, patterns growing and gone
In the tides of leaves, and the whole sky China blue.
Until I got my balance back again
All that I saw was China, China, China.

— *Richard Wilbur*

Literature Connections

Boat Ride With Lillian Two Blossom
by Patricia Polacco
Philomel/Putnam & Grosset, New York. 1988
Grades: K–4

> A wise and mysterious Native American woman takes William and Mabel on a boat ride, starting in Michigan and ranging through the sky. Explanations for the rain, the wind, and the changing nature of the sky refer to spirits such as the caribou or polar bear, which are magically shown.

Einstein Anderson Lights Up the Sky
by Seymour Simon; illustrated by Fred Winkowski
Viking Press, New York. 1982
Grades: 4–7

> In "The World in His Hands," Einstein punctures his friend Stanley's plan to build a scale model of the solar system in his basement. He discusses the relative sizes of the sun and the planets and the distances between them. In "The Stars Like Grains of Sand," Einstein enlightens his younger brother Dennis about the star population.

Einstein Anderson Tells a Comet's Tale
by Seymour Simon; illustrated by Fred Winkowski
Viking Press, New York. 1981
Grades: 4–7

> In "Tale of the Comet" there is some very interesting information about possible connections between comets, asteroids, and dinosaurs. Even though the book was published in 1981, the information is still accurate.

Follow the Drinking Gourd
by Jeanette Winter
Alfred A. Knopf, New York. 1988
Grades: K–6

> By following the hidden directions in the song "The Drinking Gourd," taught to them by an old sailor named Peg Leg Joe, runaway slaves follow the stars along the Underground Railroad and the connecting waterways to Canada and freedom. The "drinking gourd," another name for the Big Dipper, guided them north. In Activities 5 and 6, students learn how to use the Big Dipper to tell time and find the North Star.

Grandfather Twilight
by Barbara Berger
Philomel Books/Putnam & Grosset, New York. 1984
Grades: Preschool–2

> At the end of the day, as he does each day, Grandfather Twilight delivers the moon to the sky. The moon is a pearl that is removed from a strand and grows in size with each step grandfather takes. The story is portrayed simply, with few words and peaceful, yet magical illustrations.

The moon is made
of green cheese.

— *John Heywood*
Proverbs

The Heavenly Zoo, Legends and Tales of the Stars
by Alison Lurie; illustrated by Monika Beisner
Farrar, Straus & Giroux, New York. 1979
Grades: 4–8

> "Long before anyone knew that the stars were great burning globes of gas
> many millions of miles from the earth and from one another, men and
> women saw the sky filled with magical pictures outlined with points of
> light. Some of these (16) tales are heroic, some comic, some sad; but all
> are full of the wonder we still feel when we look at the sky full of stars."
> The illustrations are striking, showing each beast, bird, or fish against the
> stars that indicate its position.

How Many Stars in the Sky?
by Lenny Hort; illustrated by James E. Ransome
Tambourine Books/William Morrow, New York. 1991
Grades: K–2

> An African-American father and son set off on a journey of discovery to
> count the stars in a summer night sky. As city dwellers, they discover the
> obstacles to stargazing—city lights, for example—and end up driving to the
> country.

How to Dig a Hole to the Other Side of the World
by Faith McNulty; illustrated by Marc Simont
HarperCollins, New York. 1990
Grades: K–8

> A child takes an imaginary 8,000-mile journey through the earth and
> discovers what's inside. This activity connects beautifully with Activity 2
> of the GEMS guide, in which students are asked to imagine what might
> happen if an obect could be dropped through the center of the Earth to the
> other side. See page 344 for a more detailed description of the ways this
> book connects to *Earth, Moon, and Stars*.

In the Beginning:
Creation Stories from Around the World
by Virginia Hamilton; illustrated by Barry Moser
Harcourt, Brace, Jovanovich, San Diego. 1988
Grades: All

> An illustrated collection of 25 legends that explain the creation of the
> world, with commentary placing the myths geographically and classifying
> them by type of myth tradition such as "world parent," "creation from
> nothing," and "separation of earth and sky." Some of the selections are
> extracted from larger works such as the *Popol Vuh* or the Icelandic Eddas.
> Excellent connection with Activity 1 of the GEMS guide in which students
> learn about the ways several ancient peoples modeled how the Sun and
> Earth move.

The Magic School Bus Lost in the Solar System

by Joanna Cole; illustrated by Bruce Degen
Scholastic, New York. 1990
Grades: K–6

Ms. Frizzle and her class leave the earth and visit the moon, sun, and each planet in the Solar System, noting the temperature, color, size, and unique features. Excellent literature connection for students who want to extend their investigations on the Earth, Moon and stars to a study of the Solar System.

Many Moons

by James Thurber; illustrated by Louis Slobodkin
Harcourt, Brace & World, New York. 1943
Grades: K–5

This is the tale of a little princess who wanted the moon, and how she got it. Neither the King, the Lord High Chamberlain, the Royal Wizard, the Royal Mathematician, nor the Court Jester were able to solve the problem—it took a 10-year-old princess to figure it out. The story includes a debate about how far away the moon is.

Moon-Watch Summer

by Lenore Blegvad; illustrated by Erik Blegvad
Harcourt, Brace, Jovanovich, San Diego. 1972
Out of print
Grades: 4–6

Adam's eager anticipation of the Apollo landing and first moon walk turns to sullen resentment when he learns that he and his younger sister will be spending the summer on his grandmother's farm where there isn't even a television set. Once there, he is surprised when his grandmother confesses that she has always been "a sort of ancient moon-worshiper" and is fascinated by the Sea of Rainbows. He consoles himself with an old radio, hearing with frustration reports of the good television transmission. "Mission Control called it a 'superb' quality picture." The landing had been seen live in the United States, Japan, Western Europe, and South America, but not in Grannie's house. As the summer progresses, he makes charts and drawings summarizing the mission's progress and learns to put his family responsibilities before personal disappointments. Today's students may find it hard to imagine life without television, but will appreciate the significance of the moon walk to Adam and to society at that time.

Ancient Models of the World

EGYPT

INDIA

CHINA

GREECE

Nine O'Clock Lullaby

by Marilyn Singer; illustrated by Frane Lessac
HarperCollins, New York. 1991
Grades: Preschool–6

Children are transported through many lands showing what people might be doing on different parts of the globe at the "same" time. The pictures of the various cultures are fresh and lively, from cooking on a "barbie" in Australia to conga drumming and coconut candy in Puerto Rico. There's a brief astronomical explanation of time. This, and the idea that it is day on one part of the earth while it is night on another can be modeled for students at the beginning of Activity 3 of the GEMS guide, and is the question they explore in Activity 1. This book could be read aloud to students even as old as the sixth grade.

Planet of Exile

by Ursula LeGuin
Ace Books, New York. 1966
Grades: 6–Adult

Cooperation is the central theme of this thin but gripping book about the clash of three cultures—two that have inhabited a harsh planet for eons, and the one that has been exiled only a few generations. Difficult seasonal conditions on the planet are the result of how long it takes for the planet to revolve once around its central star. Because one "year" is equivalent to many Earth years, people only live through a very small number of winters.

The Planet of Junior Brown

by Virginia Hamilton
Macmillan Publishing, New York. 1971
Grades: 5–12

This unusual and moving book begins with three people (two students who regularly cut eighth grade classes and a school custodian who was formerly a teacher) in a secret room in a school basement with a working model of the solar system. The model has one incredible addition—a giant planet named for one of the students, Junior Brown. How can the Earth's orbit not be affected by this giant planet? Is there a belt of asteroids that balances it all out? How does this relate to equilateral triangles? From these subjects, the universe of the book expands outward into the Manhattan streets and inward into the hearts, minds, and friendship of the two students who are both African-American. After the first chapter, the solar system becomes more metaphor than scientific model, until the end of the book when the real model must be dismantled and the three must find a way to help Junior Brown and to affirm their solidarity against all odds. Powerfully and poetically written, this book humanizes the statistics about homelessness and the educational crisis in a profound and unforgettable way.

The Planets

edited by Byron Preiss
Bantam Books, New York. 1985
Grades: 8–Adult

> This extremely rich, high-quality anthology pairs a nonfiction essay with a fictional work about the earth, moon, each of the planets, and asteroids and comets. Introductory essays are by Issac Asimov, Arthur C. Clarke, and others. The material is dazzlingly illustrated with color photographs from the archives of NASA and the Jet Propulsion Laboratory, and paintings by astronomical artists such as the movie production designers of *2001* and *Star Wars*.

Quillworker: A Cheyenne Legend

adapted by Terri Cohlene; illustrated by Charles Reasoner
Watermill Press/Educational Reading Services, Mahwah, New Jersey. 1990
Grades: 2–5

> This Cheyenne legend explains the origin of the Big Dipper. Quillworker is an only child and an expert needle worker. Her dreams direct her to make seven buckskin warrior outfits for her mysterious new seven brothers. To escape the buffalo nation who want to take Quillworker, they all ride a tree up into the sky where they remain, with Quillworker as the brightest star in the dipper. Good tie to Activity 5 of the GEMS guide in which students learn about constellations and make star clocks.

Sky Songs

by Myra Cohn Livingston; illustrated by Leonard E. Fisher
Holiday House, New York. 1984
Grades: 5–12

> Fourteen poems about various aspects of the sky such as the moon, clouds, stars, storms, and sunsets. Wonderful images portray the planets as "wanderers of night," shooting stars are "bundled up in interstellar dust and bright icy jackets," and the morning sky is "earth's astrodome, floodlit."

Space Songs

by Myra Cohn Livingston; illustrated by Leonard E. Fisher
Holiday House, New York. 1988
Grades: 5–12

> Series of short poems about aspects of outer space including the Milky Way, moon, sun, stars, planets, comets, meteorites, asteroids, and satellites. The astronomy content is accurate. The black background illustrations are dynamic and involving.

The fault, dear Brutus,
is not in our stars,
But in ourselves . . .

— William Shakespeare
Julius Caeser

Star Tales: North American Indian Stories

by Gretchen W. Mayo
Walker & Co., New York. 1987
Grades: 5–12

> The nine legends in this collection explain observations of the stars, moon, and night sky. Accompanying each tale is information about the constellation or other heavenly observation and how various tribes perceived it. In *More Star Tales* the same author includes "The Never-Ending Bear Hunt" and seven other tales.

To Space and Back

by Sally Ride with Susan Okie
Lothrop, Lee, and Shepard/Morrow, New York. 1986
Grades: 4–7

> This is a fascinating description of what it is like to travel in space—to live, sleep, eat, and work in conditions unlike anything we know on Earth, complete with colored photographs aboard ship and in space. The astronauts conducted a number of scientific experiments as they observed and photographed the stars, the Earth, the planets, and galaxies. Working outside the shuttle, they feel the warmth of the sun through their gloves, but cool off on the dark side of Earth in the shade. This, and other descriptions, could lead to a better understanding of the Earth's shape and gravity (Activity 2) as well as day/night and phases of the moon.

The Truth about the Moon

by Clayton Bess; illustrated by Rosekrans Hoffman
Houghton Mifflin, Boston. 1983
Grades: K–4

> An African boy is puzzled by the changing size of the moon and asks for an explanation. His father says there is only one moon and that the moon he saw last night is the same moon he will see tomorrow. "It is growing, just as a child like you grows to be a man like me. It starts small, just a silver sliver, and every night grows bigger and bigger until it is as big as it can be, a full circle. Then, just as a man grows smaller when he is very old, so does the moon. Smaller and smaller until death." His mother explains that there is only one moon. "It is like a woman. And you know how sometimes a woman will grow larger and larger, more and more round?" The Chief tells a long tale about the sun and the moon being married and how the moon lost its heat.

The Way To Start a Day

by Byrd Baylor; illustrated by Peter Parnall
Charles Scribner's Sons, New York. 1973
Macmillan Publishing Co., New York. 1986
Grades: 3–7

> The peoples of the world have celebrated the dawn in many ways—with drum beats, ringing of bells, gifts of gold or flowers. "The way to start the day is this: Go outside and face the east and greet the sun with some kind of blessing or chant or song that you made yourself and keep for early morning." Relates well to Activity 1 of the GEMS guide, which explores ideas about the rising and setting of the Sun. Caldecott honor book.

Why Mosquitoes Buzz in People's Ears

retold by Verna Aardema; illustrated by Leo and Diane Dillon
Dial Books, New York. 1975
Grades: K–6

> This West African folk tale explains why mosquitoes buzz in people's ears, and how the owl's call is what makes the sun rise each morning! Considering ideas that explain why the sun rises each morning is related to what students do in Activity 1 of the GEMS guide.

The Year of The Comet

by Roberta Wiegand
Bradbury Press, Scarsdale, New York. 1984
Out of print
Grades: 4–9

> The first two chapters are specifically about Halley's Comet in 1910, the rumors about massive destruction that preceded it, and its actual impact on a small Nebraska town. The second chapter starts with an interesting narrative involving the theme of scale, as the heroine Sarah puts herself, like Alice going down the rabbit hole, inside a map of the United States to delve into the detail of the buildings and streets of her small town. For many students, it will be easy to continue reading of Sarah's other adventures as she gains a new maturity during "the year of the comet." Many touching and powerful passages; a good sense of the universal scope of the comet and the real-life complexity of human relations.

I have always read that the world, both land and water, was spherical, as the authority and researches of Ptolemy and all the others who have written on this subject demonstrate and prove, as do the eclipses of the moon and other experiments that are made from east to west, and the elevation of the North Star from north to south.

> — *Christopher Columbus*
> *Letter to the Sovereigns*
> *on the Third Voyage, 1498*

Editor's Note: Although many people still believe that Columbus was the first person to say the Earth is round, Columbus cites sources as far back as the third century B.C. The spherical nature of the Earth was well known by many people in Columbus's day.

The Moon in Pictures

BY JOHN ERICKSON

Where do people get their ideas about the sun and the moon? Of course, we see them in the sky, but we also see them, from a very early age, in pictures from books. Our favorite children's books have illustrations that we see over and over again. Do these pictures show things as they really are?

To some people an illustration of the night sky is hardly complete if it does not show a shining moon, yet in the real sky the moon is often absent. Most of the time the moon can be seen in the night sky it is in one of its gibbous phases. But just try to find a picture of a gibbous moon in a children's book, or in almost any other drawing or painting for that matter!

By analyzing illustrations from books of their early childhood, older students can rediscover and examine critically some of the material that has shaped their perceptions of the moon and its phases. This literary investigation will also give them practice applying what they have learned about the moon from the activities in the GEMS guide *Earth, Moon, and Stars.*

Analyzing Depictions of the Moon

You could have the class make a collection of books with illustrations showing the moon. Some of these could be selected from the library or other classrooms, but also encourage students to find books they remember from earlier in their childhood.

As a class, examine the illustrations in one or two of the books, exploring questions like: What phase of the moon is shown in the illustration? Could the moon actually appear as shown in each illustration? Take into account:

○ The shape of the moon in the illustration

○ Which side of the moon is bright and which side is dark

○ The position of the sun or the time of day. This may be shown in the picture or there may be clues in the text.

An illustration may also bring up questions about whether a phase of the moon *could* appear at a certain time of day or in a certain part of the sky. Use actual observations of the moon or the "Modeling Moon Phases" activity from *Earth, Moon, and Stars* to help find answers.

Working in Teams

Pass out two or more of the books to groups of three or four students. Tell the students that after they study the illustrations together they should choose one book and prepare to give an oral report to the rest of the class about how the moon is used in the illustrations and whether or not it is shown accurately.

If questions arise during discussion, or if students assert that the illustration(s) are inaccurate, encourage the students to refer to the moon modeling activity or their own direct observations of the moon to explain to the rest of the class why they believe a particular depiction of the moon is incorrect.

After all the teams have finished their reports, ask if there are still questions that remain. Ask students to suggest ways they could help resolve the questions through experiment or observation.

Some Examples of Children's Books

Here are some examples of children's literature with notable pictures of the moon. Many of these books are well known to students and teachers.

(*Editor's note*: Some of the comments regarding inaccuracies in the following books refer to the moon's appearance in the northern or southern hemispheres. This topic is not explored directly in the GEMS guide *Earth, Moon, and Stars*, and you should not expect your students to be aware of it. Learning more about how the moon appears in the two different hemispheres would make an excellent area of investigation for advanced students.)

Yertle The Turtle and Other Stories
by Dr. Seuss
Random House, New York. 1958

King Yertle, an overly ambitious turtle, tries to satisfy his desire for a tall throne by stacking up more and more of his turtle subjects. Before his final downfall he is distressed that the moon of the evening is rising higher than himself. A rising evening moon is never a crescent, but that is what the picture shows. (True, a thoughtful reader might suggest that it is actually a full moon undergoing an eclipse, which is perfectly possible.

However, if you accept the eclipse explanation in every case where a children's book shows a crescent moon out of its proper place then you will find that eclipses are entirely commonplace in children's literature!)

Sleep Book
by Dr. Seuss
Random House, New York. 1962

One Fish, Two Fish, Red Fish, Blue Fish
by Dr. Seuss
Random House, New York. 1960

In these books, and many other works, pictures of the moon are almost always thick crescent shapes oriented like the letter 'C'. If the evening pictures are not eclipsed full moons then they must be in the southern hemisphere. Perhaps some of them are showing the moon before dawn in the northern hemisphere. More likely Dr. Seuss just tended to draw the same kind of moon all the time. He got it right in *Dr. Seuss's ABC* (Random House, New York, 1963) on the 'M' page which shows a full moon high in the sky at midnight.

The Little House
by Virginia Lee Burton
Houghton Mifflin, Boston. 1942

A house built in the country finds a city growing up around it. The references to the difficulty of watching the sun, moon and stars from the city are entirely accurate. However, on the last page there is a reference to a new moon rising during a quiet night. This is impossible as a new moon rises when the sun rises or slightly later in the morning. The moon is pictured near the Big Dipper which is also impossible, but perhaps this can be forgiven as these illustrations often rearrange the stars for artistic effect with beautiful results. (One illustration from *The Little*

House shows a calendar with the phases of the moon as they would appear only in the southern hemisphere, yet the pages before shows the sun as it would appear in the northern hemisphere, moving across the sky from left to right.)

Mike Mulligan and His Steam Shovel
by Virginia Lee Burton
Houghton Mifflin, Boston. 1939

A sequence of pictures shows the sun rising on the right side of the scene and moving across the sky to set on the left. This implies that Popperville, the setting for the story, is in the southern hemisphere. Read or reread these stories and see what part of the world you are reminded of the most.

Goodnight Moon
by Margaret Wise Brown;
illustrated by Clement Hurd
Harper & Row, New York. 1947

This classic has recently regained popularity. It describes a young bunny's bedroom and all the things in it that get a "goodnight" before the bunny at last settles down to sleep. A full moon is shown rising at bedtime which is correct if bedtime is around sundown. As the moon rises its path slants toward the right, which puts the story somewhere in the mid-latitudes of the northern hemisphere.

Whiskers and Rhymes
written and illustrated by Arnold Lobel
William Morrow, New York. 1985

This collection of original poems has the quality and feeling of a book of favorite traditional nursery rhymes. The poem that begins "There was an old woman of long ago" is illustrated with a picture that shows the sun and moon side by side on the horizon. They have been stitched to the earth for various reasons by a seam-stress cat. Other issues aside, if the moon and the sun were really side-by-side one would expect to see half of the moon lit up, one of the quarter phases. Instead both the picture and the text of the poem make it a crescent moon. Furthermore, the bright side of the moon is pictured away from the sun. The incongruity is obvious even to someone who does not fully understand the phases of the moon. In this case, the astronomical defects of the picture add to its humorous quality.

Where The Wild Things Are
written and illustrated by Maurice Sendak
Harper & Row, New York. 1963

An evening crescent moon is visible through Max's bedroom window at suppertime. Since the part of the moon that is lit up is on the left, Max either lives in the southern hemisphere, or his family eats supper before dawn! Later in the story, after Max's return from the place where the wild things are, the moon is still seen through Max's window, but it is a full moon. If the picture with the crescent moon was in the evening, then the picture with the full moon must be just before dawn *at least ten days later*. The story itself is a bit ambiguous about how time passes, but it is noted that Max's food stayed hot while he was gone. Probably these phases of the moon, like much of the rest of this highly imaginative book, have no tame, conventional explanation.

Night and Day
by Catherine Ripley; illustrated by Debi Perna and Brenda Clark
Western Publishing, Toronto. 1985
Out of print

This nonfiction account describes the 24-hour routines in the lives of a several wild animals. The cover of the book and the first page both show a crescent moon in the morning twilight sky. The part of the moon that is lit up is away from the horizon. That's fiction!

Works of art outside children's literature can also be examined for astronomical content. The magazine *Sky and Telescope* has had some good examples.

○ "What's Wrong With a Gibbous Moon" by William Livingston examines "Sheepfold by Moonlight," an oil painting by Jean-Francois Millet, with comments about astronomy in paintings in general. (*Sky and Telescope,* February 1992, Vol. 83, No. 2, pages 159–160.)

○ "Ansel Adams' Moonrise Turns 50" by Dennis DiCicco uses the phase and position of the moon to determine when a photograph was taken. Ansel Adams remembered taking the picture, but he had forgotten exactly *when* he took it. (*Sky and Telescope,* November 1991, Vol. 82, No. 5, page 480.)

○ "Van Gogh, Two Planets, and The Moon" by Donald W. Olson and Russell L. Doescher, analyzes van Gogh's painting "Road With Cypress and Star." Given the phase of the moon in the painting and the times when van Gogh visited the setting the authors determined the most likely date for the scene. A computer program that simulates the sky allowed them to determine that the two "stars" in the picture are very possibly Venus and Mercury. (*Sky and Telescope,* October 1988, Vol. 76, No. 4, pages 406–408.)

John Erickson is a member of the Astronomy and Physics Department of the Lawrence Hall of Science, a co-author of the GEMS guide Global Warming and the Greenhouse Effect, *and the father of three young children, including twins. The bedtime stories he reads to them gave him the idea for this article.*

EARTHWORMS

Grades 6–10   **Three Sessions**

Earthworm **heartbeat patterns** are safely investigated in living earthworms in this GEMS unit, as students observe blood vessels in worms, and observe and record the pulse rates. The students experiment to discover the responses of earthworms to different temperatures, and graph the results. In discussing why earthworms respond as they do, students learn about "cold-blooded" animals, the concept of adaptation, and circulatory systems.

Session 1

Students observe earthworms, getting a sense of general activity level. Students observe the large dorsal blood vessel and are shown how to count the pulse when the worms are placed in a small amount of water. Worms with pulses that are easier to see are sorted for use in next session.

Ode to the Earthworm

A pause to thank this worm of earth
Through which life's nutrients come to birth
The butt of many jokes and hurt
Tunneler in tons and tons of dirt
The leaves that fall from autumn trees
Ingested in part by worms like these
Mixed in the soil, so new plants thrive
Earthworm labor keeps people alive
Small pinkish thing of so much worth
We thank you gently worm of earth.

— *Lincoln Bergman*

Session 2

Students observe, count, and record the pulse rates of earthworms in water of three different temperatures.

Session 3

Students analyze and graph the results of the previous session's temperature experiments. Class discusses how an earthworm's response to temperature helps it survive, and the meaning of "cold-blooded." Differences between animals whose temperature is determined by the temperature of the environment (such as earthworms) and warm-blooded animals who regulate their own body temperature are discussed.

Skills

Observing, Measuring, Experimenting, Predicting, Averaging, Graphing, Interpreting Data, Inferring

Concepts

Circulatory Systems, Pulse Rates, Cold-Blooded Animals, Effects of Temperature

Themes

Systems & Interactions, Stability, Patterns of Change, Evolution, Structure, Energy, Diversity & Unity

Notes: Background information on care of worms and additional information on worm adaptation to environment and poikilothermic response to temperature. Going Further activities include an experiment to see how extensively just a few earthworms are able to mix and re-layer the soil after only a few days. New edition includes an essay on earthworms that emphasizes the interconnectedness of all life.

MAKING CONNECTIONS

The books range from down-to-earth scientific reality to far-flung fantasy. Human involvement with worms is elaborated on in the following eclectic books related to these underground creatures. Earthworms are the focus of a fantasy story about giant extraterrestrial worms, and a novel about a team of young people who set up a worm business raising earthworms for fishing. On the other hand, one book describes a surgeon's research on the preservation of **blood plasma**, which connects to the idea of the human's blood circulatory system, following student investigations of the earthworm's. A user-friendly resource on earthworms describes **worm behavior** and simple **investigations** with worms and their **soil habitat**.

LITERATURE CONNECTIONS

Charles Drew

by Robyn Mahone-Lonesome
Chelsea House, New York. 1990
Grades: 5–8

Biography of the surgeon who conducted research on properties and preservation of blood plasma, and was a leader in establishing blood banks. Drew, the first African-American to receive a Doctor of Science degree in Medicine, even donated his own blood for a patient whose blood type he matched in the days before blood banks. There is a lot of information about blood chemistry and circulation, presented in the dramatic framework of Drew's work on the Blood for Britain effort during World War II, his many other "firsts," and personal struggles. After students have studied the circulatory system of the earthworm, this exploration of the human system is a natural extension.

Earthworms, Dirt, and Rotten Leaves

by Molly McLaughlin
Atheneum/Macmillan, New York. 1986
Avon, New York. 1990
Grades: 4–7

The earthworm and its environment are examined. Experiments to examine the survival of the earthworm in its habitat are suggested. Answers the question, "Why would anyone want to have anything to do with earthworms?" Recipient of awards for writing from Library of Congress, American Library Association, and the New York Academy of Sciences.

CROSS REFERENCES

Be Kind To Animals

How To Eat Fried Worms

by Thomas Rockwell; illustrated by Emily McCully
Franklin Watts, New York. 1973
Dell Publishing, New York. 1975
Grades: 4–7

On a dare, Billy has to eat a worm every day for 15 days. He wins the bet and even starts to like to eat them, even bringing a worm-and-egg on rye sandwich to school.

The Winter Worm Business

by Patricia R. Giff; illustrated by Leslie Morrill
Delacorte Press, New York. 1981
Grades: 4–7

Several 5th graders plan a "winter worm business" and raise 80 worms to sell to ice fishermen. However, the care of the worms is threatened by jealousies that ensue as a cousin of the main character comes to town. Worse yet, even after all the rivalries are resolved, an accident spills out all but 19 of the worms. With some quick thinking, the kids gather up some pods with grubs in them. There is significant information on how to help keep the earthworms alive, but not much other scientific information about earthworms. The main female character is positively portrayed as not afraid of either worms or grubs.

I'm Going To Pet A Worm Today

I'm going to pet a worm today.
I'm going to pet a worm. Don't say,
"Don't pet a worm"—I'm doing it soon.
Emily's coming this afternoon!
And you know what she'll probably say:
"I touched a mouse," or
"I held a snake," or
"I felt a dead bird's wing."
And she'll turn to me with a kind of smile.
"What did you do that's interesting?"
This time
I am
Going to say,
"Why, Emily, you should have seen me
Pet a worm today!"
And I'll tell her he shrank and he stretched like elastic,
And I got a chill and it felt fantastic.
And I'll watch her smile
Fade away when she
Wishes, that moment,
That she could be me!

— *Constance Levy*

The Worms of Kukumlima

by Daniel Pinkwater
E.P. Dutton, New York. 1981
Out of print
Grades: 5–8

This fantasy adventure story is about an expedition to Africa in search of giant extraterrestrial earthworms who can talk, play chess, provide everyone with crunchy granola, and who colonize elephant mice. The plot is full of fantastic twists and turns. There is only a limited amount of scientific information, but this book will capture some students' imaginations, and is a nice contrast to the very "grounded" activities in this GEMS unit.

EXPERIMENTING WITH
MODEL ROCKETS

Grades 6–10   **Seven Sessions**

Controlled experimentation is introduced in this series of exciting rocketry activities. Students experiment to see what factors influence how high a model rocket will fly by varying the number and placement of fins or the length of the body tube. Safety and teamwork are stressed. Because students use "Height-O-Meters" to measure rocket altitudes, it is necessary to complete that GEMS unit before doing these rocketry activities.

Session 1

The stages of a rocket's flight are introduced, with an optional demonstration launching. The teacher introduces the concept of controlled experimentation, illustrating the difference between a "good" and "not-so-good" experiment. The teams receive model rocket kits and design their own controlled experiments: to determine the factors that allow some rockets to fly higher than others, with the only variables being body tube length or number/placement of fins.

Sessions 2 and 3

Student teams construct rockets in careful step-by-step fashion, using a Student Experimenter's Guide.

Session 4

Student teams prepare for launching, reviewing each launch step and safety code.

Session 5

Each team launches their rockets and measures how high they fly, using Height-O-Meters.

Session 6

Student teams analyze the results from the launch, then complete a "Conclusions Poster" to graph results and draw conclusions about what their experiment revealed.

Session 7

Students hold a meeting to present their results and discuss their significance and plan possible future experiments.

Skills

Planning and Conducting Controlled Experiments, Measuring in Degrees and Meters, Graphing, Interpreting Data

Concepts

Rocketry, Space Science, Technology, Triangulation, Models

Themes

Systems & Interactions, Models & Simulations, Stability, Patterns of Change, Structure, Energy, Matter

GEMS
98

MAKING CONNECTIONS

The single most important concept in this unit is the idea of a **controlled experiment**. One excellent way to begin a Rocketry unit would be to introduce your students to this idea via literature, as it is well described in Chapter 10 of *Einstein Anderson Tells a Comet's Tale.*

The GEMS activities were carefully tested and modified to ensure the full participation of girls, and feedback from teachers confirms their "gender inclusivity." In our classes at the Lawrence Hall of Science, however, we still find that, initially at least, boys tend to be more interested in signing up for model rocketry classes, but when girls have the opportunity to build model rockets they excel at conducting their own model rocketry experiments. One way to create and foster that initial interest, or to extend both boys' and girls' excitement and knowledge about professional astronauts, is to have them read two of the books included here—one written by astronaut Sally Ride and the other a biography of her life.

If your students "take-off" on the idea of space travel, you might want to have them read *Supersuits*, about space suit design. The other books in this series are engaging science fiction stories about kids who have explored strange planets and met alien beings. There are certainly many other books about rocket ships and we would especially welcome hearing about those that also convey an accurate sense of the scientific method and controlled experimentation.

CROSS REFERENCES

Energy 322
Geometry 270
Hot Water and Warm Homes 133
Measurement 286
Models & Simulations 346
Scale 357
Structure 369

LITERATURE CONNECTIONS

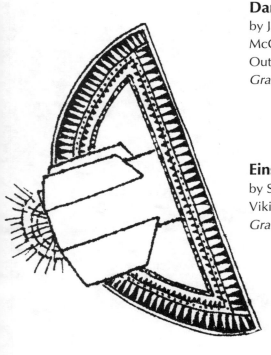

Danny Dunn and the Anti-Gravity Paint
by Jay Williams and Raymond Abrashkin; illustrated by Ezra J. Keats
McGraw-Hill, New York. 1956
Out of print
Grades: 4–8

With the invention of "anti-gravity paint," our heroes escape the Earth's gravity. Page 56 has a short explanation of the mechanics of traditional rockets.

Einstein Anderson Tells a Comet's Tale
by Seymour Simon; illustrated by Fred Winkowski
Viking Press, New York. 1981
Grades: 4–7

Chapter 10 describes a soapbox derby race in which teams have to build soapbox racing cars that weigh the same amount and are started in the same way. Our hero identifies the one test variable that allows his team to win the race. This episode is a wonderful example of a controlled experiment. You might ask your students how the soapbox derby experiment could be improved to determine whether the size of the wheels is the only important variable. (Build the racers exactly the same, except for the size of the wheels!)

From the Earth to the Moon

by Jules Verne
Airmont, New York. 1967
Grades: 8–12

The members of the Baltimore Gun Club plan to shoot a space gun to the moon. The planning, casting, and outfitting of the projectile are described in great detail. Many of Verne's ideas have come true—the site chosen for the launch is Florida! There are a few unfortunate references to the possibility of Seminole "savages" in the area, though none are encountered. Modern students could learn more about the actual achievements and way of life of the Seminole. Students could also be assigned to research which of the various scientific "facts" in the book are plausible, and which not, especially Chapter 4 which contains actual calculations of distance, velocity, coordinates, and related matters.

June 29, 1999

by David Wiesner
Clarion Books/Houghton Mifflin, New York. 1992
Grades: 3–6

The science project of Holly Evans takes an extraordinary turn—or does it? This highly imaginative and beautifully illustrated book has a central experimental component, related to controlled experimentation. Holly uses balloons rather than rockets to launch her efforts, but her planning, preparations, and analysis of unexpected results provide humorous and useful lessons.

The Paper Airplane Book

by Seymour Simon; illustrated by Byron Barton
Viking Press, New York. 1971
Grades: 4–8

Experiments with paper airplanes are described as well as explanations of the principles of aerodynamics involved.

Round the Moon

by Jules Verne
Airmont, New York. 1968
Grades: 8–12

In this sequel to *From the Earth to the Moon* the projectile (which had missed the moon) is traveling around the moon as its satellite. Describes the experiences of the three travelers and their adventures including experiencing weightlessness, narrowly missing an encounter with a meteor, and sighting a volcano. An appendix summarizes the errors in Verne's hypothesis.

Sally Ride and the New Astronauts: Scientists in Space
by Karen O'Connor
Franklin Watts, New York. 1983
Grades: 5–8

> This biography of Sally Ride is engagingly written and illustrated with many black and white photos of the Space Shuttle and dozens of training and support facilities. The book emphasizes the prejudices that women have had to overcome to be accepted as astronauts, and acknowledges the exceptional capabilities of Sally Ride and other women astronauts as scientists and engineers.

Stinker from Space
by Pamela F. Service
Charles Scribner's Sons, New York. 1988
Ballantine Books, New York. 1989
Grades: 5–8

> A girl encounters an extraterrestrial being who has had to inhabit the body of a skunk after an emergency landing. The girl and a neighbor boy help the skunk, Tsynq Yr (Stinker), to evade his enemies, the Zarnks, and get an important message to his own people. Stinker's departure from Earth involves "borrowing" the space shuttle. Rockets are mentioned during a discussion comparing the superior propulsion system used in Stinker's world to the solid and liquid-fueled rockets used to lift the shuttle.

Supersuits
by Vicki Cobb; illustrated by Peter Lippman
J.B. Lippincott, Philadelphia. 1975
Grades: 4–7

> Describes severe environmental conditions that require special clothing for survival: freezing cold, fire, underwater work, and thin or nonexistent air. Chapter 5 discusses spacecraft and the section "Why Step Outside?" looks at temperature requirements, anti-fire materials, and other design needs for pressure suits to be worn in space. Recent developments are lacking given the book's publication date.

The Time and Space of Uncle Albert
by Russell Stannard
Henry Holt, New York. 1989
Grades: 5–8

> Students who wish to go beyond the concrete experiments of the laboratory may be interested in conducting some "thought experiments" dreamed up by Albert Einstein, alias "Uncle Albert" in this whimsical story about a high school girl who gets some unusual help on her science project. Though the writing is a bit elementary for the high school level, the concepts of time and space are challenging; and accurately portray Einstein's Theory of Relativity—a cornerstone of modern physics.

To Space and Back

by Sally Ride with Susan Okie
Lothrop, Lee, and Shepard/Morrow, New York. 1986
Grades: 4–7

This is a fascinating description of what it is like to travel in space—to live, sleep, eat, and work in conditions unlike anything we know on Earth, complete with colored photographs aboard ship and in space. Details about weightlessness including gravity toilets and the 11 steps necessary to prepare lunch ("attach trays to the wall with Velcro") should fascinate students. The descriptions of what it's like to be inside the shuttle as the rockets propel it away from Earth (pages 17–18) are a great tie-in. Specifics about the spacecraft include a cross-section diagram showing the layout of the flight and mid-deck areas, a log of the countdown routine before takeoff, and a description of the space walk procedures where astronauts "become human satellites" to rendezvous with a satellite in orbit.

The Wonderful Flight to the Mushroom Planet

by Eleanor Cameron; illustrated by Robert Henneberger
Little Brown & Co., Boston. 1954
Grades: 5–8

Chuck and David respond to an advertisement from the mysterious Mr. Tyco Bass (inventor, astronomer, and mushroom grower): "Wanted: a small space ship about eight feet long, built by a boy, or by two boys." In Chapters 7 and 8, the boys meet Mr. Bass and have their spaceship outfitted and fueled by him. There are details about the rocket motor, invention of a special fuel, and the energy requirements of the space ship. This book is one of a series—all of which contain interesting scientific information in a science fiction format.

Along a parabola
life like a rocket flies,
Mainly in darkness,
now and then on a rainbow.
— *Andrei Voznesenski*
Parabolic Ballad

FINGERPRINTING

Grades 4–8   **Three Sessions**

Students explore the similarities and variations of fingerprints in these fingers-on activities. Students take their own fingerprints, devise their own classification categories, then apply their classification skills to solve a crime. The mystery scenario, "Who Robbed the Safe?" includes plot and character sketches.

Session 1

Students learn how to take their own fingerprints, using pencil, paper, and tape. They make the best clean set of their own fingerprints they can.

Session 2

Students group ten different fingerprints, according to how they look, to create their own scheme for classifying prints. They are introduced to the standard, arch-loop-whorl system, applying it to their own prints to get their "fingerprint formulas."

Session 3

Students are told of an imaginary theft from the safe of a company president. There are five suspects. A suspects sheet has the right hand prints of all suspects. After students determine fingerprint formulas and speculate on who did it, they receive a "Safe with Prints" sheet. Class evaluates evidence and determines whose prints are on safe—although that in itself does not prove who committed the crime. The ending is open-ended, asking students for any other techniques to help solve the crime.

Skills

Observing, Classifying, Drawing Conclusions, Making Inferences

Concepts

Fingerprints, Standard Fingerprint Classification System, Problem Solving

Themes

Systems & Interactions, Stability, Evolution, Diversity & Unity

Notes: Going Further activities include Fingerprint Art; using the guide as an introduction to genetics, biology, etc.; a language arts extension with students as news reporters covering the crime in Session 3, and drawing pictures of the suspects; a Find My Thumb game; classifications and mathematical exponents.

MAKING CONNECTIONS

Several books explore how **footprints** can be used as clues to the identities of the animals who left them. The rest of the books are **mysteries**, some of which involve the technique of **fingerprinting**. Like those books under *Crime Lab Chemistry*, these mysteries all involve the discovery of **evidence** and its subsequent analysis to make **inferences**.

Some teachers use newspaper articles with their students, which describe a crime (usually unsolved), the evidence and some possible inferences. There are books containing **nonfiction accounts of mysteries or scientific discoveries**, in which detective-like behavior was required. Such a mystery would be particularly apt if it involved fingerprinting or even what's currently called **DNA fingerprinting**. See also the books on **classification** in the Math Strand section on Logic.

CROSS REFERENCES

LITERATURE CONNECTIONS

Cam Jensen and the Mystery of the Gold Coins

by David A. Adler; illustrated by Susanna Natti
Viking Press, New York. 1982
Dell Publishing, New York. 1984
Grades: 3–5

Cam Jensen uses her photographic memory to solve a theft of two gold coins. Cam and her friend Eric carry around their 5th grade science projects throughout the book and the final scenes take place at the school science fair. (Other titles in the series include *Cam Jensen and the Mystery at the Monkey House* and *Cam Jensen and the Mystery of the Dinosaur Bones* in which she notices that three bones are missing from a museum's mounted dinosaur.)

Chip Rogers: Computer Whiz

by Seymour Simon; illustrated by Steve Miller
William Morrow, New York. 1984
Out of print
Grades: 4–8

Two youngsters, a boy and a girl, solve a gem theft from a science museum by using a computer to classify clues. A computer is also used to weigh variables in choosing a basketball team. Although some details about programming the computer may be a little dated, this is still a good book revolving directly around sorting out evidence, deciding whether or not a crime has been committed, solving it, and demonstrating the role computers can play in human endeavors. By the author of the Einstein Anderson series.

From the Mixed-Up Files of Mrs. Basil E. Frankweiler

by E.L. Konigsburg
Atheneum, New York. 1967
Dell Publishing, New York. 1977
Grades: 5–8

Twelve-year-old Claudia and her younger brother run away from home to live in the Metropolitan Museum of Art and stumble upon a mystery involving a statue attributed to Michelangelo. This book is a classic, and has been recommended to GEMS by many teachers.

The Great Adventures of Sherlock Holmes

by Arthur Conan Doyle
Viking Penguin, New York. 1990
Grades: 6–Adult

These classic short stories are masterly examples of deduction. Many of the puzzling cases are solved by Holmes in his chemistry lab as he analyzes fingerprints, inks, tobaccos, mud, etc. to solve the crime and catch the criminal. Nearly every Sherlock Homes story is suitable for this GEMS guide. These stories are available from many different publishers and in many editions.

Let's Go Dinosaur Tracking

by Miriam Schlein; illustrated by Kate Duke
HarperCollins, New York. 1991
Grades: 2–5

The many different types of tracks dinosaurs left behind and what these giant steps reveal is explored. Was the creature running … chasing a lizard … browsing on its hind legs for leaves … traveling in pairs or in a pack … walking underwater? At the end of the book, you can measure your stride and compare the difference when walking slowly, walking fast, and running. The process involved in attempting to draw conclusions about an animal's behavior or movement patterns from its tracks is similar to the way inferences are drawn from evidence in the GEMS mystery-solving activities. You could discuss with your students how they would weigh the evidence and consider the suspects if, for example, muddy shoeprints of a suspect had also been found at the scene of the crime.

The Mystery of the Stranger in the Barn

by True Kelley
Dodd, Mead, & Co., New York. 1986
Grades: K–4

A discarded hat and disappearing objects seem to prove that a mysterious stranger is hiding out in the barn, but no one ever sees anyone. A good opportunity to contrast evidence and inference.

The One Hundredth Thing About Caroline

by Lois Lowry
Houghton Mifflin, Boston. 1983
Dell Publishing, New York. 1991
Grades: 5–9

Fast-moving and often humorous book about 11-year-old Caroline, an aspiring paleontologist, and her friend Stacy's attempts to conduct investigations. Caroline becomes convinced that a neighbor has ominous plans to "eliminate" the children and Stacy speculates about the private life of a famous neighbor. Due to hasty misinterpretations of real evidence, both prove to be wildly wrong in their inferences. Gathering evidence, weighing it, and deciding what makes sense are good accompanying themes. A somewhat inaccurate portrayal of "color blindness" is a minor flaw.

Susannah and the Blue House Mystery

by Patricia Elmore

E.P. Dutton, New York. 1980

Scholastic, New York. 1990

Grades: 5–7

> Susannah (an amateur herpetologist) and Lucy have formed a detective
> agency. They check into the death of a kindly old antique dealer who
> lived in the mysterious "Blue House." They attempt to piece together clues
> in hopes of finding the treasure they think he has left to one of them. The
> detectives evaluate evidence, work together to solve problems, and prevent
> a camouflaged theft from taking place.

Susannah and the Poison Green Halloween

by Patricia Elmore

E.P. Dutton, New York. 1982

Scholastic., New York. 1990

Grades: 5–7

> Susannah and her friends try to figure out who put the poison in their
> Halloween candy when they trick-or-treated at the Eucalyptus Arms
> apartments. Tricky clues, changing main suspects, and some medical
> chemistry make this an excellent choice, with lots of inference and mys-
> tery.

Who Really Killed Cock Robin?

by Jean Craighead George

HarperCollins, New York. 1991

Grades: 3–7

> A compelling ecological mystery examines the importance of keeping
> nature in balance, and provides an inspiring account of a young environ-
> mental hero who becomes a scientific detective.

Whose Footprints?

by Masayuki Yabuuchi

Philomel Books, New York. 1983

Grades: K–4

> A good guessing game for younger students that depicts the footprints of a
> duck, cat, bear, horse, hippopotamus, and goat.

FROG MATH

PREDICT, PONDER, PLAY

Grades K–3   **Six Sessions**

*T*his series of lively mathematics activities jumps off from one of the well-known Frog and Toad stories. Although the title suggests this book is just about frogs, it also includes many activities unrelated to frogs. From free exploration of buttons, sorting and classifying, and a "Guess the Sort" game, students go on to design their own buttons, and use a graphing grid to organize data. Students "guesstimate" the number of small plastic frogs in a jar and the number of lima beans in a handful to develop the valuable life skill of estimating. A "Frog Pond" board game helps students develop strategic-thinking skills. The "Hop to the Pond Game" is an experiment in probability and statistics. Emphasis throughout is on student cooperation. Many extension and age modification suggestions are included.

Session 1
Students hear and interact with "The Lost Button" story by Arnold Lobel, then freely explore buttons. Younger students play the "Button Up" game while older students play "Button Factory."

Session 2
Students sort and classify buttons working with partners. They begin sorting by color, then devise new ways to sort and play a "Guess the Sort" game.

Session 3
Students create and decorate paper buttons of many shapes and sizes. Their buttons are then sorted and classified on a class graphing grid.

Session 4
Students "guesstimate" the number of small plastic frogs in a jar. Estimates are recorded and revised as the frogs are poured out and counted. Older students also estimate how many lima beans in a handful.

Session 5
In the "Frog Pond" game, students play an adaptation of the ancient Chinese logic game of NIM, as they try to take the last, or "magic" frog. When students have refined their strategies, the game can be changed to try to avoid taking the last, or "poison" frog.

Session 6
Students play the "Hop to the Pond "game, as 12 frogs race across the pond according to throws of the dice. You can read "The Swim" story from *Frog and Toad Are Friends.*

Skills
Observing, Cooperating, Logical Thinking, Problem Solving, Collecting and Interpreting Data, Sorting and Classifying, Using Geometric Vocabulary, Noticing and Articulating Patterns, Graphing, Estimating, Predicting, Developing and Testing Strategies

Concepts
Literature and Art Connections to Mathematics, Classification, Number, Place Value, Pattern Recognition, Estimation, Probability and Statistics

Themes
Diversity & Unity, Models & Simulations, Systems & Interactions, Structure, Scale

Math Strands
Number, Geometry, Patterns and Functions, Statistics and Probability, Logic

MAKING CONNECTIONS

Several books featuring **buttons** are included to accompany the **sorting and classifying** activities in the first three sessions that involve buttons. The other sessions use plastic frogs for a variety of activities that involve **estimation, counting, organizing data, logical thinking skills, probability** and **statistics**. There are books listed on several of those topics though many more can be found in the cross-reference section under the individual math strands. In addition, this list includes a variety of books on frogs from storybooks to inviting informational books.

CROSS REFERENCES

Toad stood on his head
for a long time.
But he could not think
of a story to tell Frog.
— *Arnold Lobel*
Frog and Toad Are Friends

LITERATURE CONNECTIONS

Alligator Shoes

by Arthur Dorros
E.P. Dutton, New York. 1982
Grades: Preschool–K

Being locked in a shoe store by mistake is Alvin alligator's dream come true. He tries on an endless variety of footwear, but his decision about what footgear he prefers will surprise and delight your students. Great to read before having students use their own shoes to sort and use as the data for a real-life graph. Ties in with Session 2: Sort, Classify and "Guess the Sort."

The Button Box

by Margarette S. Reid; illustrated by Sarah Chamberlain
Dutton's Children's Books, New York. 1992
Grades: Preschool–2

As a young child explores and examines the buttons in his Grandma's special button box, he sorts and classifies the buttons. Many attributes of buttons are illustrated including color and size, materials from which buttons are made, and buttons for different articles of clothing. This serves as a great introduction to Session 2. In *The Button Box*, buttons are also used to make a toy and create eyes on a puppet. A brief history of buttons for adults is included.

An old pond—
A frog leaping in—
The sound of water.
— *Matsuo Bashō*

A Chair For My Mother
by Vera B. Williams
Greenwillow Books/William Morrow, New York. 1982
Grades: K–3

A child, her waitress mother, and her grandmother save coins to buy a comfortable armchair after all their furniture is lost in a fire. The accumulation of coins of various denominations in a jar grows to a significant amount and is exchanged for dollars at the bank. Ties in with estimation activities in Session 4. See the article on Vera Williams on page 121 and her other books—*Something Special For Me* and *Music, Music for Everyone*. All the stories involve the money jar.

Corduroy
by Don Freeman
Viking Press, New York. 1968
Grades: Preschool–2

Corduroy, an adventurous teddy bear, explores his department store home one night in search of a button lost from his overalls. After he is returned to the toy display, an African-American girl named Lisa buys him with the pennies that she has saved and sews a new button on his overalls as soon as she gets him home. Nice connection to the lost button discussion and the button sorting and classifying activities in *Frog Math*.

Count & Find 100 Frogs & 10 Flies
by Polly Jordan
McClanahan Book Company, New York. 1992
Out of print
Grades: Preschool–2

Count by tens all the way to one hundred with this delightful, whimsical frog counting book. Added attraction is a hidden fly to find on each page. Ties in with Session 4: Frog Guesstimation.

The Frog
by Angela Royston; illustrated by Bernard Robinson
Ideals Children's Books, Nashville, Tennessee. 1989
Grades: K–3

The life cycle of a frog through the seasons, beginning with mating in spring, is explained. Predator and prey relationships among animals in a pond habitat are illustrated. A small glossary and scientific information on frogs are provided. A good sampling of the diversity of frogs around the world.

The Frog Alphabet Book
by Jerry Pallotta; illustrated by Ralph Masiello
Charlesbridge Publishing, Watertown, Massachusetts. 1990
Grades: K–3

A beautifully illustrated book that shows the diversity of frogs and other "awesome amphibians" from around the world.

Frog and Toad Are Friends
by Arnold Lobel
Harper & Row, New York. 1970
Grades: K–2

This well-known collection of stories provides the "jumping off place" for *Frog Math* activities. The story "A Lost Button" begins the button sorting and classification session and "The Swim" is often read with the "Hop to the Pond" game. The entire book is a strong example of how mathematics and literature can be "leapfrogged" together in fun and stimulating ways.

How Much Is A Million?
by David M. Schwartz; illustrated by Steven Kellogg
Lothrop, Lee & Shepard Books, New York. 1985
Grades: K–5

With detailed, whimsical illustrations that include children, goldfish, and stars, this book leads the reader to conceptualize what at first seems inconceivable—a million, a billion, and a trillion. Gives young children as concrete a representation of large numbers as possible. An adult-level explanation to calculate the numbers is included. Ties in with Session 4: Frog Guesstimation.

It's Mine!
by Leo Lionni
Alfred A. Knopf, New York. 1985
Grades: K–2

Three selfish frogs quarrel over who owns their pond and island until a storm makes them value the benefits of sharing. As children work as partners in many of the activities in *Frog Math*, the delightful frog friends in this book can be used to illustrate the benefits of the cooperative spirit.

Jumanji
by Chris Van Allsburg
Houghton Mifflin, Boston. 1981
Scholastic, New York. 1988
Grades: 3–Adult

A bored brother and sister left on their own find a discarded board game (called Jumanji), which turns their home into an exotic jungle. A final roll of the dice for two sixes helps them escape from an erupting volcano. Goes along well with Session 6: Hop to the Pond Game with 12 frogs racing to the pond.

Kimako's Story
by June Jordan; illustrated by Kay Burford
Houghton Mifflin, Boston. 1981
Grades: 2–3

A little girl works poetry puzzles indoors and has outdoor adventures with a dog. The story includes poem puzzles for the reader to complete and a map of Kimako's city paths. Promotes logical-thinking skills.

The Magic Bubble Trip

by Ingrid & Dieter Schubert

Kane/Miller Book Publishers, New York. 1981

Grades: K–3

> James blows a giant bubble that carries him away to a land of large hairy frogs where he has a fun, fanciful adventure.

The Stupids Die

by Harry Allard and James Marshall

Houghton Mifflin, Boston. 1981

Grades: K–3

> This book about the Stupid family—along with *The Stupids Step Out* and *The Stupids Have a Ball*—has children laughing with delight as they see the Stupid children do such silly things as slide up a banister or take a bath fully clothed in an empty bath tub. These books promote logical-thinking skills as kids find all the outrageous things and suggest ways to correct them. Ties in with Session 5, as logical thinking helps catch the magic frog!

Tuesday

by David Wiesner

Clarion Books, New York. 1991

Grades: K–Adult

> Frogs rise on their lily pads, float through the air, and explore the nearby houses while the inhabitants sleep in this beautifully illustrated and almost wordless book.

Whose Hat Is That?

by Ron Roy; photographs by Rosemarie Hausherr

Clarion Books/Ticknor and Fields, New York. 1987

Grades: Preschool–2

> Text and photographs portray the appearance and function of eighteen types of hats including a top hat, a jockey's cap, and a football helmet. The children and adults modeling the hats represent a rainbow of peoples. Makes a nice connection to classification activities in Session 2: Sort, Classify and "Guess the Sort." Children can bring in hats and sort them! His other book, *Whose Shoes Are These?*, shows nineteen different types of shoes and works equally well with the same session.

The Yellow Button

by Anne Mazer; illustrated by Judy Pedersen

Alfred A. Knopf, New York. 1990

Grades: K–8

> There is a button in a pocket, worn by a girl, lying on a couch, in a living room, in a house, at the edge of a field, near a mountain range, in a country, on a planet, in a universe. This book communicates the relationship of small to big in a very simple fashion.

Global Warming

& Greenhouse Effect
the

Grades 7–10

Eight Sessions

Students explore this controversial topic through a wide variety of formats, from hands-on science activities and experiments, to a simulation game, analysis of articles, a story about an island threatened by rising sea levels, and a world conference on global warming. This GEMS guide has two major aims: to present the scientific theories and evidence behind the environmental problem of global warming, and to help students see environmental problems from different points of view—from people who live on islands in the Pacific, to those who work in the lumber and auto manufacturing industries.

Session 1

Students discuss what they have already heard about the greenhouse effect. They examine two graphs that show global temperature records for the past 110 years and the climatic history of Earth for the past 160,000 years.

Session 2

Students perform a "greenhouse" experiment in which they compare the heating of an open container with the heating of a closed container. This experiment teaches the concepts of heating and cooling and equilibrium, and students discover that a closed container of air is able to trap heat more effectively than an open one.

Session 3

Students play a simulation game that approximates the greenhouse effect. They find out that photons of light energy from the sun are absorbed by the Earth and emitted as infrared (heat) photons; these photons are absorbed by carbon dioxide in the Earth's atmosphere; the more carbon dioxide, the more heat is trapped; and that is why carbon dioxide is called a "greenhouse gas." Several different greenhouse gases are discussed.

Sessions 4 and 5

Students conduct a series of experiments in which they compare the relative concentration of carbon dioxide in gas samples from four different sources: ambient air, human breath, car exhaust, and a chemical reaction between vinegar and baking soda. As homework, the students find out how much of the carbon dioxide in the atmosphere is contributed by various nations.

Session 6

Students consider the social and ecological consequences that might result if the Earth's climate should increase by 5° Fahrenheit, including both positive and negative effects. A story about a youngster's life on an island threatened by rising sea levels is read and discussed.

Sessions 7 and 8

Students participate in a "world conference" debate about the nature of the global warming problem, what can be done to avoid it, or reduce its detrimental effects, if climate change turns out to be unavoidable.

Skills

Observing, Measuring, Recording Data, Interpreting Graphs, Experimenting, Drawing Conclusions, Synthesizing Information, Role Playing, Using Simulation Games, Problem Solving, Brainstorming Solutions, Critical Thinking

Concepts

The Atmosphere, Visible and Infrared Photons, The Interaction of Light, Heat, and Matter, The Greenhouse Effect, Sources of Carbon Dioxide, Climate and Weather, The Effects of Climate Change

Themes

Systems & Interactions, Models & Simulations, Stability, Patterns of Change, Evolution, Scale, Structure, Energy, Matter

MAKING CONNECTIONS

There are some excellent books available to help your students better understand the global warming dilemma, including scientific aspects of the problem, what we can do about it as individuals, and the ethical and moral dimensions of the problem. These books, like GEMS activities, seek to help students comprehend differing points of view and consider ways to resolve conflicts and find solutions. Both the *Global Warming* and the *Acid Rain* GEMS units also endeavor to foster a sense of student empowerment in the face of environmental difficulty, and books like *Kid Heroes of the Environment* can certainly aid in that effort.

Several books are about deforestation and biological diversity—issues that are closely related to global warming. Others are science fiction that will help your students imagine what might happen in the future if our misuse of the environment goes unchecked.

CROSS REFERENCES

The Great Global Warming Limerick Debate

"The topic's a hot one at that,"
Said the first, putting on his straw hat,
"Evidence it is forming
That this globe is warming"
And the sweat dribbled down his cravat.

His opponent, unflappably cool,
Said, "Please don't take *me* for a fool—
If the temperature's rising
It isn't surprising
It goes up, then goes down, as a rule."

"Sure, cycles exist," said the first,
Gulping water to stave off his thirst,
"But our excess pollution's
A new contribution
For this reason, I fear the worst!"

"Just put all your worries on ice,"
Said the other, tossing some dice,
"I'd much rather wait
Take my chances with fate
Till the bill is due, why pay the price?"

The outcome my friend's up to you
To find out which side is more true
Before more time passes
Analyze all the gases
And help figure out what to do.

— *Lincoln Bergman*

LITERATURE CONNECTIONS

The Day They Parachuted Cats on Borneo: A Drama of Ecology
by Charlotte Pomerantz; illustrated by Jose Aruego
Young Scott Books/Addison-Wesley, Reading, Massachusetts. 1971
Out of print
Grades: 4–7

> This cautionary verse, based on a true story, explores how spraying for
> mosquitoes in Borneo eventually affected the entire ecological system,
> from cockroaches, rats, cats, and geckoes to the river and the farmer. A
> good example of the interacting elements of an ecosystem. The strong,
> humorous text makes the book successful as a read-aloud or as a play to be
> performed. Could be accompanied by dramatic presentation of the
> "Global Warming" limericks featured in the GEMS guide.

The Earth is Sore: Native Americans on Nature
adapted and illustrated by Aline Amon
Atheneum, New York. 1981
Out of print
Grades: 4–Adult

> This collection of poems and songs by Native Americans celebrates the
> relationship between the Earth and all creatures and mourns abuse of the
> environment. Could provide an introduction to the ecological and spiritual
> beliefs and traditions of many Native American (and other Indigenous)
> peoples, which have had a profound influence on the modern environ-
> mental movement. Illustrated with black and white collage prints made
> from natural materials.

The Endless Pavement
by Jacqueline Jackson and William Perlmutter; illustrated by Richard Cuffari
Seabury Press, New York. 1973
Out of print
Grades: 5–8

> Josette lives in a strange, bleak future where people are the servants of
> automobiles, ruled by the Great Computermobile. One night the "Screen"
> goes blank, and the father reminisces about what it was like before pave-
> ment when there was grass, "a soft green blanket that people used to walk
> on." Josette is inspired to free herself. She sabotages the Computermobile,
> starting a mass pedestrian movement towards the chain-link fence bound-
> ary. Automobile manufacturers are one of the interest groups represented
> at the World Conference on Global Warming students hold as part of the
> GEMS activities.

> Unless we slow and reverse the greenhouse gas buildup, we may have to adapt to such dramatic temperature shifts and ecological chaos in decades instead of millennia.
>
> — *Sierra Club "Global Warming," 1989*

The Faces of Ceti
by Mary Caraker
Houghton Mifflin, Boston. 1991
Grades: 6–12

Colonists from Earth form two settlements on adjoining planets of the Tau Ceti system. One colony tries to survive by dominating the natural forces that they encounter, while those who land on the planet Ceti apply sound ecological principles and strive to live harmoniously in their new environment. Nonetheless, the Cetians encounter a terrible dilemma—the only edible food on the planet appears to be a species of native animals called the Hlur. Teen-age colonists Maya Gart and Brock Magnus risk their lives in a desperate effort to save their fellow colonists from starvation without killing the gentle Hlur.

The Great Kapok Tree: A Tale of the Amazon Rain Forest
by Lynne Cherry
Harcourt, Brace, Jovanovich, San Diego. 1990
Grades: K–4

Animals that live in a great kapok tree in the Brazilian rain forest try to convince a man with an ax of the importance of not cutting down their home. While he is asleep, the animals, including a boa constrictor, bee, monkeys, toucan, macaw, tree frog, jaguar, porcupines, anteaters, and a sloth, all try to influence his dreams by conveying the beauty and utility of the rain forest. At the end of the story, the man puts down his ax and walks away. The preservation of the rain forest, sometimes called "the lungs of the Earth," has a direct bearing on the other factors that contribute to possible global warming, and deforestation is discussed in more detail in the background section of the GEMS guide.

The Greenhouse Effect: Life on a Warmer Planet
by Rebecca L. Johnson
Lerner Publications, Minneapolis. 1990
Grades: 5–9

Recent research on the causes and probable impact of the greenhouse effect is presented. There are chapters on carbon dioxide and deforestation, other greenhouse gases, the effects of global warming and modeling changing climates, and a scenario for life on a warmer planet. The last chapter "What Can We Do?" looks at reducing carbon dioxide, exploring new energy sources, and planning and working together. A good deal of scientific information is presented and made accessible along with accompanying graphs, charts, and photographs. This is a good scientific introduction to the issues raised by "global warming."

Just A Dream

by Chris Van Allsburg

Houghton Mifflin, Boston. 1990

Grades: 1–6

> When he has a dream about a future Earth devastated by pollution, Walter begins to understand the importance of taking care of the environment. Planning and taking part in the world conference on global warming in the GEMS guide encourages a similar awareness.

Kid Heroes of the Environment:
Simple Things Real Kids Are Doing To Save the Earth

edited by Catherine Dee; illustrated by Michele Montez

Earth Works Press, Berkeley, California. 1991

Grades: 3–12

> Twenty-nine profiles of individuals and organizations working at home, at school, locally, and nationally, to impact the environment through "kid power." In addition to specific projects about pollution, recycling, and endangered species, many profiles deal with collecting and publicizing information such as a "green" yearbook, fundraising for adoption of rain forest acreage, and organizations like "Just Say Yes" and "Out of the Ozone," which deal with legislation and public opinion. An encouraging and excellent connection to ideas of environmental responsibility embedded in the GEMS activities.

The Lorax

by Dr. Seuss

Random House, New York. 1971

Grades: Preschool–8

> Humans are destroying a beautiful forest because it contains "truffula" trees, needed to make "thneeds." The lorax, who speaks for the environment, explains that the deforestation has affected not only Brown Bar-ba-loots who eat truffula fruits, but also the swans, fish, and other creatures. Ironically, at the end of the book, the thneeds factory owner is placed in charge of the last truffula tree seed. Deforestation relates to the "global warming game" in Session 2 and to many other activities in this GEMS guide.

One Day in the Tropical Rain Forest

by Jean C. George; illustrated by Gary Allen

HarperCollins, New York. 1990

Grades: 4–7

> When a section of rain forest in Venezuela is scheduled to be bulldozed, a young boy and a scientist seek a new species of butterfly for a wealthy industrialist who might preserve the forest. As they travel through the ecosystem rich with plant, insect, and animal life, everything they see on this one day is logged beginning with sunrise at 6:29 a.m. They finally arrive at the top of the largest tree in the forest and fortuitously capture a specimen of an unknown butterfly.

Restoring the Earth: How Americans Are Working to Renew Our Damaged Environment

by John J. Berger
Doubleday, New York. 1987
Grades: 7–12

> This book profiles individuals and groups active in conservation. It is written at the adult level and does not include any photographs or illustrations. Especially relevant is the first section which deals with restoration of a polluted lake, a trout stream, and a dead marsh. "Mother Nashua" is the true story of the clean-up effort portrayed in Lynne Cherry's picture book *A River Run Wild*. Other sections relate to land use and waste disposal, human settlements and their environmental impact, and wildlife preservation.

Sweetwater

by Laurence Yep; illustrated by Julia Noonan
Harper & Row, New York. 1973
Grades: 5–8

> The sea level around a star colony rises year after year as the inhabitants attempt to save their way of life and protect the city from flooding. They investigate why the sea level continues to rise and the tides become increasingly larger.

Who Really Killed Cock Robin?

by Jean C. George
HarperCollins, New York. 1991
Grades: 3–7

> This compelling ecological mystery examines the importance of keeping nature in balance, and provides an inspiring account of a young environmental hero who becomes a scientific detective.

Will We Miss Them?

by Alexandra Wright; illustrated by Marshall Peck
Charlesbridge Books, Watertown, Massachusetts. 1992
Grades: 2–5

> A sixth-grader writes about "some amazing animals that are disappearing from the earth." On each double-paged spread, the question is asked "Will we miss . . . ?" and gives basic information on 13 animal species and how their habitats may be threatened. The illustrations are strong, a simple map shows approximate locations of threatened species, and the book presents the hopeful message that we don't have to miss these animals, we can still save them!

GROUP SOLUTIONS

Cooperative Logic Activities

for Grades K-4

The more than fifty highly involving cooperative logic activities in this guide are designed for groups of four students. Each student receives a clue to a problem and needs to share the information with all other group members to find the solution. The entire group is responsible for finding the solution, and it can **ONLY** be figured out by connecting the information from **ALL** the clues. The games, puzzles, and problems are clearly stated and cleverly illustrated, with easy-to-use, fun manipulatives. Introductory sections explain how to use the book and discuss cooperative learning and logic in the classroom. Many key elements in mathematics and science are naturally explored, leading to student questions such as, "Are we really doing math? Can we do this again, soon?"

Skills

Visual Discrimination, Counting, Sorting, Classifying, Using the Process of Elimination, Using Deductive Reasoning, Communicating, Sequencing, Spatial Visualization, Recognizing Shapes and Colors, Using Charts, Comparing Amounts, Map Reading, Finding Locations on Maps, Using Map Legends

Concepts

Numeration, Computation (Fractions), Directionality, Ordinal Numbers, Money, Mapping

Themes

Patterns of Change, Models & Simulations, Systems & Interactions, Stability, Structure, Scale

Math Strands

Number, Measurement, Geometry, Pattern, Functions, Statistics and Probability, Logic, Algebra

MAKING CONNECTIONS

There are a great many children's literature books that convey positive lessons relating to cooperation. You and your students no doubt have your own favorites. Reading such books can help set the stage for children to work together on cooperative logic activities, and in many other ways as well. Several books that center on **cooperation** are listed. In addition, there are books to go along with the activities about **money, maps** and **number**. A delightful version of the classic fairy tale "Goldilocks and the Three Bears" is included to complement the many activities in this GEMS guide that include small bear figures, such as "Bear Line-Ups" and "Bear Park Map."

CROSS REFERENCES

LITERATURE CONNECTIONS

Alexander Who Used to Be Rich Last Sunday
by Judith Viorst; illustrated by Ray Cruz
Atheneum, New York. 1978
Grades: K–3

A humorous look at how Alexander spends the dollar that his grandparents give him on a Sunday visit. Though Alexander would like to save the money for a walkie-talkie, saving money is hard! He and his money are quickly parted on such items as bubble gum, bets, a snake rental, and a garage sale.

Annabelle Swift, Kindergartner
by Amy Schwartz
Orchard Books/Franklin Watts, New York. 1988
Grades: K–2

Although some of the things her older sister taught her at home seem a little unusual at school, other lessons help make Annabelle's first day in kindergarten a success. The lesson that ties in beautifully with the Coin Count activity involves counting pennies and nickels. Annabelle surprises her classmates with her expertise in counting money. This book will especially appeal to kindergarten kids, but older students will also identify with it.

As the Crow Flies: A First Book of Maps
by Gail Hartman; illustrated by Harvey Stevenson
Bradbury Press, New York. 1991
Grades: Preschool–2

Different maps chart the worlds and favorite places of an eagle, rabbit, crow, horse, and seagull, each from its own perspective. A large map on the last two pages incorporates all the geographical areas—the mountains, meadow, lighthouse and harbor, skyscrapers, hot dog stand, etc. A wonderful book with which to introduce or reinforce an appreciation of scale. Connects very well with the map activities in *Group Solutions.*

Babushka's Doll
by Patricia Polacco
Simon & Schuster, New York. 1990
Grades: K–3

Natasha is a demanding and rambunctious little girl who borrows a doll that turns out to be even more demanding than she is. Natasha learns something about herself—and that playing with Babushka's doll once is enough! A good book to start a discussion about cooperative behavior.

A Chair for My Mother

by Vera B. Williams

Greenwillow Books/William Morrow, New York. 1982

Grades: K–3

> A child, her waitress mother and her grandmother save coins to buy a comfortable armchair after all their furniture is lost in a fire. The accumulation of coins of various denominations in a jar grows to a significant amount and is exchanged for dollars at the bank. Ties in with the "Coin Count" activities 1–6 in which students put coins in a cup and count them.

Goldilocks and the Three Bears

retold and illustrated by Jan Brett

Dodd, Mead & Company, New York. 1987

Grades: K–3

> This classic tale introduces a consistent and predictable scale comparison, as Goldilocks encounters the three bowls of porridge, the three chairs, the three beds, and finally, the three bears themselves. Pay attention to the gorgeous illustrations in this version and notice the caterpillars changing to butterflies, the many varieties of birds' eggs, seeds and leaves, and forest scenes, which show a system of interacting plants and animals. Your students will miss none of these details. A fun way to build on the bear motif.

How Many Snails?

by Paul Giganti, Jr.; illustrated by Donald Crews

Greenwillow Books, New York. 1988

Grades: Preschool–3

> A young child takes walks to different places and wonders about the amount and variety of things seen on the way, from fish to fire trucks to cupcakes. This book invites the reader to actively participate and count meaningfully by attributes.

It's Mine!

by Leo Lionni

Alfred A. Knopf, New York. 1985

Grades: K–2

> Three quarrelsome frogs quibble over ownership of their pond, the island, and even the air! A storm makes them value the benefits of sharing when they must share the last rock rising above the flooded waters. This is a helpful story to use to introduce the merits of cooperation before beginning cooperative activities.

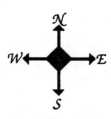

Music, Music For Everyone

by Vera B. Williams

Greenwillow Books, New York. 1984

Grades: K–4

> Rosa's grandmother is sick and likes to listen to Rosa and her three friends play their musical instruments. They form the Oak Street Band and Rosa contributes her share of the money to the "money jar" in the living room. Great example of four young girls of diverse cultures cooperating and working together to help out with a family crisis.

The Purse

by Kathy Caple

Houghton Mifflin, Boston. 1986

Grades: K–2

> Katie keeps her money in a band-aid box until her older sister convinces her to buy a purse. Because she uses all her money for the purse, she has nothing left to put in it! Katie does earn more money and the way she spends it provides a novel twist to the end of the story.

The Secret Birthday Message

by Eric Carle

Harper & Row, New York. 1986

Grades: Preschool–3

> Instead of a birthday package, Tim gets a mysterious letter written in code. Full-color pages, designed with cut-out shapes, allow children to fully participate in this enticing adventure and follow the trail of shapes in search of the birthday gift.

Stone Soup

by Marcia Brown

Charles Scribner's Sons, New York. 1947

Grades: K–3

> Three hungry soldiers come marching into a French village in search of a bit of food. Not until the soldiers begin to make a pot of stone soup do the villagers begin to share their food. Each family contributes a bit of vegetable, meat, grain, milk, or spice to make a soup that the whole village sits down to eat. Though the villagers are tricked into sharing, the benefits of working together are wonderfully illustrated.

The Wolf's Chicken Stew

by Keiko Kasza

G.P. Putnam's, New York. 1987

Grades: K–3

> A hungry wolf's attempts to fatten a chicken for his stew pot have unexpected results. A delightful book to begin an investigation of 100 as the delectable items that the wolf brings to the chicken's doorstep come in quantities related to or of 100. When he arrives at the chicken's house to capture her for his stew, he has quite a surprise, and the unexpected ending is very touching.

The "Money Jar" Trilogy

By Jaine Kopp

One of my favorite authors, Vera B. Williams, has written a trilogy of books about real-life situations that—as so many situations do—involve money. Money is of high interest to most children, and I use these books to introduce activities about the monetary system to my primary students. All three stories feature a young girl named Rosa, her mother, who works as a waitress, and her grandmother. Strong values of social cooperation and caring for others resound throughout these three books as well.

The first story, *A Chair For My Mother*, introduces us to their money jar. Rosa and her family are accumulating coins in the large jar to purchase a new comfortable armchair as they lost all their furniture in a fire. As the story progresses, the jar is filled and then the coins are exchanged for dollars at the bank. And, of course, the perfect armchair is purchased with the money!

After this story, I take out a small jar filled with coins (appropriate to the level of the students I am currently teaching). The children make guesses about the amount of money in the jar and what they could buy with that amount of money. Then we pour out the money and guess again about the amount since we can see the coins more clearly now. (This ties in nicely with the "Frog Guesstimation" activity in the GEMS guide *Frog Math*.) I ask the children for ways to count the money. We usually end up sorting the coins by value and then counting. It is interesting to compare our guesses with the actual amount and even more interesting to see how many of the items can be purchased with that amount of money! Sometimes, for homework, I have students make a list of things they could buy for that amount. I also like to use the same small jar and fill it with coins of the same value (e.g., all nickels) and do a follow-up estimation with the class on a subsequent day.

In the next story, *Something Special For Me*, the money from the once again partially-filled jar is to be used to purchase a special birthday present for Rosa. There isn't enough money to purchase all the things Rosa might like, and she has a hard time deciding on the one gift she wants the most—which turns out to be an accordion. I particularly like this real-world situation which most of us face—how to decide what to buy with a limited amount of money! In connecting to this story, I do an art project: students have to "purchase" their supplies with a set amount of play money. I give each student the same "allowance." They use the money at the "Art Store" where special supplies, such as construction paper, pipe cleaners, tissue paper, cotton balls, straws, stickers, glitter, etc. are available at various prices. It is interesting to see how students spend their money. The imaginative array of art pieces they produce is a wonderful example of the diverse uses that can be made of similar materials!

Rosa's accordion becomes the lead-in to the third story, *Music, Music For Everyone*.

In this story Rosa wants to help earn money to pay for added family expenses due to her grandmother's illness. She forms a band with three of her friends and collectively they make beautiful music at a local celebration. Rosa puts her portion of the money they earn into the jar. To help children understand the value of class materials, I have children work cooperatively to purchase one new item for the classroom. First, we brainstorm a list of things they would like for the class. Then, we research the cost of items and decide upon one to purchase. As a group, we think about fund raising efforts we can do collectively as well as how students can earn money individually. We get a large money jar and watch the money grow! It is important for the children to see that it takes time and planning to acquire new special things. It also gives them a greater appreciation for the items we already have in our classroom. Sometimes we have special fund raisers for field trips as well. (*Music, Music For Everyone* is also a great book to read with the GEMS *Group Solutions* guide. As in the cooperative logic activities in *Group Solutions*, the four girls in the book work together to solve a problem. In addition, *Group Solutions* includes some challenging "Coin Count" games.)

There are many more activities with coins and money that I do with my students as part of units on the monetary system. The activities above are only those inspired by the wonderful stories by Vera B. Williams! In fact, our class publishes its own books and we have dedicated one to Vera B. Williams and another to Rosa!

Jaine Kopp is a teacher in the Mathematics Education Department at the Lawrence Hall of Science.

HEIGHT·O·METERS

Grades 6–10

Four Sessions

Students are introduced to the principle of triangulation by making simple cardboard devices called Height-O-Meters. Students measure angles to determine the height of the school flagpole, and compare how high a Styrofoam and rubber ball can be thrown. *Height-O-Meters* is a prerequisite for GEMS rocketry activities.

Session 1

Each students constructs a Height-O-Meter from cardboard, push pin, and a provided template.

Session 2

Students calibrate Height-O-Meters with a paper clip on disk part of instrument.

Session 3

Students measure the angular height of an object, such as the school flagpole, and then learn how to calculate the actual height in meters.

Session 4

After predicting what they think will be the result, students use Height-O-Meters to measure the height of two objects thrown into the air by the teacher: a Styrofoam ball and a rubber ball.

Skills

Predicting, Estimating, Making and Calibrating Scientific Instruments, Measuring in Degrees, Graphing, Calculating, Interpreting Data

Concepts

Angular and Linear Measurement, Triangulation with Scale Drawings, the Metric System

Themes

Systems & Interactions, Models & Simulations, Stability, Scale

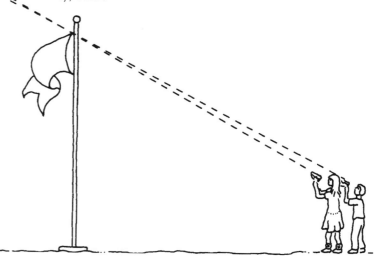

Notes: Going Further activities relate triangulation to the real-life activities of forest rangers and astronomers and also introduce the tangent function of trigonometry.

Challenge to Teachers

We're stumped! Any ideas of literature extensions for this activity guide involving angular and linear measurement and triangulation with scale drawings? Possible tie-ins would be books relating to forest rangers, model rocketry, angles, the metric system, calibration, helium balloons, and the distances to the stars. And don't forget those favorite novels about flagpole sitting! We did find one with a flagpole motif.

CROSS REFERENCES

LITERATURE CONNECTIONS

M.C. Higgins, The Great

by Virginia Hamilton
Macmillan, New York. 1974
Grades: 7–12

This strange and moving slice-of-life tale features teen-aged Mayo Cornelius Higgins who sits on a 40-foot pole and surveys the denizens of Sarah's Mountain. M.C.'s life is changed forever when a wandering young woman spends time in the area, and a man from the city comes to record his mother's folk singing. The early chapters, particularly the end of Chapter 1 and all of Chapter 2, are concerned with the view and perspective from the pole's height, and so connect most directly with the flagpole measurement activity. Eventually the pole becomes a symbol, a marker of the family's multi-generational connection to their home. There is also information on the way that strip mining has ravaged the land, is threatening their house, and affecting the lives of their unusual neighbors and the local animals. The real "measurement" in this award-winning novel is of maturity, responsibility, and a growing appreciation of family and an understanding of self.

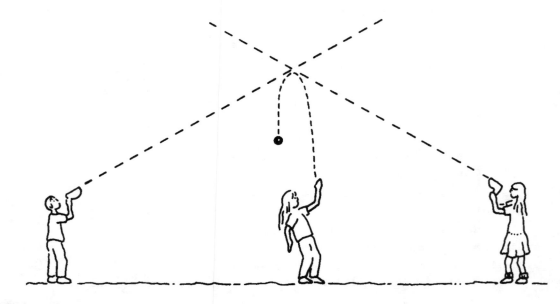

HIDE A BUTTERFLY

Preschool–K   **Three Sessions**

Children create camouflaged butterflies, hungry birds, and a meadow of flowers to dramatize a familiar scene from nature.

Session 1

The concept of a mural is introduced. The teacher demonstrates gluing flowers and grass; children identify parts of a flower, make flowers and grass for the mural, then talk about small animals they've seen in real grass or on flowers.

Session 2

Students learn about a meadow. They decorate paper butterflies and make bird puppets, later acting out the behavior of birds and butterflies. Camouflage, feeding on nectar, and other refinements are introduced as children enact "The Butterfly Play."

Session 3

Children learn more about real butterflies and their various means of protective coloration. Teacher displays pictures of different kinds of butterflies, including use of false eyes on wings, and takes children for a walk to observe butterflies in nature.

Skills

Observing, Communicating, Comparing, Matching, Role Playing, Using Scissors (optional)

Concepts

Biology, Entomology, Butterflies, Protective Coloration, Predator/Prey Relationships

Themes

Systems & Interactions, Models & Simulations, Scale, Energy, Diversity & Unity

Notes: Has modifications for Grades 1–3 and special instructions for younger preschoolers. Has art mural of meadow with flowers, a great deal of artwork, paper bag bird-puppet, camouflage.

MAKING CONNECTIONS

The following books look at a butterfly's **lifestyle** and **life cycle**, as well as showing how many animals use **camouflage** and other **defensive coloration** for **protection**. Language is encouraged in stories and rhymes that challenge children to go on a nature hunt with animals that are hidden in realistic and fantasy settings. To introduce the topic of butterflies, you may want to use the books that present the flower or meadow habitat of a butterfly through beautiful illustrations and verse. Comparing defenses of other animals is highlighted in two fantasy books about chameleons who use their ability to change color to their advantage.

CROSS REFERENCES

LITERATURE CONNECTIONS

The Butterfly Hunt
by Yoshi
Picture Book Studio, Saxonville, Massachusetts. 1990
Grades: Preschool–2

A boy pursues and captures elusive butterflies, but decides it is more fun to carry home memories than a trophy. He sets the butterfly free, "forever and ever the butterfly was his very own." Beautiful full-color illustrations of a wide variety of butterflies.

Chameleon Was a Spy
by Diane R. Massie
Scholastic, New York. 1979
Grades: K–5

Chameleon uses his ability to change colors to help the Pleasant Pickle Company get back their secret recipe from the competitors who stole it!

The Flower Alphabet Book
by Jerry Pallotta; illustrated by Leslie Evans
Quinlan Press, Boston. 1988
Grades: Preschool–2

Beautiful, scientifically precise alphabet picture book showing many varieties of flowers and plants. A good early primary accompaniment to the meadows ideas and the activities relating to nectar, pollen, and flowers in *Buzzing a Hive.*

The Girl Who Loved Caterpillars
adapted by Jean Merrill; illustrated by Floyd Cooper
Philomel Books/Putnam & Grosset, New York. 1992
Grades: 2–6

Izumi loves caterpillars but wonders "Why do people make such a fuss about butterflies and pay no attention to the creatures from which butterflies come? It is caterpillars that are really interesting!" Izumi is interested in the "original nature of things," and in doing things naturally. This book, at a somewhat higher age level than the *Hide a Butterfly* activities, could be read out loud to younger children, adapted as needed to their vocabulary and level of understanding. An excellent portrayal of an independent-thinking female role model.

How to Hide a Butterfly and Other Insects
by Ruth Heller
Grosset & Dunlap, New York. 1985
Grades: Preschool–3

Written in rhyme, this beautifully illustrated book describes and shows how insects camouflage themselves and are often "out of view, although they're right in front of you."

How to Hide a Polar Bear and Other Mammals
by Ruth Heller
Grosset & Dunlap, New York. 1985
Grades: Preschool–3

> Go on a nature hunt to find the camouflaged polar bear, deer, zebra and other handsome mammals hiding in the brilliantly illustrated pages of this book.

The Lamb and the Butterfly
by Arnold Sundgaard; illustrated by Eric Carle
Orchard Books, New York. 1988
Grades: K–3

> A protected lamb and an independent butterfly discuss their different ways of living. Spirited introduction to the concept of diversity and acceptance of differences.

Lizard in the Sun
by Joanne Ryder; illustrated by Michael Rothman
William Morrow, New York. 1990
Grades: Preschool–2

> A friendly narration guides you through your day as a lizard: you are camouflaged from hungry birds and hidden from insects that become your next meal. Children enjoy seeing the natural world from a lizard's view-point, and learn interesting facts about the lizard's lifestyle.

The Mixed-Up Chameleon
by Eric Carle
Harper & Row, New York. 1975
Grades: Preschool–2

> A bored chameleon wishes it could be more like all the other animals it sees, but soon decides it would rather just be itself. Protective coloration (the chameleon changes color according to the surface on which it rests) and energy (when the chameleon is warm and full, it turns one color, when cold and hungry, it turns another) are woven into the story, as are a discussion of the attributes of various other animals.

The Rose in My Garden
by Arnold Lobel; illustrated by Anita Lobel
Greenwillow Books, New York. 1984
Grades: Preschool–2

> Each page adds a new rhyming line to a poem as a beautiful garden of flowers, insects, and animals grows. A surprise interaction among the garden residents takes place at the end of the book. Young readers will enjoy the repeated patterns in the story.

Float like a butterfly,
Sting like a bee.

— *Muhammad Ali*

The Very Hungry Caterpillar

by Eric Carle
Philomel Books, New York. 1969
Grades: K–3

> Follow the progress of a hungry little caterpillar as it eats its way through a varied and very large quantity of food until, full at last, it forms a chrysalis around itself and goes to sleep. A good opportunity to correct the common misuse of the word "cocoon" (moths emerge from cocoons), with the correct term "chrysalis" for butterflies.

We Hide, You Seek

by Jose Aruego and Ariane Dewey
Greenwillow, New York. 1979
Grades: Preschool–2

> Young readers, led by a bumbling rhinoceros, try to find animals that are hidden in their natural environment. A delightful introduction to the concept of camouflage.

Where Butterflies Grow

by Joanne Ryder; illustrated by Lynne Cherry
Lodestar Books/E.P. Dutton, New York. 1989
Grades: K–5

> Here's a delightful description of what it might feel like to change from a caterpillar into a butterfly. Structure, metamorphosis, locomotion, camouflage, and feeding behaviors are all described from the point of view of the butterfly. Also included are gardening tips on how to attract butterflies. This is a beautiful book and offers unusually detailed drawings of metamorphosis.

Where Does the Butterfly Go When It Rains?

by May Garelick; illustrated by Leonard Weisgard
Scholastic Book Services, New York. 1961
Out of print
Grades: Preschool–3

> The child narrator describes, in simple rhyming text, how various insects and animals might respond to rain and questions what he doesn't understand, "Where does the butterfly go when it rains?" Good model of questioning process, ending with the idea to go out and find a butterfly and observe its behavior. A good tie in to Session 3 where students observe butterflies in nature.

LHS GEMS

Who's Hiding Here?

by Yoshi
Picture Book Studio, Saxonville, Massachusetts. 1987
Grades: Preschool–3

This magnificently illustrated book is about animals that use camouflage to protect themselves. Each page has a full-color batik-style illustration as well as a riddle that ends with the predictable question, "Who's hiding here?" Two pages of information on camouflage and mimicry are included at the end for older readers.

Wild Wild Sunflower Child Anna

by Nancy W. Carlstrom; illustrated by Jerry Pinkney
Macmillan Publishing Co., New York. 1987
Grades: Preschool–1

This poetic and vividly illustrated story is about a young child's playful adventures with nature and shows potential butterfly habitats. Young listeners will enjoy the lively verse; they can also recite the poem aloud.

It is eternity now. I am in the midst of it. It is about me in the sunshine; I am in it, as the butterfly in the light-laden air. Nothing has to come; it is now. Now is eternity; now is the immortal life.

— *Richard Jefferies*
The Story of My Heart

Dear GEMS

I felt compelled to write and share some of the ways that I have incorporated literature with the *Hide A Butterfly* GEMS unit. I have been doing this unit for the past three years with my first grade class. They love the art projects and drama activities.

I must confess, however, that science is not my strong point. My first love is literature and language arts. Therefore, I wanted to incorporate some of the wonderful, colorful and language-rich books about butterflies into this unit. Enclosed is the result of my endeavors.

My classroom has a "poetry wall" and I change the poems to relate to the theme we are currently studying. Poems about butterflies that I have used are listed on a separate page. For choral reading, the children read the poetry wall together on an almost daily basis. They also copy the poems in their handwriting book using their best handwriting. In this way the children have a record of all the poems that we have studied at the end of the year.

I hope you find some ideas that might be useful for your handbook.

Jackie Corsetti
First Grade Teacher
St. Agnes School
Concord, California

Here is an edited version of the lesson plan.

Lesson 1

Objective:
Understanding the concept of a meadow.

1) Read *Over in the Meadow*.
2) Read book a second time, but leave off the rhyming word at the end of second line so children can join in. Children enjoy joining in this predictable rhyme and it keeps their attention focused on the language.
3) Read *Life in the Meadow*.
4) Prepare children for an observation walk by asking them to list the animals they might see on a walk. Explain how the area where they will walk is the same or different from a meadow.
5) Go for a walk. Have the children bring a notebook and write down each animal and insect they see as well as how many.
6) Back in the classroom, discuss what was seen on the walk. Record the information on a large chart or divide the class into small groups and have each group create a chart. This information can be compared, categorized, counted, summarized, and the chart displayed throughout the time you work on *Hide A Butterfly*.

Lesson 2

Objective

Awareness of the different kinds of butterflies, moths, and flowers.

1) Read *I Like Butterflies*.
2) Have children recall butterflies they have seen and where. You list them on a chart as a language experience story.
3) Explain concept of a meadow and have children discuss what should be included on a mural of a meadow the children will make.
4) Have the children make flowers for their meadow mural.

Lesson 3

Objective

Describe colors and behavior of butterflies, and recall facts about migration of monarch butterfly.

1) Read *The Butterflies Come*.
2) Have children retell story and record main events on a chart.
3) Re-read sections of story describing monarch butterflies.
4) Have children describe butterflies using descriptive language. Refer to children's story "Butterflies We Have Seen."
5) Have children make the butterfly for their mural. Butterfly color should match flower color.

Lesson 4

Objective

Understanding the four stages of a butterfly's life.

1) Read *The Very Hungry Caterpillar*.
2) Re-read story. On the "Saturday" page, have children name all kinds of "junk food" the caterpillar ate, and imagine even more.
3) Re-read story. Cut pieces of "fruit" out of felt (don't forget the hole), and have children volunteer to stand up with a piece of felt fruit when that type of fruit is mentioned in the story. The children place the felt fruit on their shirts, which act as a flannel board. (We usually have to do this many times as it is so much fun.)
4) Have children act out four stages of a butterfly's life—a grassy area outside is best.
Stage 1: Have children get down on their hands and knees and curl up in a round ball to demonstrate the egg stage.
Stage 2: Have children uncurl and crawl on hands and knees demonstrating the caterpillar stage.
Stage 3: Children roll over and over, with their bodies elongated, demonstrating the chrysalis stage.
Stage 4: Children show emergence from chrysalis stage by getting on their feet, flapping their arms, and "flying" around the play area demonstrating the butterfly stage.

Dear GEMS

Lesson 5

Objective

Understanding the butterfly's behavior and need for protection from predators.

1) Read *The Lamb and the Butterfly*.
2) Discuss with children the different lives of a lamb and a butterfly.
3) Ask children how a butterfly protects itself from the weather and from predators such as birds. Discuss the concept of protective coloration (refer to the book *The Butterflies Come*).
4) Have children make bird puppets from paper bags and construction paper.
5) Have children enact "The Butterfly Play" using bird and butterfly puppets. Try to get the children to use descriptive language to describe what is happening as they act out the play.

The following books were used as literature connections to the *Hide A Butterfly* GEMS guide. The starred books are part of the lesson plan. All of the books, plus a pictorial reference book on butterflies, were kept available for the children to look at while *Hide A Butterfly* was being presented. Children are encouraged to share other books and pictures about butterflies.

Backyard Insects
   by Millicent E. Selsam and Ronald Goor
Chickens Aren't the Only Ones
   by Ruth Heller

*I Like Butterflies by Gladys Conklin
*Life in the Meadow
   by Eileen Curran
*Over in the Meadow
   (Traditional Rhyme; illustrated by Ezra Jack Keats)
*The Butterflies Come by Leo Politi
*The Lamb and the Butterfly
   by Arnold Sundgaard
*The Very Hungry Caterpillar
   by Eric Carle
Where Does the Butterfly Go When It Rains? by May Garelick

POEMS ABOUT BUTTERFLIES AND CATERPILLARS

Hurt No Living Thing

Hurt no living thing;
Ladybird, not butterfly,
Nor moth with dusty wing,
Nor cricket chirping cheerily,
Nor grasshopper so light of leap,
Nor dancing gnat, nor beetle fat,
Nor harmless worms that creep.

— *Christina Rossetti*

Butterfly

What is a butterfly? At best
He's but a caterpillar dressed.

— *Benjamin Franklin*

HOT WATER AND WARM HOMES FROM SUNLIGHT

Grades 4–8

Five Sessions

Students build model houses and hot water heaters to discover more about solar power. They conduct experiments to determine the effects of size, color, and number of windows on the amount of heat produced from sunlight. Information on the "greenhouse effect" connects to the GEMS guide on *Global Warming and the Greenhouse Effect.*

The Introductory Activity: "Controlled Experimentation" is an on-paper experiment about growing plants that defines controlled experiment, variable, and outcome, with data sheet.

Session 1

Students discuss the ways energy is used in homes. Students build model houses out of paper or cardboard (from the template provided with the guide) to find out how sunlight can be used to heat homes.

Session 2

In teams of four, students conduct the "Solar House Experiment" with the variable being whether or not the window on the model house is covered. The outcome is temperature change in the house after 11 minutes. Results are graphed by minute.

Session 3

Students calculate net temperature change for their house and share with other members of their team. Class results are summarized and the "greenhouse effect" explained.

Session 4

Students discuss conventional home water heaters, then conduct an experiment to heat water using aluminum pie pans and sunlight. The variable is whether or not the pan is enclosed in a plastic bag.

Session 5

Students calculate the net temperature change of their solar water heater experiment, share results with other team members and the class. The class discusses the results, reviews the greenhouse effect, and considers other possible experiments.

Skills

Experimenting, Controlling Variables, Measuring and Recording Data, Graphing, Drawing Conclusions

Concepts

Models and Systems, Solar Energy, Using Sunlight to Heat Homes and Water, the Greenhouse Effect, Home Energy Use, Conducting Science Experiments

Themes

Energy, Models, Systems & Interactions, Equilibrium

Notes: Good for discussions on general environmental issues, especially energy and alternative energy resources; good lead-in to global warming discussions; excellent introduction to controlled experimentation.

MAKING CONNECTIONS

Books that include information on solar and other alternative energies and/or the energy crisis would make good connections to this guide. We welcome your suggestions.

A good way to introduce your students to the idea of controlled experiments is through Chapter 10 of *Einstein Anderson Tells a Comet's Tale.*

Although not addressed specifically by this GEMS guide, the concept of **insulation** is an important extension. Once we've warmed our homes and hot water reservoirs with solar energy, we need to find ways to keep that energy until it's needed. Insulation presents some difficult intellectual challenges for students, and many have misconceptions about it. For example, many students believe that putting a blanket or sweater over an object will make the object get hotter (as opposed to cool down more slowly). Several books listed here may help your students overcome some of their misconceptions, and learn more about the important phenomenon of insulation.

CROSS REFERENCES

LITERATURE CONNECTIONS

A Chilling Story: How Things Cool Down

by Eve and Albert Stwertka; illustrated by Mena Dolobowsky

Julian Messner/Simon & Schuster, New York. 1991

Grades: 5–8

> Humorous black and white drawings show a family and its cat testing out the principles of how refrigeration and air conditioning work, with sections on heat transfer, evaporation, and expansion. The explanations are simple and easy to understand.

Einstein Anderson Lights Up the Sky

by Seymour Simon; illustrated by Fred Winkowski

Viking Press, New York. 1982

Grades: 4–7

> In Chapter 4, "The Snow Job," the properties of snow as an insulator figure in the solution of the mystery.

Einstein Anderson Shocks His Friends

by Seymour Simon; illustrated by Fred Winkowski

Viking Press, New York. 1980

Grades: 4–7

> In "The Case of the Snow Sculpture Contest" Einstein Anderson saves the 6th grade snow sculpture from melting before the judging by insulating it with blankets.

Einstein Anderson Tells a Comet's Tale

by Seymour Simon; illustrated by Fred Winkowski

Viking Press, New York. 1981

Grades: 4–7

> Chapter 10 describes a soapbox derby race in which both teams have to build soapbox racing cars that weigh the same amount and are started in the same way. Our hero identifies the one test variable that allows his team to win the race.

June 29, 1999

by David Wiesner

Clarion Books, Houghton Mifflin, New York. 1992

Grades: 3–6

> The science project of Holly Evans takes an extraordinary turn—or does it? This highly imaginative and humorous book has a central experimental component, and conveys the sense of unexpected results.

Why does this magnificent applied science, which saves work and makes life easier, bring us so little happiness? The simple answer runs: Because we have not yet learned to make sensible use of it.

— *Albert Einstein*

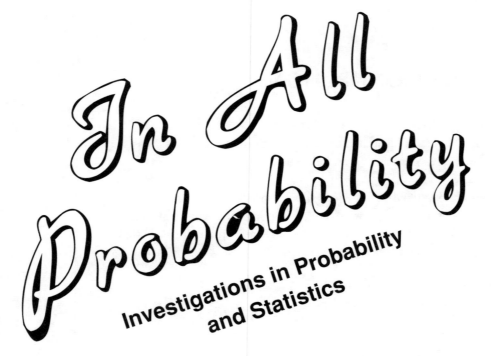

In All Probability

Investigations in Probability and Statistics

Grades 3–6

Students play games that involve coins, spinners, dice, and Native American gaming sticks. They investigate chance and probability with concrete materials, learn how to gather and analyze data, make predictions, and draw conclusions. As they gain direct experience, they also build confidence in their ability to explore probability and statistics. These activities provide a solid basis for the development of much-needed (and often neglected) real-life understandings and skills. Cooperation is stressed and students learn that mathematics is fun!

Activity 1

Students play the "Penny Flip" game, investigating the probability of getting heads or tails, and developing ideas about how to describe their results.

Activity 2

Students play the "Track Meet" game with two different spinners, one fair, the other unfair. They record, compare, and discuss their results.

Activity 3

Students roll one die and consider the outcome. They generate data, then analyze and discuss the results.

Five Activities

Activity 4

Students play the "Horse Race" game, in which 12 horses compete to cross the finish line first, based on a roll of the dice. Students make predictions, cheer on their favorites, generate data and discuss results.

Activity 5

Students create their own decorated game sticks to play "Native American Game Sticks," a complex experiment in probability that uses six two-sided sticks. The outcome of the game and the ways that probability theory relates to it are then discussed.

Skills

Predicting, Describing, Comparing, Estimating, Drawing Conclusions, Working with a Partner

Concepts

Probability, Statistics, Prediction

Themes

Patterns of Change, Models & Simulations

Math Strands

Number, Pattern, Statistics and Probability, Logic

Notes: Some additional background on Native American games is provided.

MAKING CONNECTIONS

There are several books in which **statistics** are generated that also involve collecting, organizing, and recording data. Though some of these books may not be at the exact reading level of your students, they can still provide the basis of a statistics activity. In addition, books about **prediction, chance** and **probability** as well as **number** are included. One session in the guide includes a **Native American game** from California and to complement that session there is a book of California Indian legends.

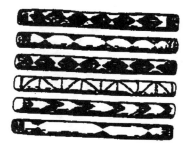

LITERATURE CONNECTIONS

Alice

by Whoopi Goldberg; illustrated by John Rocco
Bantam Books, New York. 1992
Grades: 2–6

This book not only highlights the comedic skills of its author, it contains a lesson about friendship and some statistical wisdom relating to sweepstakes and their deceptive enticements. Alice enters "every sweepstakes, every giveaway, every contest," because she wants more than anything else to be rich. She lives in New Jersey and one day is notified she has won a sweepstakes. Alice convinces her friends to go with her on an odyssey to New York City to collect the prize. What happens makes for a rollicking adventure, which your students will enjoy, as they realize the probability and statistics lessons they are learning in the classroom have lots of applications in the real world!

Maybe Yes, Maybe No, Probably So

Prediction's the art of horse before cart
A race against time; a flip of the dime,
Probably yes, or possibly no,
Perhaps it will happen—maybe not so.
The things that we think most likely to be,
May or may not come to pass you see,
The chance of a snowball that falls toward the Sun
Or the order in which a race will be won,
The roll of a die or how game sticks will fall,
Probability helps us look at it all—
To estimate what we think will occur
Then try it out to find out more sure.
A crystal ball could cloud over my friend,
But on probability, you can depend,
Not every time, it's not that reliable,
But in general, it's undeniable,
From a flip of a coin to a throw of the dice,
Grains of sand on a beach or a bowl of rice
Yes, 9 out of 10 math *funsters* agree
You'll get into *In All Probability*!

— *Lincoln Bergman*

Back in the Beforetime: Tales of the California Indians

retold by Jane L. Curry; illustrated by James Watts

Macmillan Publishing Co., New York. 1987

Grades: 2–6

A retelling of 22 legends about the creation of the world from a variety of California Indian tribes. In the myth "The Theft of Fire," the animal people spend an evening gambling with the people from the World's End. After the animal people lose all they can gamble, Coyote wagers the animal people's fire stones in a final bet. The outcome of that bet is the basis for the mythic explanation of how the animal people got fire. The story ties in with Session 5 "Game Sticks," a California Native American gambling game.

Cloudy With a Chance of Meatballs

by Judi Barrett; illustrated by Ron Barrett

Atheneum, New York. 1978

Grades: K–3

A hilarious look at weather conditions in the town of Chewandswallow, which needs no food stores because daily climatic conditions bring the inhabitants food and beverages, such as a storm of giant pancakes or an outpouring of maple syrup. This book presents a non-threatening way to look at predictions. Students can follow up the story by listening to weather reports and charting the accuracy of meteorologists. They can also use *The Cloud Book* by Tomie dePaola to observe and chart clouds, one aspect of weather patterns.

Jumanji

by Chris Van Allsburg

Houghton Mifflin, Boston. 1981

Scholastic Books, New York. 1988

Grades: K–5

A bored brother and sister left on their own find a discarded board game (called Jumanji) which turns their home into an exotic jungle. A final roll of the dice for two sixes helps them escape from an erupting volcano. The story complements the horse racing game in the GEMS guide where the roll of the dice also determines an important outcome.

People

by Peter Spier

Doubleday, New York. 1980

Grades: Preschool–6

Here's an exploration of the differences between (and similarities among) the billions of people on earth. It illustrates different noses, different clothes, different customs, different religions, different pets, and so on. This is a great book to use in collecting statistics and creating graphs about characteristics of people. Pairs of students can investigate the occurrence in their class of a physical feature (hair type, eye color, etc.), preference (types of food), or other distinguishing attribute (where one lives), and report their findings to the class.

Investigating Artifacts

Making Masks, Creating Myths, Exploring Middens

Grades K–6   **Six Sessions**

This guide weaves together three activities related to anthropology and archaeology and to diverse Native American and world cultures. Students sort and classify natural objects found on a class walk, then make their own masks. They create their own stories to explain natural phenomena and learn how ancient peoples evolved myths to explain and represent the natural world. Students learn that a midden is an archaeological term for deposits earlier peoples left behind, including utensils, garbage, and other artifacts, then teams of students carefully sift through "artifacts" in shoebox-middens.

Possible extensions raise important social issues. A major scientific thread in all three activities concerns inferences that can be drawn from varying evidence. Resource sheets provide information on Native American and world masks, myths, and archaeological sites. An annotated listing of related young people's literature is included.

Session 1

Students gather natural materials outside, then in the classroom explore, sort and classify what they've found. They play a "Secret Sort" game and make inferences about the neighborhood.

Session 2

Students use the natural materials they've gathered to create masks, then sort and discuss the masks they've made, learning how anthropologists make inferences about cultures from masks and other artifacts.

Session 3

Students are asked to imagine themselves living 2,000 years ago. After hearing several Native American myths or stories that explain a natural phenomena, students make up their own myths.

Session 4

Student storytellers share their stories, using their masks if they wish. After hearing "How the Stars Came to Be" story, students make inferences about the people who originated the story.

Session 5

After an introductory role play, students work in teams to carefully investigate shoebox "middens" with simulated artifacts.

Session 6

Students list and describe the objects they uncovered in their middens, then brainstorm ways the objects might have been used by an ancient culture.

Skills

Observing, Recording, Sorting and Classifying, Finding and Making Patterns, Mapping and Diagramming, Making Inferences, Designing Models, Writing, Relating, Communicating, Drama and Role Playing, Working Cooperatively, Analyzing Data

Concepts

Anthropology; Archaeology; Cultural Diversity and Similarity; Art, Language, and Culture; Myths, Legends, and Storytelling

Themes

Patterns of Change, Models & Simulations, Systems & Interactions, Structure, Scale, Diversity & Unity

Math Strands

Functions, Measurement, Number, Pattern, Logic

Notes: Resource pages provide information on Native American and world myths, North American and world archaeological sites, and a map showing tribal locations in North America.

MAKING CONNECTIONS

The annotations include quite a few books of **Native American myths**. Many of these myths focus on explaining natural phenomena, as do the myths created by students in Sessions 3 and 4. Others provide a window into the culture of specific tribes or peoples. Providing students with a sense of these differences is an important step towards reducing the stereotypical misrepresentation of a uniform Native American culture.

There are several books that present **other aspects of Native American cultural issues**. They range from a description of the structure and uses of tipis, a collection of poems and songs by Native Americans celebrating the relationship between earth and all creatures, to an account of the uprooting of Native American life told by a young Navajo girl. These are important in that they provide rich detail of past and current Native American peoples, their values and their struggles.

There are also several books with **myths from other world cultures**, including African, Greek, and Jewish folktales, legends, and stories. These can be used to show how all cultures leave a trail of stories from which we can learn.

In Sessions 1 and 2, students make their own masks and discuss their role in ritual and celebration. There are a couple of books which address **celebration and ritual** in general and one which focuses on the security and continuity provided by a **tradition** of a quilt that is passed down through several generations. A biography of Diego Rivera sheds light on the **relationship between art and society**, which becomes apparent to students as they learn what masks can tell us about the people who made them.

There are several books that relate to Sessions 5 and 6, where students excavate pretend middens. These books involve **archaeology** and archaeological techniques.

One of the unit's Going Further activities involves students gathering oral histories of their families. Several books have to do with **exploring one's own roots**, from tracing family records to historical research.

Some of the books cross referenced under *Crime Lab Chemistry* and *Fingerprinting* deal with **evidence** and **inference**. They can be used with *Masks, Myths, and Middens* to extend the experience students have of examining evidence (in this case, masks, myths, and the artifacts in middens) and making inferences about the people who produced them.

Of course, many teachers have their own favorite books on Native American life, and there is at least some inclusion of information about local or regional tribes in many elementary school curricula. Combining this unit with a focus on learning more about the indigenous peoples of your region is an excellent way to extend learning and emphasize cultural diversity.

Especially noteworthy are two books by Michael J. Caduto and Joseph Bruchac. They combine clearly written nature-based and science activities for children with beautiful, sensitively told, and stirring stories that are identified as to tribal origin. Published by Fulcrum Publishing, Golden, Colorado, we highly recommend, as resources of the highest quality, *Keepers of the Earth: Native American Stories and Environmental Activities for Children* and *Keepers of the Animals: Native American Stories and Wildlife Activities for Children*. An expanded look at those books is on page 397.

CROSS REFERENCES

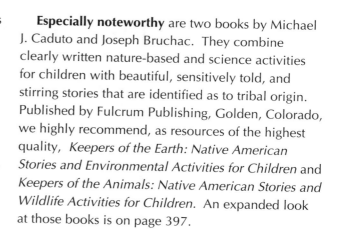

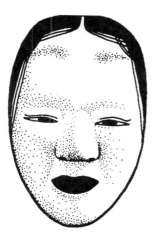
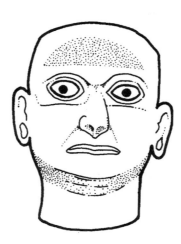

LITERATURE CONNECTIONS

The book listings *for Investigating Artifacts* are divided into three sections:

➤ **Native American Myths, Legends, Stories**
➤ **World Culture**
➤ **Archaeology/Exploring Your Roots**

Screaming the night
away, with his great
wing feathers
swooping the
darkness up;
I hear the Eagle Bird
pulling the blanket
back off from the
eastern sky.

— *Iroquois Invitation Song*

NATIVE AMERICAN MYTHS, LEGENDS, STORIES

Arrow to the Sun
by Gerald McDermott
Viking Press, New York. 1974
Grades: K–3

> This Pueblo Indian myth explains how the spirit of the Lord of the Sun was brought to the world through "the Boy." The bold, angular illustrations in bright orange tones are uniquely suited to the tale. After passing through the trial of the Kiva of Lightning on a quest for his father, a boy is transformed and is filled with the power of the sun. Winner of the Caldecott award.

Boat Ride With Lillian Two Blossom
by Patricia Polacco
Philomel/Putnam & Grosset, New York. 1988
Grades: K–4

> A wise and mysterious Native American woman takes William and Mabel on a boat ride, starting in Michigan and ranging through the sky. Explanations for the rain, the wind, and the changing nature of the sky refer to spirits such as the caribou or polar bear, which are magically shown.

Corn is Maize: The Gift of the Indians
by Aliki
Thomas Y. Crowell, New York. 1976
Grades: 3–5

> This book tells how corn was first cultivated, stored, and used by Native Americans and how it came to be a main food source all over the world.

Dancing Teepees: Poems of American Indian Youth
Selected by Virginia Driving Hawk Sneve; illustrated by Stephen Gammell
Holiday House, New York. 1989
Grades: 3–8

> A selection of chants, lullabies, prayers, and poems from Native American oral tradition. These celebrate rites of passage and other symbolic events such as a buffalo hunt or corn ceremony. The muted color illustrations of beadwork, petroglyphs, and motifs derived from nature are quite beautiful and seem well matched to the verses.

Dragonfly's Tale

by Kristina Rodanas

Clarion Books/Houghton Mifflin, New York. 1991

Grades: K–4

Based on a Zuni legend, this story tells of the origin of the dragonfly with a secondary theme about appreciating one's blessings. After the people foolishly waste food in a festive "food fight," the Corn Maidens teach them a lesson by sending a famine and drought. A little boy fashions a toy insect from a cornstalk for his sister. The toy comes to life, secures the Corn Maidens help in providing a harvest, and can be seen every summer humming among the corn as a dragonfly. The illustrations depict the landscape, honeycombed lodgings, and ceremonial and daily clothing particular to the culture with great warmth and detail.

Dream Wolf

by Paul Goble

Bradbury Press/Macmillan, New York. 1990

Grades: 1–4

A Plains Indian boy and his sister wander away from a berry-picking expedition and are lost in the hills as night falls. A wolf comes to their aid, leading them back to their home. There has been close kinship with the Wolf People since then. But wolves are no longer heard in the evenings at berry-picking time because they have been killed or driven away. The wolves will return only when we have them in our hearts and dreams again, the People say. Goble has written and illustrated many Native American legends including the Iktomi series of humorous tales about the Trickster. This book was originally published in 1974 as *The Friendly Wolf*.

The Earth is Sore: Native Americans on Nature

by Aline Amon

Atheneum, New York. 1981

Out of print

Grades: 4–Adult

Collection of poems and songs by Native Americans that celebrates the relationship between the earth and all creatures and mourns abuse of the environment. Illustrated with black and white collage prints made from natural materials.

Earthmaker's Tales:
North American Indian Stories About Earth Happenings
by Gretchen W. Mayo
Walker and Co., New York. 1989
Grades: 5–7

This is a particularly appropriate book providing many specific examples of myths or stories that explain natural events (the same thing children do in the "Myths" section of the GEMS activities). Among the natural events that the stories relate to are: earthquakes, floods, night and day, storms, thunder and lightning, fog, volcanoes, etc. A teacher could choose two or three of her favorites to widen student acquaintance with Native American cultures and/or orient them to their task. A second volume, *More Earthmaker's Tales*, includes eight more.

I'm in Charge of Celebrations
by Byrd Baylor; illustrated by Peter Parnall
Charles Scribner's Sons, New York. 1986
Grades: 4–9

A Native American girl, who might seem lonely to others, celebrates things such as a triple rainbow, a meteor shower, and a chance encounter with a coyote, delighting in her surroundings. Vivid pictures of the world of the desert.

Iroquois Stories: Heroes and Heroines, Monsters and Magic
as told to Joseph Bruchac; illustrated by Daniel Burgevin
Crossing Press, Freedom, California. 1985
Grades: 3–8

Wonderful collection of over 30 stories by a leading storyteller. Good for reading out loud to younger children and for students in third or fourth grade to read by themselves. The book is in fairly large type and written in clear, focused language. Opens with "The Coming of Legends," which tells how legends came into the world, and a creation poem about life springing from a handful of seeds dropped from the Sky-World.

The Indian knows his village and feels for his village as no white man for his country, his town, or even his own bit of land. His village is not the strip of land four miles long and three miles wide that is his as long as the sun rises and the moon sets. The myths are the village, and the winds and rains. The river is the village, and . . . the talking bird, the owl who calls the name of the man who is going to die.

— *Margaret Craven*
I Heard the Owl Call My Name

Ladder to the Sky:
How the Gift of Healing Came to the Ojibway Nation

retold by Barbara J. Esbensen; illustrated by Helen K. Davie
Little, Brown & Co., Boston. 1989
Grades: K–4

This legend tells of the time when all people were healthy. When they grew old a "shining spirit-messenger" carried them up a magic vine to the sky where they lived forever. After a grandmother climbs up the forbidden vine in pursuit of her grandson, the Great Spirit punishes the people by sending sickness and death, but blesses them with the gift of healing. There is a very strong message here about the negative value of disobedience in the culture. The villagers call the old woman a witch and resent her for bringing "shame and disaster" to the people. The first section has good detail on how the Chippewa integrated the plant world into their culture, weaving rushes and reeds into mats, incorporating flower designs into clothing, making birch bark containers, and gathering milkweed down for bedding.

The Legend of the Bluebonnet

retold and illustrated by Tomie dePaola
G.B. Putnam's Sons, New York. 1983
Grades: K–3

Set in what is now Texas, this legend tells of a young Native American girl, She-Who-Is-Alone, whose people are desperately praying to the Great Spirits to end a drought. Only when the orphan girl sacrifices her treasured and only possession, a warrior doll with blue jay feathers, do the spirits send the rain. She is renamed One-Who-Dearly-Loved-Her-People and the bluebonnet flower comes every spring, as blue as the feathers of the blue jay.

The Legend of the Indian Paintbrush

retold and illustrated by Tomie dePaola
G.P. Putnam's Sons, New York. 1988
Grades: K–4

After a Dream Vision, the Plains Indian boy Little Gopher is inspired to paint pictures as pure as the colors in the evening sky. He gathers flowers and berries to make paints but can't capture the colors of the sunset. After another vision, he goes to a hilltop where he finds brushes filled with paint that he uses and leaves on the hill. The next day, and now every spring, the hills and meadows are ablaze with the bright color of the Indian Paintbrush flower.

Lightning Inside You and Other Native American Riddles

edited by John Bierhorst; illustrated by Louise Brierley

William Morrow, New York. 1992

Grades: All

> This intriguing book divides its more than 140 riddles into categories such as the human body, animals, things made to be used, etc. The author discusses various riddling situations: hunter's riddles (out of respect or fear, avoid calling game animals by their actual names); dream guessing (other people try to interpret a person's troubling dreams); initiation riddles; riddle contest (form of gambling); riddle dance; courtship riddling; and riddles in stories. Here's a Comanche riddle: "What is there inside you like lightning?" "Meanness."

A people without history is like the wind on the buffalo grass.

— *Sioux saying*

Mama, Do You Love Me?

by Barbara M. Joosse; illustrated by Barbara Lavallee

Chronicle Books, San Francisco. 1991

Grades: K–4

> Set in the Arctic, a child tests the limits of her independence. Whales, wolves, puffins, sled dogs, and Inuit culture are depicted in stunning, fresh illustrations. The answer to the title's question is always, "yes," even, "if you put lemmings in my mukluks." A glossary lists some animals and objects which may not be familiar (such as "ptarmigan"), noting their particular significance in the culture.

The Moon, the Sun, and the Coyote

by Judith Cole; illustrated by Cecile Schoberle

Simon & Schuster, New York. 1991

Grades: K–4

> Written in a traditional, folkloric style by a modern author, the amusing descriptions of the rivalry between the sun and the moon could trigger some imaginative student ideas as they work on the "creating myths" portion of the GEMS activities. The story also "explains" why coyotes look like they do, thus providing a model for what students are asked to do: create a story that explains a natural phenomena. The never-satisfied Coyote of the story has all-too-human elements, and discussion of his foibles could help students make inferences about the values the story projects, especially in connection with Session 4 of this GEMS guide.

Moon Was Tired of Walking on Air

by Natalia M. Belting; illustrated by Will Hillenbrand

Houghton Mifflin, New York. 1992

Grades: 3–5

> Fourteen creation myths from tribes of South America that explain the worlds above and below the earth. The dynamic illustrations combine a sense of magic and power with very literal depiction of the characters. The moon walks on earth among the squash vines with giant feet; fox opens a bottle tree and is pursued by rushing waters; the sculpted muscular figures of North and South wrestle and bend the rainbow; and armadillo plunges from the sky, making a hole in it.

Native Dwellings

by Bonnie Shemie
Tundra Books, Montreal, Canada. 1991
Grades: 2–6

This series of books deals with the structure of Native American dwellings, including how they are lived in, and the building materials, techniques, and tools used. Included in this series are *Houses of hide and earth* (on plains dwellings), *Houses of bark* (on woodland dwellings), and *Houses of snow, skin and bones* (on Northern dwellings). The books are designed with a number of double-paged spreads showing a panoramic view and with smaller insets showing details such as tools. The more recently published *Houses of wood* (on Northwest Coast dwellings) has especially effective artwork. The next volume scheduled in the series is *Houses of straw and mud* about Southeastern dwellings.

You have noticed that everything an Indian does is in a circle, and that is because the Power of the World always works in circles, and everything tries to be round. In the old days when we were a strong and happy people, all our power came to us from the sacred hoop of the nation, and so long as the hoop was unbroken, the people flourished. The flowering tree was the living center of the hoop, and the circle of the four quarters nourished it. The east gave peace and light, the south gave warmth, the west gave rain, and the north with its cold and mighty wind gave strength and endurance. This knowledge came to us from the outer world with our religion. Everything the Power of the World does is done in a circle. The sky is round, and I have heard that the earth is round like a ball, and so are all the stars. The wind, in its greatest power, whirls. Birds make their nests in circles, for theirs is the same religion as ours. The sun comes forth and goes down again in a circle. The moon does the same, and both are round. Even the seasons form a great circle in their changing and always come back again to where they were. The life of a man is a circle from childhood to childhood, and so it is in everything where power moves.

— *Black Elk*
Black Elk Speaks

The Night of the Stars
by Douglas Gutierrez and Maria F. Oliver
Kane/Miller Book Publishers, New York. 1988
Grades: 3–5

There was a man (long, long ago) who did not like the night and the dark sky. During the day he worked weaving baskets and watching over his animals, but at night he shut himself in his house and lit his lamp. One night he went up to the highest point on the mountain, stood on his tiptoes and with his finger poked a hole in the black sky. A pinprick of light appeared through the hole so he poked more holes all over the sky. Then he poked a really big hole for the moon. That night no one slept and everyone stayed up late looking at the moon and stars. The story is a good accompaniment to Session 4 of the GEMS guide where students hear and discuss "How the Stars Came to Be." A Spanish edition of *The Night of the Stars* is also available.

Quillworker: A Cheyenne Legend
adapted by Terri Cohlene; illustrated by Charles Reasoner
Watermill Press, Mahwah, New Jersey. 1990
Grades: 2–5

A Cheyenne legend that explains the origin of the Big Dipper constellation. Quillworker is an only child and an expert needle worker. Her dreams direct her to make seven buckskin warrior outfits for her mysterious new seven brothers. To escape the buffalo nation who want to take Quillworker, they all ride a tree up into the sky where they remain, with Quillworker as the brightest star in the dipper. A reference section at the back includes a short glossary and brief overview of the Cheyenne people and some of their customs.

Rainbow Crow: A Lenape Tale
by Nancy Van Laan; illustrated by Beatriz Vidal
Alfred A. Knopf, New York. 1989
Grades: K–3

When the weather brings a long period of snow, the animals become worried and decide to send a messenger to the Great Sky Spirit to ask him to stop it. The most beautiful bird, brightly colored Rainbow Crow (also known as Raven), offers to make the long journey and is rewarded with the gift of fire which he carries in his beak. Forever after, he has a hoarse cry and blackened feathers, but with tiny rainbows of color. The outstanding illustrations are perfectly wedded to the text.

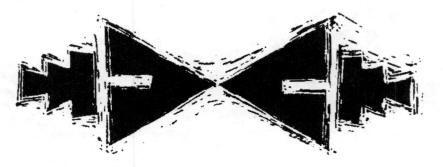

Sing Down the Moon

by Scott O'Dell

Houghton Mifflin, Boston. 1970

Dell Publishing, New York. 1976

Grades: 6–12

Through the eyes of 14-year-old Bright Morning, a Navajo girl who lives in Canyon de Chelly, Arizona in 1863, we see the uprooting of Native American life, first by Spanish slavers and then by U.S. soldiers who lead the Navajos on the 300-mile Long Walk to Fort Sumner. The major events are tragic and cruel, but Bright Morning survives to return to her home and her sheep. Accounts of daily life include construction of huts and shelters, preparations for the marriage and Womanhood ceremonies, making garments, planting seed, and the logistics of travel. Newbery honor book.

Sparrow Hawk

by Meridel LeSueur; illustrations by Robert Desjarlait

Holy Cow Press, Stevens Point, Wisconsin. 1987

Grades: 4–12

A beautifully written and moving novel about a young Sauk boy growing to manhood and his white friend Huck, both caught in the midst of the Black Hawk War. The images of the closeness to the land and connection to the life-giving corn are poetic and powerful.

The Star Maiden

retold by Barbara J. Esbensen; illustrated by Helen K. Davie

Little, Brown & Co., Boston. 1988

Grades: K–4

A star grows tired of wandering the sky and wants to live on earth. She tries becoming a rose and then a prairie flower, but finally finds her "place on earth" as a water lily. The watercolor illustrations with their unique borders are by the same artist as *Ladder to the Sky*.

The Story of Jumping Mouse

by John Steptoe

Lothrop, Lee & Shepard, New York. 1984

Grades: 3–6

In this Native American legend, the smallest and humblest of creatures (a mouse) becomes the noblest (the eagle). In a spirit of hope, compassion, and generosity, a young mouse gives away his sense of sight and smell to other needy animals and is rewarded by a transformation. Beautiful wash drawings capture the world of forest and desert from a low-on-the-ground perspective. The story strongly relates to the cultural component of the GEMS guide, and places a high value on compassion and cooperation. The class can consider cultural values as it discusses "How the Stars Came To Be" in Session 4.

> [The Indian] sees no need for setting apart one day in seven as a holy day, since to him all days are God's.
>
> — *Ohiyesa*
> *The Soul of the Indian*

> Nearness to nature
> . . . keeps the spirit
> sensitive to
> impressions not
> commonly felt,
> and in touch
> with the unseen
> powers.
>
> — *Ohiyesa*
> *The Soul of the Indian*

Thirteen Moons on Turtle's Back:
A Native American Year of Moons

by Joseph Bruchac and Jonathan London; illustrated by Thomas Locker
Philomel Books/Putnam & Grosset, New York. 1992
Grades: 3–7

Stories from different Native American cultures about the 13 moons—
believed to correspond to the scales on the shell of the turtle's back—
explain the changes in the seasons. A numbered drawing shows the
location of each of the 13 moons on the turtle's shell. The names of the
moons are evocative and the poems that accompany them closely intertwined with respect for nature's ways: Baby Bear Moon, Moon When
Wolves Run Together, Moon of Popping Trees, Moose-Calling Moon, and
Wild Rice Moon. Locker's rich landscape paintings are powerful and
distinctive.

The Tipi: A Center of Native American Life

by David and Charlotte Yue
Alfred A. Knopf, New York. 1984
Grades: 5–8

This excellent book describes not only the structure and uses of tipis, but
the Plains Indians social and cultural context as well. Some of the cultural
language and oversimplifications are less vital than they might be, but it is
written in an accessible style. There are good charts, exact measurements,
and information on the advantages of the cone shape. The central role
played by women in constructing the tipi and in owning it are discussed.
While this book includes some mention of the negative consequences of
European conquest, noting that in some places tipis were outlawed, it is
weak in this important area, and should be supplemented with other
books.

Totem Pole

by Diane Hoyt-Goldsmith; photographs by Lawrence Migdale
Holiday House, New York. 1990
Grades: 3–5

A Tsimshian Indian boy describes how his father carved a totem pole for
the Klallam tribe in Washington. The complete process is shown through
color photographs, from finding a straight tree through raising the pole and
the ceremonies accompanying it. Drawings show the figures of
Thunderbird, Killer Whale, Bear, Raven and a Klallam Chief carved on the
pole, explaining their mythic significance. The pride of the young boy in
his father, the dedication to the craft, and the passing on of tradition are
well conveyed. "Like my father, I look for the animal shapes hidden inside
the wood."

The Village of Blue Stone

by Stephen Trimble; illustrated by Jennifer Dewey and Deborah Reade
Macmillan, New York. 1990
Grades: 5–8

Badger Claws (the Sun Watcher), Turquoise Boy, Dragonfly, and baby Blue Feather are involved in the full range of events of one year in the life of an imaginary Anasazi pueblo in 1100 in what is now New Mexico. Their daily life, including pottery-making, a wedding, the Harvest Dance, illness, and a death, is wonderfully depicted. Illustrates the connections between the land and architecture, work and art, material culture and spiritual beliefs.

When Clay Sings

by Byrd Baylor;
illustrated by Tom Bahti
Charles Scribner's Sons/
Macmillan, New York. 1972
Grades: 1–6

Prose poem retraces the daily life and customs of prehistoric Southwestern tribes from designs in the remains of their pottery. The striking illustrations in black, brown, and ochre tones include design motifs from the Anasazi, Hohokam, Mogollon, and Mimbres cultures. In the text, parents tell their children to treat each fragment with respect because "every piece of clay is a piece of someone's life," and each has its own song. Excellent connection to the "Middens" activities in the GEMS guide, in which students carefully uncover simulated artifacts, including pieces of broken pottery.

Turtle Island Mother Earth

To begin with there was no earth, only water,
With animals who swam or flew nearby–
Then there came a wondrous human Daughter
Who fell down from a torn place in the sky.
She needed soil of earth, or she would die.

Two Loons cried out and caught her as she fell,
Called for the other animals to lend a paw,
Set her to rest upon a giant Turtle's shell,
Began to dive into the Sea's vast craw,
To seek the earthly soil with tooth and claw.

The Beaver tried, his broad tail slapped the tide,
A Muskrat lent his whiskers to the search,
When they came up the Turtle looked inside
Their mouths to see if they had captured any earth.
Others tried and failed, returned to their sad perch.

Poor Toad stayed down so long he almost died,
But in his mouth, so deep the Toad had dived,
Turtle found a mound of earth (Toad glowed inside)
The wondrous Daughter of the sky survived!
She patted earth around the Turtle's shell, revived!

The soil began to grow and grow, becoming land,
Earth grew and grew upon Great Turtle's shell,
From clumps and countries continents expand.
Then Woman brought forth Children, with their Truths to tell,
Maize, beans, and pumpkins sprout from Her as well.

So Children of today, please listen as we say
That this Daughter from the sky created our Life's way
And were it not for the animals, the Loons and the Toad,
Who knows what might have happened on Life's long road
It's still a question, a mystery—
If not for that Turtle, where would we be?

— *Lincoln Bergman*

WORLD CULTURE

> Science and art belong to the whole world, and before them vanish the barriers of nationality.
>
> — *Goethe*
> *Truth and Poetry*

The All Jahdu Storybook
by Virginia Hamilton; illustrated by Barry Moser
Harcourt, Brace, Jovanovich, San Diego. 1991
Grades: K–6

These stories about the folkloric trickster hinge on no specific traditions, but aim to express the timelessness of folklore. The diverse illustrations reflect the changes in the trickster: one minute he is in the jungle, the next in a taxi in Harlem. Characters he meets include animals (Bandicoot Rat or the chicken Cackle G.), or are abstract (Shadow, Thunder, or Grass). The author uses the generic name Jadhu for the trickster who appears in the folklore of various cultures.

Ashanti to Zulu: African Traditions
by Margaret Musgrove; illustrated by Leo and Diane Dillon
Dial Books, New York. 1976
Grades: 3–7

Beautifully illustrated and well-researched alphabet book that describes African ceremonies, celebrations, and day-to-day customs as well as reflecting the richness and diversity of the peoples and cultures. A man, woman, child, an artifact, a local animal, and living quarters are depicted in most of the paintings so that each page is quite detailed, even though all these elements might not ordinarily be seen together. The border design is based on the Kano Knot, a seventeenth-century design that symbolizes endless searching. Caldecott award winner.

Cornrows
by Camille Yarbrough; illustrated by Carole Byard
Coward, McCann Inc., New York. 1979
Grades: K–5

This powerful and tender book recounts a family story that Mama and Great-Grammaw tell as they braid intricate cornrow patterns into the children's hair. This book blends poetic accounts of African traditions, brutal slavery, cultural heritage, and the achievements of many famous African-Americans, with a strong and loving sense of family. It could be read as part of "Masks, Myths, and Middens" to introduce African-American contributions in general, and more specifically, to discuss the way the braiding of cornrows, the telling of stories, and the depiction of masks and sculptures connects to modern children's understanding of their own culture.

Diego

by Jeanette Winter

Alfred A. Knopf, New York. 1991

Grades: K–4

> Story of the great Mexican muralist Diego Rivera with special attention to his childhood and how it influenced his art. Important themes include the relationship between art and society and the importance of direct experience for an artist. Winter's vibrant miniature paintings seek to convey Rivera's spirit but do not attempt to copy his work. The book has been criticized for not including any mention of Diego's wife, the artist Frida Kahlo, although she does appear in one illustration. The artistic theme of this book could make a nice connection to the creative "Masks" activities in this GEMS guide.

Elinda Who Danced in the Sky

adapted by Lynn Moroney; illustrated by Veg Reisberg

Children's Book Press, San Francisco. 1990

Grades: K–4

> Estonian folk tale about the sky goddess Elinda who overcomes her disappointment at losing her fiancé Prince Borealis, whose land would not let him leave. There are clever explanations of why Elinda turned down previous suitors: the North Star would be distant and unmoving, the moon always takes the same narrow path, and the sun's light too harsh and overpowering. Elinda returns to her vocation of guiding the birds in their migrations, putting them on the right path. Her wedding veil, woven from dewdrops and dragonfly wings, is the Milky Way.

How Many Spots Does a Leopard Have? and Other Tales

by Julius Lester; illustrated by David Shannon

Scholastic, New York. 1989

Grades: 6–10

> A fine collection of African folktales, including two Jewish tales and one African-Jewish hybrid. Written for somewhat older students, these stories can be read out loud very sucessfully to younger students. We learn why the sun and moon live in the sky, why monkeys live in trees, and why dogs chase cats, but no one ever finds out how many spots the leopard really has!

In the Beginning: Creation Stories from Around the World

by Virginia Hamilton; illustrated by Barry Moser

Harcourt, Brace, Jovanovich, San Diego. 1988

Grades: All

> An illustrated collection of 25 legends that explain the creation of the world. The myths are placed geographically and by type of myth tradition such as "world parent," "creation from nothing," and "separation of earth and sky." Some of the selections are extracted from larger works such as *Popol Vuh* or the Icelandic Eddas. Newbery honor book.

> The people are a story that never ends.
>
> — *Meridel Le Sueur*
> *North Star Country*

Just So Stories

by Rudyard Kipling

Viking Penguin, New York. 1987

Grades: 2–6

These amusing tales of how things came to be—such as how the elephant got his long nose, how the leopard got his spots, and how the camel got his hump—are inspired classics. The stories captivate and provide a perfect springboard to the question, which in any exploration of myths you may wish to ask: "Could that be true?"

The Keeping Quilt

by Patricia Polacco

Simon & Schuster, New York. 1988

Grades: K–5

A homemade quilt ties together the lives of four generations of an immigrant Jewish family. Made from their old clothes, it helps them remember back home "like having the family in Russia dance around us at night." The quilt is used in marriage ceremonies, as a tablecloth, and as a blanket for a newborn child, symbolizing the family's enduring love and faith. In the GEMS guide, students practice making inferences about a culture based on artifacts. What does the quilt described in the story tell us about the culture of the people who created it? A resource to begin a quilt project. Sidney Taylor Award winner.

Land of the Long White Cloud: Maori Myths, Tales, and Legends

by Kiri Te Kanawa; illustrated by Michael Foreman

Arcade Publishing/Little, Brown & Co., Boston. 1989

Grades: K–5

Stories from the Maoris, the indigenous people of New Zealand, about the trickster and mischief maker Maui; the woman in the moon, the birds, the lakes, rivers and trees; and assorted fairies and monsters. These exciting tales reflect the life of a people whose survival depended on their close knowledge of the sea in all its aspects.

Legend of the Milky Way

retold and illustrated by Jeanne M. Lee

Henry Holt & Co., New York. 1982

Grades: K–5

The weaver Princess came down from heaven to marry a mortal; but her mother objects and punishes them by making them stars separated by the Silver River (the Milky Way). The Chinese celebrate this story on the seventh day of the seventh Chinese month. If it rains that night, they say the princess is crying because she must say good-bye to her husband. The last page explains which familiar stars and constellations represent the characters in this legend.

Nine-in-One Grrr! Grrr!

by Blia Xiong; illustrated by Nancy Hom
Children's Book Press, San Francisco. 1989
Grades: K–3

In this folktale from Laos, when the first female tiger asks the kind and gentle God Shao how many cubs she will have, he tells her she will have nine cubs a year, if she remembers his words. Tiger does not have a great memory, so she makes up a little song to remember: "Nine-in-One, Grr! Grr!" When the other animals find this out, they are worried because that many tigers could eat all of them. A clever bird succeeds in distracting the tiger long enough to make her forget the song, then convinces her that the song was, "One-in-Nine, Grr! Grr!" (one cub born every nine years). That is why, the Hmong people of Laos say, "we don't have too many tigers on earth today." This direct and compelling explanatory myth could open a basic discussion of the balance of nature as well as the tricks that people use to remember things! This is an excellent book to select in connection with the world cultural aspects of the GEMS guide, especially if you wish to highlight Southeast Asian cultures.

The Patchwork Quilt

by Valerie Flournoy; illustrated by Jerry Pinkney
Dial/Dutton, New York. 1985
Grades: K–5

Tanya, an African-American child, and her grandmother make a quilt using scraps cut from the family's old clothing including her African princess Halloween costume. The grandmother becomes ill and the whole family is involved with completing the quilt of memories. Referring to her "master-piece," the grandmother says "A quilt won't forget. It can tell your life story." The GEMS guide encourages students to explore their own cultural roots and gain an appreciation for the peoples of the past. The process of unearthing and analyzing artifacts in the middens activity is compared to trying to put together puzzle pieces of the past. Might this also be compared to making a patchwork quilt?

A Promise to the Sun

by Tololwa M. Mollel; illustrated by Beatriz Vidal
Joy Street Books/Little, Brown & Co., Boston. 1992
Grades: K–4

This African tale explains why bats fly only at night. In a time of great drought, the birds draw lots to see who will journey to seek rain and the lot falls to a visiting bat. The bat successfully persuades the sun to bring about rain, but is left holding the bag when the birds don't follow through on a promise to the sun. To avoid the sun's wrath, the bat hides in a cave and lives there to this day. The theme of an individual acting to benefit the group and solve a problem is also found in *Rainbow Crow*, which is illustrated by the same artist.

The Truth About the Moon

by Clayton Bess; illustrated by Rosekrans Hoffman

Houghton Mifflin, Boston. 1983

Grades: K–4

An African boy is puzzled by the changing size of the moon and asks for an explanation. His father says there is only one moon and that the moon he saw last night is the same moon he will see tomorrow. "It is growing, just as a child like you grows to be a man like me. It starts small, just a silver sliver, and every night grows bigger and bigger until it is as big as it can be, a full circle. Then, just as a man grows smaller when he is very old, so does the moon. Smaller and smaller until death." His mother explains that there is only one moon. "It is like a woman. And you know how sometimes a woman will grow larger and larger, more and more round?" The Chief tells a long tale about the sun and the moon being married and how the moon lost its heat.

The Turtle and the Island:
A Folktale from Papua New Guinea

by Barbara K. Wilson; illustrated by Frane Lessac

J.B. Lippincott, New York. 1990

Grades: K–4

In this creation myth, New Guinea was made by a great sea turtle, the mother of all sea turtles. The turtle makes the island by adding more sand and rocks to a high hill. Then she brings the lonely sole man in the ocean from his cave to the island together with a lonely weeping woman. They have beautiful children whose children have more children. The vibrant illustrations show the lovable sea turtle and an island teeming with plant and marine life.

Why Mosquitoes Buzz in People's Ears

by Verna Aardema; illustrated by Leo and Diane Dillon

Dial, New York. 1975

Grades: K–6

This West African folk tale is a clever story of why mosquitoes buzz in people's ears. Devising explanations for things in nature by creating myths is what students do in Sessions 3 and 4 of the GEMS activities.

Why Rat Comes First: A Story of the Chinese Zodiac

retold by Clara Yen; illustrated by Hideo C. Yoshida

Children's Book Press, San Francisco. 1991

Grades: K–4

The Jade King invites all the animals to a feast, but only 12 show up. He rewards them by naming a year after each animal, starting with the rat whose quick thinking wins him first place. Even more fun comes after the story is done, when each person can look up her/his birth year and the corresponding animal and characteristics. This book could also be used to discuss the lunar calendar, and the different ways that world cultures keep track of time.

Why the Sky Is Far Away

retold by Mary-Joan Gerson; illustrated by Hope Meryman
Harcourt, Brace, Jovanovich, San Diego. 1974
Grades: K–4

Nigerian folk tale about the sky that offers a strong moral message about squandering natural resources. It tells of a time when the sky was so close to the earth that anyone who was hungry just cut off a piece of sky and ate it. The king even had a special team of servants whose only job was to cut and shape the sky for ordinary meals and for special ceremonies. But the sky was getting tired of being wasted. When one woman throws away a leftover piece saying "What does it matter? ... one more piece on the rubbish heap," the sky finally moves away. Ever since then people have had to work very hard to grow their own food. In addition to an imaginative explanation for why the sky is far away (similar to the stories explaining natural phenomena that students create and share in Sessions 3 and 4 of this GEMS guide) the story connects to the "Going Further" activities for Sessions 5 and 6 in which students are asked to discuss or prepare a written assignment on the question: "What does our garbage tell us about our culture?"

Why the Sun and the Moon Live in the Sky

by Elphinstone Dayrell; illustrated by Blair Lent
Houghton Mifflin, Boston. 1968
Grades: K–4

In this Nigerian folk tale, the sun and his wife, the moon, built a large house for entertaining the water. By the time the water and all his people have flowed in and over the top of the roof, the sun and moon are forced to go up into the sky. The main characters and the fish and water animals are all represented as African people in tribal costumes and masks in brown, green, blue and gold-patterned drawings. Caldecott Honor book.

Why the Tides Ebb and Flow

by Joan C. Bowden; illustrated by Marc Brown
Houghton Mifflin, Boston. 1979
Grades: K–4

A feisty old woman bargains with the Sky Spirit, finally gaining a hut, a daughter and son-in-law, and the loan of a very special rock to beautify her yard. She borrows the rock twice each day from the hole in the bottom of the sea, and that is why the tides ebb and flow. The tale is not attributed to any specific culture, but the design motifs seem African inspired.

ARCHAEOLOGY AND EXPLORING YOUR ROOTS

The best of prophets
of the future
is the past.

— *Lord Byron*
Journal

Do People Grow on Family Trees?: Genealogy for Kids and Other Beginners
by Ira Wolfman; illustrated by Michael Klein
Workman Publishing Co., New York. 1991
Grades: 5–12

The chapter "Ancestor Detector" tells how to trace family records, find documents, and includes general background material on American immigration and family names. With many photographs and short articles, this book is slickly designed to grab the reader. In the "Paper Chase" chapter, some of these short articles are useful such as "keeping dates straight," "that old-time handwriting," "the mystery of the missing days," and "using the sounds of Soundex" about an indexing code for names used by the government.

Mitzi and Frederick the Great
by Barbara Williams; illustrated by Emily A. McCully
E.P. Dutton, New York. 1984
Grades: 5–9

Humorous fictional account of the summer Mitzi spends with her mother and brother Frederick on an archaeological dig in Chaco Canyon, one of the most important Native American historical sites. While much of the book is about the family dynamics of Mitzi and her brother, there is accurate information on archeology and its techniques.

Motel of the Mysteries
by David Macaulay
Houghton Mifflin, Boston. 1979
Grades: 6–Adult

Presupposing that all knowledge of our present culture has been lost, an amateur archeologist of the future discovers clues to the lost civilization of "Usa" from a supposed tomb, Room #26 at the Motel of the Mysteries, which is protected by a sacred seal ("Do Not Disturb" sign). For older students, this cleverly illustrated archaeological satire is a particularly apt accompaniment to the midden activities. Students construct their own inferences based on evidence from their middens, myths, and masks. Motel of the Mysteries is an elaborate and logically constructed train of inferences based on partial evidence, within a pseudo-archaeological context. Reading this book, whose conclusions they know to be askew, can encourage students to maintain a healthy and irreverent skepticism about their own and other's inferences and conclusions, while providing insight into the intricacies and pitfalls of the reasoning involved. An extended review of *Motel of the Mysteries* is on page 160.

My Backyard History Book

by David Weitzman; illustrated by James Robertson
Little, Brown & Co., Boston. 1975
Grades: 4–12

> A do-it-yourself history primer with activities and projects for tracing your
> own roots. Create a birthday time capsule, be creative with family photo-
> graphs using a photocopy machine, make a family map, record family
> activities and memories through photography, oral history, or gravestone
> rubbings. Emphasizes the theme that the past is all around you and history
> is more than just dates.

Skara Brae: The Story of a Prehistoric Village

by Olivier Dunrea
Holiday House, New York. 1985
Grades: 4–8

> This book describes a stone age settlement preserved almost intact in the
> sand dunes of one of the Orkney Islands, how it came to be discovered in
> the mid-nineteenth century, and what it reveals about the life and culture
> of this prehistoric community. Learning about the archaeological aspects
> of this discovery sheds light on archaeological techniques that students
> practice in the GEMS midden activities.

Who Do You Think You Are?: Digging for Your Family Roots

by Suzanne Hilton
Westminster Press, Philadelphia. 1976
Out of print
Grades: 6–Adult

> Describes how to do primary and secondary research, to construct a family
> tree, and to find "problem" records for immigrants, adopted children,
> Native Americans or black slave ancestors, with the example of Alex Haley
> and his research for the "Roots" series. The author advocates looking at
> history in a new way, history made by your own people.

Who Put the Cannon in the Courthouse Square: A Guide to Uncovering the Past

by Kay Cooper; illustrated by Anthony Accardo
Walker and Co., New York. 1985
Grades: 5–Adult

> How to research local history, not just people, but
> landmarks, battles, accidents and natural disasters,
> cemeteries, and other secret places. Chapters describe
> how to do research at libraries, museums, and interview
> people. An appendix includes a summary of three
> secondary school local history projects.

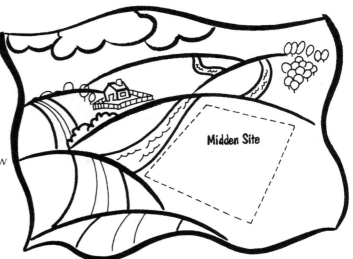

Motel of the Mysteries

BY LINCOLN BERGMAN

*M*otel of the Mysteries by David Macaulay is an excellent literary and scientific accompaniment to any unit (especially *Investigating Artifacts*) that focuses on the collection of clues and evidence from which to make inferences, draw conclusions, or solve mysteries. *Motel of the Mysteries* could also be used with older students to great educational advantage in connection with the GEMS units *Crime Lab Chemistry*, *Mystery Festival*, and *Fingerprinting*, both of which use classic mystery story plots to engender interest and controversy.

In brief, this brilliantly conceived and carefully designed book begins with the proposition that late 20th century civilization has been buried under an avalanche of paper, mostly direct mail advertising fliers, and much about the life of this lost era is not known until 4022. Only scanty evidence has been pieced together, including some remnants of the "imposing Temples of Bigapple," until a previously obscure individual named Howard Carson accidentally stumbles upon the "Motel of the Mysteries," whose contents and interpretations of their roles in the culture are displayed in the book.

The incredible and bizarre archaeological twists put upon interpretations of common household items, including toilet seats and other bathroom accessories, are stated in an academic, logical tone, as if they make perfect sense as final conclusions, yet we know that they are wildly and sometimes hilariously off the mark. The toilet bowl becomes the "Sacred Urn" of the "inner chamber," thought to be "carved from a single piece of porcelain and then highly polished." The archaeologists of the future believe that "the Urn was the focal point of the burial ceremony. The ranking celebrant, kneeling before the Urn, would chant into it while water from the sacred spring flowed in to mix with sheets of Sacred Parchment." The toilet seat itself is believed to have been a "sacred collar" worn by the "ranking celebrant" and *secured* to the urn after the ceremony. As for the "Sacred Parchment …"

Your students will no doubt appreciate both the down-to-earth humor and linguistic sophistication of this book. You could consider having students list or make a chart showing a possible inference process that the scholars of the future could have used to come to their conclusions. Ask them to think of reasons why it was believed a "burial chamber" was discovered. What was the evidence? Are there other interpretations? How does this process compare to the one students used to try to solve the mysteries in the GEMS activities? Is it possible that their proposed solution to the mystery is just as far afield as the interpretations in *Motel of the Mysteries*? Why or why not? Are there other possible reasons why an individual's fingerprints might be on the safe? What other factors need to be taken into account besides which pen was used to write the ransom note?

You could also have your students do some research on a famous archaeological discovery, such as the ancient tombs of Egyptian kings or Chinese emperors, or the numerous Native American sites closer to home. Have any early assump-tions or conclusions about these cultures been shown to be false? Are there differences in interpretation or controversies about the meaning of the findings? For a creative finale to this investigation, why not have your students work in teams to write and illustrate their own versions of a book similar to *Motel of the Mysteries*, focusing perhaps on the future discovery of a school and a classroom a lot like your own!

Lincoln Bergman is the principal editor of the GEMS series as well as co-author of Investigating Artifacts.

INVOLVING DISSOLVING

Grades 1–3　　　　　　　　　　　　　　**Four Activities**

Students learn about the concepts of dissolving, evaporation, and crystallization. Using familiar substances, they create homemade "gel-o," colorful disks, and crystals that emerge on black paper to make a "starry night."

Activity 1

Students receive a "mystery solid" (gelatin) and a "mystery liquid" (apple juice) to observe, examine, and describe. Students then mix the solid with hot juice, discuss where the solid went, and learn the meaning of the word "dissolve." The next day students discuss, then eat, the gelled results.

Activity 2

Students discuss what might happen with a little hot water and lots of gelatin, then make stained-glass-like hanging ornaments or gelatin disks.

Activity 3

Students measure, mix, and observe one substance (salt) that dissolves in water and another that does not (pepper). They pour solutions onto black paper, form crystals, and create a "starry night."

Activity 4

Students observe, over a one- to two-week period, the results when eggs, both raw and hard-boiled, are placed in vinegar and water. Eggs in vinegar begin to lose their shells. When the solution is poured off and evaporates, the calcium from the eggshells recrystallizes.

Skills

Observing, Comparing, Describing, Measuring, Recording, Predicting, Drawing Conclusions

Concepts

Chemistry, Liquids and Solids, Dissolving, Solutions, Evaporation, Crystals

Themes

Systems & Interactions, Patterns of Change, Structure, Matter

Notes: Much questioning as to whether or not things that dissolve "disappear." Starry Night is reminiscent of Van Gogh. Background information includes more on how the "gel-o" forms, crystals, and mold growth (which can adversely affect the gelatin disks). Has modifications for kindergarten students and "Helpful Hints for Hands-On Science in the Classroom." Art-related projects, experimenting with eggs, and chicken bones in vinegar.

MAKING CONNECTIONS

Two books depict **crystals dissolving**, one from an ant's perspective and another as it would appear through a microscope. A picture book featuring a deer licking **salt** provides a nice literary extension to those activities in this unit that involve salt. Another book deals with **evaporation** in the context of the water cycle and water purification.

Keep on the lookout for books which help children understand more about the **nature of crystals** and **dissolving**, or **the use of filtration or crystallization as a methods of purification or separation**. Let the GEMS project know about them!

LITERATURE CONNECTIONS

Greg's Microscope
by Millicent E. Selsam; illustrated by Arnold Lobel
Harper & Row, New York. 1963
Grades: 2–4

> Greg's father buys him a microscope and he finds an unlimited array of items around the house to observe, even the hair of Mrs. Broom's poodle. The illustrations show the salt and sugar crystals, threads, hair, and other material as it appears to him magnified. Solutions of salt and sugar give him a chance to see crystals dissolve. Although this is not a high tech, state-of-the art representation, the fun and empowering experience of playing with scale are well portrayed.

The Magic School Bus at the Waterworks
by Joanna Cole; illustrated by Bruce Degen
Scholastic, Inc., New York. 1986
Grades: K–6

> When Ms. Frizzle, the strangest teacher in school, takes her class on a field trip to the waterworks, everyone ends up experiencing the water purification system from the inside. Evaporation, the water cycle, and filtration are just a few of the concepts communicated in this whimsical fantasy field trip.

CROSS REFERENCES

Cake Poem

I feel good about cooking.
The planning, arranging
Of pots like waterbirds, pans like boats
Oars of spoons, lakes of boiling water,
Nobody watches me.
Flour collapses into milk,
A tree disintegrating into earth.
Egg yolks float
And there is the peculiar pleasure
Of breaking them, the yellow blood.
I pursue recipes
Like a child who must pick all the flowers.
Best of all is baking,
Putting the brown unformed mud
With all those beaten cells
The vanilla from South America,
The milk from how many cows,
The magic powder to make it rise,
Shutting it into warm gestation
The hot spirit-boat
The directions have been observed.
Later it appears
Swollen, with a steaming crack.
I have given my fingerprints
To its invisible swirls,
My cake.
I have fashioned a planet.
And then the sharing.
Cutting a map
I give it out like countries—
And it feeds me.

— *Katy Akin*
Impassioned Cows by Moonlight

Mystery Day

by Harriet Ziefert; illustrated by Richard Brown

Little, Brown & Co., Boston. 1988

Grades: 1–4

Mystery Day is a school day full of surprises for Mr. Rose's students. They have to guess the identity of five mystery powders. The students test their guesses with simple experiments as they look at the powders, touch them, taste them, and mix them with various liquids to see what happens. Once they are correctly guessed, several powders are mixed together and the investigation process starts all over again.

Salt

by Harve Zemach; illustrated by Margot Zemach

Farrar, Straus & Giroux, New York. 1977

Grades: 2–4

This Russian tale tells of a rich merchant's third son, Ivan the Fool, who discovers an island with a mountain of salt. To market his ship's cargo of salt to a foreign king, he secretly adds salt to the food cooking in the royal kitchen. The story could introduce a discussion of how and why salt enhances the flavor of food. The rest of the story involves a beautiful princess, his evil brothers, and a helpful giant.

Salt Hands

by Jane C. Aragon; illustrated by Ted Rand

E.P. Dutton, New York. 1989

Grades: Preschool–2

On a moonlit summer night, a young girl awakens to find a deer in her yard. She sprinkles salt in her hands and goes out to stand near it. The deer moves closer, and finally licks the salt from her hand until it is all gone. Lends a nice extension to the activities involving salt by providing an opportunity to discuss the need that animals (and people!) have for salt in their diets.

Two Bad Ants

by Chris Van Allsburg

Houghton Mifflin, Boston. 1988

Grades: Preschool–4

When two curious ants set off in search of beautiful sparkling crystals (sugar), it becomes a dangerous adventure that convinces them to return to the former safety of their ant colony. Illustrations are drawn from an ant's perspective, showing them lugging individual sugar crystals and other views from "the small." Good extension to those activities that deal with sugar, dissolving, and crystals.

Ladybugs

Preschool–1

Five Activities

Children engage in a series of activities that help them learn more about ladybug body structure and symmetry, life cycle, defensive behavior, and foods. Math is an integral part of this unit and role playing is interwoven in the activities. Colorful posters of ladybugs are included. In the last activity, "Ladybugs Rescue the Orange Trees," children learn important lessons about the environmental role of ladybugs and the interdependence found in nature.

Activity 1

Children are introduced to ladybugs using larger-than-life posters and live ladybugs. They watch ladybugs fly and count the spots on ladybugs. They also role play being ladybugs. In the Symmetry session, the children compare themselves to ladybugs and make a symmetrical drawing to examine the placement of the spots. They make their own paper ladybug.

Activity 2

Ladybugs and aphids are observed (either live animals or posters and drawings). The children examine the aphid posters and count feelers and legs of the aphids and ladybugs. The children pretend to be ladybugs, wear paper wings, and eat "popsicle aphids." Paper aphids are made, as are paper leaves, as homes for aphids and ladybugs.

Activity 3

With their paper leaves and ladybugs, the children role play a ladybug egg-laying drama. The children examine the baby ladybugs posters or, if available, live eggs and larvae. They go outside to look for ladybugs. A drama about egg hatching, baby ladybugs crawling, and eating aphids is presented. The children count the legs of the larvae on the posters, and examine the colors of the baby ladybugs.

Activity 4

Children are introduced to pupae and larvae using live animals or posters. A drama about a baby ladybug growing up, turning into a pupa, and eating aphids is presented. With a paper cutout of a pupa, the children pretend to be pupae. Using the ladybug life cycle poster, the life stages of a ladybug are examined. A drama of the life cycle is presented using paper cutouts. The children role play the life cycle.

Activity 5

Using the scale poster, the children discuss how scale harms plants, and how ladybugs eat scale. A drama is presented about how ladybugs saved the California orange trees. Children eat orange slices. They review what they have learned about ladybugs.

Skills

Observing, Identifying, Creative and Logical Thinking, Communicating, Comparing, Matching, Role Playing

Concepts

Ladybugs (Body Structure, Life Cycle, Defenses), Symmetry, Predator/Prey, Environmental Role of Ladybugs

Themes

Systems & Interactions, Patterns of Change, Models & Simulations, Evolution, Scale, Structure, Diversity & Unity

Math Strands

Geometry (Symmetry), Pattern, Number, Measurement, Logic

MAKING CONNECTIONS

Although we list only two books, both are excellent. Other books that discuss the life cycle of ladybugs would relate well to this GEMS guide. Books about insects and animals that play a positive environmental role would also make excellent connections.

CROSS REFERENCES

LITERATURE CONNECTIONS

The Grouchy Ladybug

by Eric Carle

Harper & Row, New York. 1986

Grades: Preschool–2

The grouchy ladybug and the friendly ladybug want to eat all the aphids on a leaf. The grouchy ladybug challenges the friendly ladybug to a fight, and then challenges every other animal it meets regardless of the animal's size or strength. For young children, this book is a wonderful springboard to measurement activities involving size. There is a clock on each page to chronicle the day in hours for older children. This book works especially well with Activity 2.

Ladybug, Ladybug

by Ruth Brown

E.P. Dutton, New York. 1988

Grades: Preschool–1

Based on the nursery rhyme "Ladybug, Ladybug, Fly Away Home," this beautifully illustrated story captures a ladybug's flight home, where she finds her children safely sleeping. While this book can be read anytime when presenting *Ladybugs*, it works especially well with Activity 1.

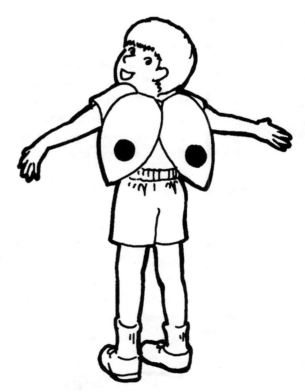

LIQUID EXPLORATIONS

Grades 1–3

Five Activities

*I*n this series of fun and engaging activities, students explore the properties of liquids. They play a classification game, observe how food coloring moves through different liquids, then create secret salad dressing recipes and an "ocean in a bottle." The "rain drops and oil drops" activity can prompt discussion of environmental issues such as oil slicks.

Activity 1

Teacher introduces a classification game with a secret rule that students guess. Using 5–15 liquids, students are asked to classify, and eventually devise their own secret rules, at various classification levels.

Activity 2

Students add a drop of food coloring to a glass of water, observing and recording the results, then compare a drop of food coloring in salt water and in seltzer water. They draw what happens in each case. The activity includes making a class container of pink lemonade.

Activity 3

Students use straw-droppers to investigate water drops and oil drops. Based on their observations, they discuss questions such as "What shape is an oil drop? Does an oil drop look like a water drop? What happens when an oil drop and a water drop are put together?"

Activity 4

The concept of mixing is introduced, as students either directly make or assist the teacher in demonstrating an "ocean in a bottle."

Activity 5

While investigating two liquids that do not mix, students create their own secret salad dressings. Using a color-coded system for recording an ingredient formula, students can then re-create the salad dressing at home.

Skills

Observing, Comparing, Describing, Classifying, Recording, Drawing Conclusions

Concepts

Properties of Liquids, Simple Definitions, Classifying, Observing Attributes, Comparing, Mixtures

Themes

Systems & Interactions, Stability, Patterns of Change, Structure, Matter

Notes: Has modifications for kindergarten, and a special section: "Helpful Hints for Hands-On Science in the Classroom." The "Raindrops and Oil Drops" activity was used in the Alaska schools as part of their efforts to educate students in relation to the 1989 Exxon oil spill.

MAKING CONNECTIONS

Several books focus on **water**—what you can do with it, where it can be found, its properties, its different phases (fog, snow, steam, etc.), the water cycle, and how water is purified so we can drink it.

Other books focus on **classification**, relating to Activity 1 in which students classify different liquids. There is one book with a section about **salad dressings** that relates nicely to Activity 5.

Finally, we have included a book in which various liquids (with various attributes) are mixed to create a mysterious liquid potion with magical powers. This and other books with a fantasy twist are great ways to unleash your students' imaginations as they use their understanding of **liquids and liquid properties** to weave and follow stories.

CROSS REFERENCES

LITERATURE CONNECTIONS

About Water

by Laurent deBrunhoff
Random House, New York. 1980
Out of print
Grades: Preschool–2

Babar the elephant finds a world of water in this tiny book—water to drink, to bathe in, to boat on, to dive in, to feed a fountain, and even to use as a mirror.

Elliot's Extraordinary Cookbook

by Christina Bjork: illustrated by Lena Anderson
R&S Books/Farrar, Straus & Giroux, New York. 1990
Grades: 4–7

With the help of his upstairs neighbor, Elliot cooks wonderful food, and investigates what's healthy and what's not so healthy. He finds out about proteins, carbohydrates, and the workings of the small intestine. He learns about the history of chickens, how cows produce milk, and how live yeast is used in rye bread. His friend shows him how to grow bean sprouts, and he sews an apron. On page 26 are two recipes for salad dressing that relate to Activity 5.

Everybody Needs a Rock

by Byrd Baylor; illustrated by Peter Parnall
Aladdin Books, New York. 1974
Grades: K–5

> What are the qualities to consider in selecting the perfect rock for play and pleasure? The properties of color, size, shape, texture, and smell are discussed in such an appealing way you'll want to rush out and find a rock of your own. This book makes a nice introduction or follow-up to a discussion of the properties of solids.

Gorky Rises

by William Steig
Farrar, Straus & Giroux, New York. 1980
Grades: 2–5

> When Gorky's parents leave the house, he sets up a laboratory at the kitchen sink and mixes up a liquid mixture with a few secret ingredients, drops of his mother's perfume, and his father's cognac! The liquid proves to have magical properties that allow him to fly over the world. Although the format is a picture book, the content makes it usable for older students. Nice connection to the attributes and properties of liquids, with a fantasy twist.

Harriet's Halloween Candy

by Nancy Carlson
Puffin/Penguin, New York. 1982
Grades: Preschool–3

> Harriet learns the hard way that sharing her Halloween candy makes her feel much better than eating it all herself. In the process, she sorts, classifies, and counts her candy. Fun activity to do at Halloween or with any food items. A good connection to the sorting and classifying game in Activity 1.

The Magic School Bus at the Waterworks

by Joanna Cole; illustrated by Bruce Degen
Scholastic, New York. 1986
Grades: K–6

> Ms. Frizzle, the "strangest teacher in school," takes her class on a field trip to the waterworks. First, they journey to the clouds where the class rains, each kid inside his own raindrop. Then they end up experiencing the water purification system from the inside, traveling through the mixing basin, settling basin, filter, and through the pipes to emerge from a faucet. Evaporation, the water cycle, and filtration are just a few of the concepts explored in this whimsical fantasy field trip.

> Water, water, everywhere,
> Nor any drop to drink.
>
> — *Samuel Taylor Coleridge*
> *The Ancient Mariner*

Mystery Day

by Harriet Ziefert; illustrated by Richard Brown

Little, Brown & Co., Boston. 1988

Grades: 1–4

Mystery Day is a school day full of surprises for Mr. Rose's students. They have to guess the identity of five mystery powders. The students test their guesses with simple experiments as they look at the powders, touch them, taste them, and mix them with various liquids to see what happens. Once they are correctly guessed, several powders are mixed together and the investigation process starts all over again.

Rain Drop Splash

by Alvin Tresselt; illustrated by Leonard Weisgard

Lothrop, Lee & Shepard, New York. 1946

Mulberry Books/William Morrow, New York. 1990

Grades: K–3

Raindrops fall to make puddles. Puddles become larger and larger to form ponds. Ponds overflow into brooks that lead to lakes. The rainstorm continues, falling on plants and animals, making mud, flooding a road. The last scene leads to the ocean, when at last the rain stops and the sun emerges.

Shoes

by Elizabeth Winthrop; illustrated by William Joyce

Harper & Row, New York. 1986

Grades: K–2

A survey of the many kinds of shoes in the world concludes that the best of all are the perfect natural shoes that are your feet. Great to read before doing a survey of shoes or sorting and classifying a group of real shoes.

The Snowy Day

by Ezra Jack Keats

Viking, New York. 1962

Grades: Preschool–2

Peter goes for a walk on a snowy day. He makes different patterns in the snow with his feet, a stick, and then his whole body. He tries to save a snowball in his pocket but is disappointed when it melts. That night Peter dreams that the sun melted all the snow outside, but when he wakes up, it's snowing again!

Splash! All About Baths

by Susan K. Buxbaum and Rita G. Gelman; illustrated by Maryann Cocca-Leffler
Little, Brown & Co., Boston. 1987
Grades: K–6

Before he bathes, Penguin answers his animal friends' questions about baths. "What shape is water?" "Why do soap and water make you clean?" "What is a bubble?" "Why does the water go up when you get in?" "Why do some things float and others sink?" and other questions. Answers to questions are both clear and simple. Received the American Institute of Physics Science Writing Award.

Very Last First Time

by Jan Andrews; illustrated by Ian Wallace
Atheneum/Macmillan, New York. 1985
Grades: 2–4

This entrancing book tells the story of the Inuit girl Eva who walks for the first time in a sea-floor cavern under the frozen ocean ice. When the tide is out she and her mother come to gather mussels and Eva goes below the ice. When she stumbles, her candle goes out and the tide starts to come in, roaring louder, while the ice shrieks and creaks. Terrified at first, Eva recovers, and eventually finds her way to the surface and her waiting mother and the moonlight. Although the book does not scientifically explain the freezing of the top of the sea or the action of the tides, you and your class may want to discuss these questions: "Why does only the top part of the water freeze?" "Why does the ice stay intact even when the water underneath it goes out with the tide?" The images of Eva on the sea floor beneath the ice are unique and fascinating, enhanced by the eerie purple tones of the illustrations. The descriptive language and Eva's intense interest in nature exemplify excellent scientific observation skills.

Water's Way

by Lisa W. Peters; illustrated by Ted Rand
Arcade Publishing/Little, Brown & Co., New York. 1991
Grades: K–3

"Water has a way of changing" inside and outside Tony's house, from clouds to steam to fog and other forms. Innovative illustrations show the changes in the weather outside while highlighting water changes inside the house.

Whose Hat Is That?

by Ron Roy; photographs by Rosemarie Hausherr
Clarion Books/Ticknor and Fields, New York. 1987
Grades: Preschool–3

Text and photographs portray the appearance and function of eighteen types of hats including a top hat, a jockey's cap, and a football helmet. The children and adults modeling the hats represent a rainbow of peoples. Makes a nice connection to classification activities in Session 1 of the GEMS guide.

> Water can both float and sink a ship.
>
> — *Chinese proverb*

THE "MAGIC" OF ELECTRICITY

Grades 3–6

This assembly presenter's guide offers exciting demonstrations of electrical phenomena and introduces important concepts of static electricity, and how current electricity is produced through batteries, generators, and solar cells. This is one of two GEMS Assembly Presenter's guides, which provide a script, detailed instructions, and materials lists for the presentation of a program to a school assembly audience of several hundred. In an hour-long session, students take part in impressive demonstrations by the "Wizard of Electricity." They distinguish magic from science, as volunteers help the Wizard create electricity by immersing strips of metal in lemon juice; moving a giant magnet inside a coil of wire; turning a pinwheel with steam, water, and wind; and using light to spin a propeller on a solar beanie.

Skills

Observing, Comparing, Predicting, Visualizing, Relating, Modeling

Concepts

Atom, Electron, Circuit, Cell, Battery, Meter, Solar Cell, Transformer, Generator, Static Electricity, Voltage

Themes

Systems & Interactions, Models & Simulations, Stability, Patterns of Change, Scale, Energy, Matter

MAKING CONNECTIONS

Literature is a good way to expand your students' understanding of electrical phenomena. Two selections listed are drawn from the Einstein Anderson series. They are short chapters about static electricity that can be read aloud in class and end with a challenging puzzle that students can work on for homework.

Your students can also explore electrical currents in the context of several fun stories listed here—as well as several biographies of Michael Faraday that offer a clear discussion of his important discoveries.

CROSS REFERENCES

LITERATURE CONNECTIONS

Coils, Magnets, and Rings: Michael Faraday's World

by Nancy Veglahn; illustrated by Christopher Spollen
Coward, McCann & Geoghegan, New York. 1976
Out of print
Grades: 5–8

This delightful biography of Michael Faraday, best known for his discovery of the electric generator, captures the spirit of the questioning scientist and the atmosphere of the Royal Institution, a nineteenth century institution dedicated to scientific research in all fields. The book portrays exciting participatory lectures given by Faraday. An excellent assignment for upper elementary students is to read the book to find out who invented the electric generator, or "dynamo."

Danny Dunn and the Swamp Monster

by Jay Williams and Raymond Abrashkin; illustrated by Paul Sagsoorian
McGraw-Hill, New York. 1971
Out of print
Grades: 4–8

Danny and friends discover a superconductor, which they use in an adventure in Africa involving an electric fish. On pages 24–26 is an excellent explanation of electrical current, resistance and magnetic fields.

Dear Mr. Henshaw

by Beverly Cleary
William Morrow, New York. 1983
Grades: 4–6

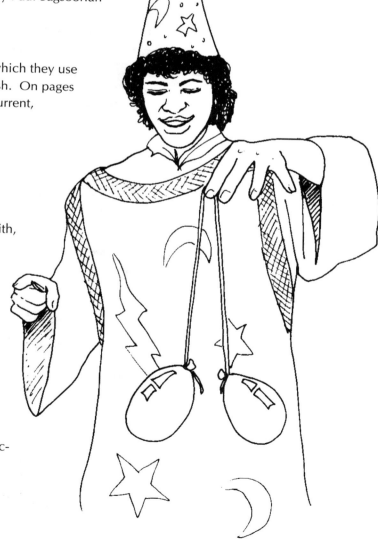

A 10-year-old boy writes to his favorite author with, over time, a growing sense of identity and self-esteem, and enhanced writing ability. A portion of the book describes how the boy sets out to catch a thief by rigging a battery-powered burglar alarm to his lunch box, and how this invention gains him respect.

Einstein Anderson Lights Up the Sky

by Seymour Simon; illustrated by Fred Winkowski
Viking Press, New York. 1982
Grades: 4–7

In Chapter 10, "The Spring Festival," static electricity helps the decorating committee with their balloon placement.

The smell of burning
fills the startled air—
The Electrician
is no longer there!
— *Hilaire Belloc*
Newdigate Poem

Einstein Anderson Shocks His Friends

by Seymour Simon; illustrated by Fred Winkowski
Viking Press, New York. 1980
Grades: 4–7

> In the first chapter, "The Electric Spark," Einstein uses static electricity to scare off his nemesis, Pat Burns.

Everything Happens to Stuey

by Lilian Moore; illustrated by Mary Stevens
Random House, New York. 1960
Out of print
Grades: 4–7

> After smelling up the refrigerator with his secret formula, turning his sister's doll green with a magic cleaner, and having his invisible ink homework go awry, budding chemist Stuey is in trouble. In the end, he uses his knowledge to rescue his sister by fabricating a homemade flashlight. The illustrations and depiction of family life and sex roles are dated, but the spirit of discovery is timeless.

The Pet of Frankenstein

by Mel Gilden; illustrated by John Pierard
Avon Books, New York. 1988
Grades: 5–7

> This book is one in an entertaining series of adventures involving Danny Keegan and his monstrous classmates. Frankie's particular expertise is electronics, so there is a lot of it in this book and some nicely interwoven technical language. From video games and computers to Tesla coils and robotic animals, the bizarre narrative jumps amusingly along. In the end, Frankie succeeds in creating a robotic dog from an old design of his namesake Baron Frankenstein. Definitely not heavy reading, the book will tickle some students' imagination and funny bones.

MAPPING ANIMAL MOVEMENTS

Grades 6–10   **Four Sessions**

Students apply field biology techniques, using a sampling and mapping system, to quantify and compare the movements of hamsters and crickets. Students plan and conduct experiments, graphing changes in movement patterns when food and shelter are added to the environment. This guide makes an excellent extension to the GEMS *Animals in Action* activities.

Session 1

Students discuss the value of and reasons for observing an animal's movements. They learn a time sampling system by helping the teacher map a student's movements in the classroom.

Session 2

Students discuss feeding and humane animal care, then observe, touch, and hold the animals. The class, working in teams of four with assigned roles, writes numbers on adhesive dots for use in next session.

Session 3

Students use a sampling system to map movements of a hamster and a cricket. Animals are observed in an empty container first, then food and shelter are added and movements compared.

Session 4

Students construct a bar graph to analyze data, then compare the two trials (box empty, box with food and shelter) and discuss differences observed by various teams.

Skills

Observing, Classifying, Mapping, Analyzing Data, Experimenting, Making Inferences

Concepts

Biology, Animals, Insects (Crickets), Hamsters, Treatment of Animals, Habitat Requirements, Research Techniques of Biologists

Themes

Systems & Interactions, Stability, Patterns of Change, Evolution, Structure, Diversity & Unity

Notes: There are sections on mapping the movements of Tule Elk, animal care, food, housing, handling, and NSTA Code of Practice on Use of Animals in Schools. Going Further includes mapping movements of a pet or family member, and other neighborhood observations.

MAKING CONNECTIONS

Books that focus on animal behavior, particularly regarding their movements and ways of observing, tracking, and recording them make excellent connections to this GEMS guide. In the following books, connections are made to real research conducted by Charles Darwin, James Audubon, and a naturalist studying wolf behavior. Readers are able to follow the true-life movements of a snapping turtle's journey on the Mississippi River with detailed descriptions and illustrations. In another book, a young girl takes a hike in search of a mysterious bird, and learns much about the woodland habitat along the way.

CROSS REFERENCES

LITERATURE CONNECTIONS

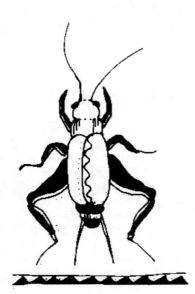

The Beagle and Mr. Flycatcher: A Story of Charles Darwin

by Robert Quackenbush
Prentice Hall, Englewood Cliffs, New Jersey. 1983
Grades: 4–8

This biography tells how Charles Darwin began his career as an unpaid naturalist aboard the brig H.M.S. Beagle on a five-year voyage around South America and began to formulate his revolutionary theory of evolution. There are brief descriptions of Darwin's specimen collecting (cuttlefish, shellfish fossils, sloth jawbone), scientific observation (finches, an ostrich skeleton) and of a subsequent eight-year study of barnacles. Although the illustrations are in a cartoon style and the writing uses a humorous approach, quite a bit of information is conveyed. Might be followed up with the fictional diary of a cabin boy in *Darwin and the Voyage of the Beagle*.

The Black Pearl

by Scott O'Dell; illustrated by Milton Johnson
Houghton Mifflin, Boston. 1967
Grades: 5–12

Ramon's dream is to dive for pearls in the waters of Baja, California, and to one day find the great pearl—the magnificent Pearl of Heaven. But to do so, Ramon must confront the giant manta ray Diablo who guards the pearl. An absorbing story of a young boy's quest for the pearl and his manhood, with the Vermilion Sea as the beautiful backdrop. The movements and interaction of underwater animals are wonderfully described.

A Caribou Alphabet

by Mary Beth Owens
Dog Ear Press, Brunswick, Maine. 1988
Farrar, Straus & Giroux, New York. 1990
Grades: K–5

> An alphabet book depicting the characteristics and ways of caribou. While at first glance, this book may seem a primary-level "A, B, C" book, it includes a compendium of information about caribou, including intricacies of their behavior, habitat requirements, and physical features.

Chipmunk Song

by Joanne Ryder
E.P. Dutton, New York. 1987
Grades: K–5

> A lyrical description of a chipmunk as it goes about its activities in late summer, prepares for winter, and settles in until spring. You are put in the place of a chipmunk—through food gathering, hiding from predators, hibernation, and more. Detailed illustrations show everything from roots invading the chipmunk's hole, to underground stashes of acorns.

Darwin and the Voyage of the Beagle

by Felicia Law; illustrated by Judy Brook
Andre Deutsch, Great Britain. 1985
(Distributed by E.P. Dutton, New York)
Grades: 4–8

> A cabin boy goes along on Charles Darwin's five-year voyage. He assists Darwin with his collection of insect, bird, and marine life specimens all the while learning about their habits and habitats. On one occasion they return with 68 different species of one beetle. They collect fossils in the Andes, straddle Galapagos tortoises, discover the skeleton of the Megatherium, and get to know Fuegan natives. The format is oversized, with many drawings, charts and maps.

Frogs, Toads, Lizards, and Salamanders

by Nancy W. Parker and Joan R. Wright; illustrated by Nancy W. Parker
Greenwillow, New York. 1990
Grades: 3–6

> The physical characteristics, habits, and environment of 16 creatures are encapsulated in rhyming couplets, text, and anatomical drawings, plus glossaries, range maps, and a scientific classification chart. A great deal of information is presented, the rhymes are engaging and humorous, and the visual presentation terrific. "A slimy Two-toed Amphiuma/ terrified Grant's aunt from Yuma" (she was picking flowers from a drainage ditch).

The Fly

Little Fly,
Thy summer's play
My thoughtless hand
Has brushed away.

Am not I
A fly like thee?
Or art not thou
A man like me?

For I dance
And drink and sing,
Till some blind hand
Shall brush my wing.

— *William Blake*
Songs of Experience

Hatchet

by Gary Paulsen
Bradbury Press/Macmillan, New York. 1987
Puffin/Viking Penguin, New York. 1988
Grades: 6–12

> After the pilot of a single-engine plane dies of a heart attack, the sole passenger, 13-year-old Brian, is left to survive alone in the Canadian wilderness. He gets the worst of encounters with a porcupine, a skunk, and a moose. He slowly learns how to provide fire, shelter, and food for himself. His motivation for observing animal, bird, and fish habits is simply to track and hunt them for food. He makes a spear and bow and arrow and learns to store and cook food. "It had always been so simple at home. He would go to the store and get a chicken and it was all cleaned and neat, no feathers or insides, and his mother would bake it in the oven."

A Bird Came Down the Walk

A bird came down the walk:
He did not know I saw;
He bit an angleworm in halves
And ate the fellow raw.

And then he drank a dew
From a convenient grass,
And then hopped sidewise to the wall
To let a beetle pass.

He glanced with rapid eyes
That hurried all around;
They looked like frightened beads, I thought;
He stirred his velvet head

Like one in danger; cautious,
I offered him a crumb,
And he unrolled his feathers
And rowed him softer home

Than oars divide the ocean,
Too silver for a seam,
Or butterflies, off banks of noon,
Leap, plashless, as they swim.

— *Emily Dickinson*

The Island

by Gary Paulsen
Orchard/Franklin Watts, New York. 1988
Bantam Doubleday Dell, New York. 1990
Grades: 6–12

> Wil Neuton seeks solace from family problems on an island in a lake in Wisconsin. He finds solitude and discovers a new interest in the bird, fish, and animal life whose behavior he observes, describes, and paints. A loon and her chicks first attract his interest, then a blue heron swallowing a frog, ants scavenging leftovers from a tin can, and a snapping turtle's bloody conquest of a sunfish. "I could see the heron, finally, with every one of his gray blue-feathers shining on the edge of purple, with his tapered beak and crest and curved-over neck and clean lines and by not looking at the heron I could see me. At the end, I could see myself in the heron."

Minn of the Mississippi

by Holling Clancy Holling
Houghton Mifflin, Boston. 1951
Grades: 5–9

> The journey of Minn, a snapping turtle, is followed from northern Minnesota to the bayous of Louisiana. Her adventures with people, animals, and the changing seasons are vividly described. Wonderful drawings and maps of her travels accompany the engaging true-life story on the Mississippi River.

Mrs. Frisby and the Rats of NIMH

by Robert C. O'Brien; illustrated by Zena Bernstein
Atheneum, New York. 1971
Grades: 4–12

A mother mouse learns that the rat colony near her home is actually a group of escapees from an NIMH research institute. These rats, injected with DNA and other substances, have acquired great intelligence, learned to read and write, and are planning to develop their own civilization. In addition to offering a great plot, the book helps us to visualize nature from the scale of a small animal, to imagine communications between birds and rodents, to consider the impact of animal experimentation, and to comment on the technological top-heaviness of modern day human society. Newbery award winner.

Never Cry Wolf

by Farley Mowat
Atlantic Monthly/Little, Brown & Co., New York. 1963
Bantam Books, New York. 1984
Grades: 6–Adult

Wolves are killing too many of the Arctic Caribou, so the Wildlife Service assigns a naturalist to investigate. Farley Mowat is dropped alone onto the frozen tundra of Canada's Keewatin Barrens to live among the wolf packs to study their ways. His interactions with the packs, and his growing respect and understanding for the wild wolf will captivate all readers.

Nicky The Nature Detective

by Ulf Svedberg; illustrated by Lena Anderson
R & S Books/Farrar, Straus & Giroux, New York. 1983
Grades: 3–8

Nicky loves to explore the changes in nature. She watches a red maple tree and all the creatures and plants that live on or near the tree through the seasons of the year. Her discoveries lead her to look carefully at the structure of a nesting place, why birds migrate, who left tracks in the snow, where butterflies go in the winter, and many many more things. This book is packed with information.

On the Frontier with Mr. Audubon

by Barbara Brenner
Coward, McCann & Geoghegan, New York. 1977
Grades: 6–9

Based on an unedited diary of 1820-26, this fictionalized journal tells of Joseph Mason, a 13-year-old assistant who really traveled with John J. Audubon for 18 months. The work gives a detailed account of their daily life, hunting, drawing birds and their habitats, a stay at a plantation, and travel by flatboat, keelboat, and steamer. Black and white illustrations include reproductions of paintings and drawings by Audubon and other artists of his day.

One Day in the Tropical Rain Forest

by Jean C. George; illustrated by Gary Allen
HarperCollins, New York. 1990
Grades: 4–7

> A section of rain forest in Venezuela is scheduled to be bulldozed as young boy and a scientist seek a new species of butterfly for a wealthy industrial-ist who might preserve the forest. As they travel through the ecosystem rich with plant, insect, and animal life, everything they see on this one day is logged, beginning with sunrise at 6:29 a.m. They finally arrive at the top of the largest tree in the forest and fortuitously capture a specimen of an unknown butterfly.

One Day in the Woods

by Jean C. George; illustrated by Gary Allen
Thomas Y. Crowell, New York. 1988
Grades: 4–7

> On a day-long outing in a forest, Rebecca, a "ponytailed explorer," searches for the elusive ovenbird. Wonderful detail on her observation and interaction with the plant and animal life enhanced by realistic black and white drawings.

Owl Moon

by Jane Yolen; illustrated by John Schoenherr
Philomel/Putnam, New York. 1987
Grades: Preschool–5

> On a moonlit winter night, a father and daughter go searching for the elusive Great Horned Owl. The suspense of the hunt, along with the lyrical language and stunning illustrations of a rural scene at night make one feel a part of the expedition. They seek the owl in its habitat, observe its behavior, and imitate its call.

How doth the little crocodile
Improve his shining tail,
And pour the waters of the Nile
On every golden scale.

How cheerfully he seems to grin,
How neatly spreads his claws,
And welcomes little fishes in,
With gently smiling jaws!
— *Lewis Carroll*
Alice's Adventures in Wonderland

The Roadside

by David Bellamy; illustrated by Jill Dow
Clarkson N. Potter/Crown, New York. 1988
Grades: K–5

> The construction of a six-lane highway in a wilderness area disrupts the balance of nature and forces animals there to struggle for existence.

The Song in the Walnut Grove

by David Kherdian; illustrated by Paul O. Zelinsky
Alfred A. Knopf, New York. 1982
Grades: 4–6

> A curious cricket meets a grasshopper. Together they learn of each other's daytime and nighttime habits while living in a herb garden. Their friendship grows as they learn to appreci-ate each other's differences. This story weaves very accurate accounts of insect behavior and their contributions to the ecology of Walnut Grove.

Urban Roosts:
Where Birds Nest in the City
by Barbara Bash
Sierra Club/Little, Brown & Co., Boston. 1990
Grades: 1–6

The inventive places birds make their homes in the heart of the city and how they adjust to such a harsh urban environment is fascinating. Some of the birds examined include pigeons, barn owls, nighthawks, and peregrine falcons. Habitats include traffic lights, tile roofs, and train trestles. This is a particularly good way to show urban students that observation of wildlife need not be confined to the countryside.

Gumbo Limbo
and You

As can be surmised from the multiple listings in this handbook, *The Missing 'Gator of Gumbo Limbo: An Ecological Mystery* by Jean C. George is, from a "GEMSian" point of view, exemplary in many respects. Annotations for Gumbo Limbo in this handbook point out specific connections to several diverse GEMS guides and science themes. Some of these annotations contain a few sentences from the book to give you a sense of the richness of the writing, and the way in which the scientific information is interwoven with the prose in relevant and educational ways. The book is as timely as homelessness and ecological devastation, and is written in a way that communicates the integration of science/mathematics with real-life issues. It also features an independent-thinking young heroine who, far from being intimidated when it comes to technical matters, pursues scientific and ecological knowledge with a fierce (and believable) intensity, as she comprehends how important that knowledge can be in preserving a delicate and threatened environment.

Several GEMS guides, notably *Acid Rain* and *Global Warming and the Greenhouse Effect* seek, through hands-on experiments and a variety of other activities, to communicate this same sense of environmental empowerment and responsibility. *River Cutters* includes sections on erosion and toxic waste. The entire GEMS series is, of course, dedicated to linking science and mathematics with real-life experience and helping young people think for themselves. The heroine of Gumbo Limbo provides a great role model for all people who have become more aware of the interconnectedness of the Earth's environment.

Mapping FISH HABITATS

Grades 6–10

Four Activities

Students learn about and apply the field-mapping techniques of aquatic biologists as they chart the movements of fish in a classroom aquarium. Students plan experiments to determine the effects of an environmental change on the home ranges of the fish. An aquarium becomes a habitat for a variety of fish that move in different areas of the tank. Students use a simple technique to map and **observe** the movements of individual fish as they **feed, defend** themselves, and establish their **territory** within the aquarium.

Activity 1

Students design the environment for their fish-mapping activities by placing rocks, plants, and fish in an aquarium. Students also measure temperature, pH, water level, and algae growth. The concept of ecosystem is introduced.

Activity 2

Students compare different fish to be able to identify each individual during mapping activities. The concept of habitat is introduced and habitat requirements discussed. A fish-mapping system, using colored dots, is demonstrated. A team of students makes a home range map.

Activity 3

The class analyzes the home range map to learn about home ranges and sampling systems. Students become familiar with the basic procedure and decide on a change to make in aquarium environment to determine its effect on the home range or other fish behaviors.

Activity 4

The second team observes and maps the fish after the one change is made, and the class analyzes and discusses. Teacher introduces concept of animal territories. Class selects another variable to change, and a third team conducts that experiment. Experiments continue until all teams have made a home range map.

Skills

Observing, Classifying, Mapping, Analyzing Data, Experimenting, Making Inferences

Concepts

Biology, Fish, Habitat Requirements, Home Range, Animal Territories, Research Techniques of Biologists, Monitoring An Aquarium Ecosystem, Humane Treatment of Animals

Themes

Systems & Interactions, Stability, Patterns of Change, Evolution, Structure, Diversity & Unity

Notes: Background information on how to set up an aquarium, and Fish Facts sheets are provided for guppy, rosy barb, and bronze catfish.

MAKING CONNECTIONS

Four books in particular describe true historical adventures involving ocean creatures and the native peoples who live nearby. The stories point out the importance of observation and **tracking** of marine life to people dependent on the sea for their survival. Other stories bring out the role of fishing in the relationships between a girl and her father, and a young boy and his grandfather. Both of these books have young people investigating the living and non-living world of **aquatic habitats**. An "ecological mystery" provides a compelling look at how tracking, observation, and scientific testing can help understand and protect the environment.

CROSS REFERENCES

LITERATURE CONNECTIONS

The Black Pearl
by Scott O'Dell
Houghton Mifflin, Boston. 1967
Dell Publishing, New York. 1977
Grades: 5–12

Ramon's dream is to dive for pearls in the waters of Baja, California, and to one day find the great pearl—the magnificent Pearl of Heaven. But to possess the pearl, Ramon must confront the giant manta ray Diablo, which guards the pearl. An absorbing story of a young boy's quest for the pearl and his manhood, with the Vermilion Sea as the beautiful backdrop. Newbery honor book.

Call it Courage
by Armstrong Sperry
Macmillan, New York. 1940
Collier Books/Macmillan, New York. 1971
Grades: 6–12

A young Polynesian chief's son is scorned for his fear of the sea, which took his mother's life. He goes on a difficult and dangerous quest in a canoe with his little dog and a pet albatross. Learning to survive on his own, he becomes fascinated by the undersea world of a barrier reef. There is a detailed section on the marine life, including a life-threatening battle with an octopus, and a struggle with a hammerhead shark. His escape from cannibals might be balanced by a discussion that modern anthropology suggests that many accusations of cannibalism are unfounded. It might be interesting to have older students analyze the cultural sensitivity of the book from today's perspective. Newbery award winner.

> Only the gamefish
> swims upstream,
> But the sensible fish
> swims down.
>
> — *Ogden Nash*
> *When You Say That, Smile*

Go Fish

by Mary Stolz; illustrated by Pat Cummings
HarperCollins, New York. 1991
Grades: 4–6

Eight-year-old Thomas and his grandfather go fishing in the Gulf of Mexico. Grandfather is a collector of shells, petrified wood, and even a sandstone with a fossil fish. In Chapter 2, their fishing gear is listed, including a record book for noting large specimens caught. They observe herons, pelicans, minnows and jellyfish, as well as the blowfish, flounder, and other fish they seek. Back at grandfather's they share a card game of Go Fish and an African folk tale. The book provides a wonderful model of friendship between generations and affirms the value of observing and questioning the world around us. The regional focus is refreshing, offering African-American characters in a non-urban setting.

Island of the Blue Dolphins

by Scott O'Dell
Houghton Mifflin, Boston. 1960
Dell Publishing, New York. 1987
Grades: 5–12

A Native American girl grows to womanhood by herself on the outermost island of the Channel Islands—about 75 miles southwest of Los Angeles. Interwoven are descriptions of the island, of fish and ocean vegetation, animals and plants. The way she interacts with nature to survive, hunt, build shelter, and design clothing, both as she has been taught by her people and as she develops her own technological and artistic skills, is a particularly strong aspect of the book. She has a wild dog she raises from a pup, an otter, and trained birds for company. A main plot line in Chapters 16 and 19 concerns the stalking of a giant devilfish, or manta ray. Students could read these chapters and then discuss or make a list of what they have learned about the habits and habitat of the devilfish. Sentences like the following give a sense of the possibilities of this assignment, and how it connects to the GEMS activities: "Seldom do you see any devilfish here, for they like deep places, and the water along this part of the reef is shallow. Perhaps this one lived in the cave and came here only when he could not find food."

June Mountain Secret

by Nina Kidd
HarperCollins, New York. 1991
Grades: 4–6

Jen and her father go fly fishing for rainbow trout and she learns about the different types of insects that attract trout, eventually using a mayfly replica as bait. When they finally set the trout they caught free, the issue of conservation regarding sport fishing is raised. A page of facts and some very specific graphic materials enable this book to be used for 6th grade even though the format is a picture book.

Minn of the Mississippi

by Holling Clancy Holling
Houghton Mifflin, Boston. 1951
Grades: 5–9

The journey of Minn, a snapping turtle, is followed from northern Minnesota to the bayous of Louisiana. Her adventures with people, animals, and the changing seasons are vividly described. Wonderful drawings and maps of her travels accompany the engaging true-life story on the Mississippi River.

The Missing 'Gator of Gumbo Limbo: An Ecological Mystery

by Jean C. George
HarperCollins, New York. 1992
Grades: 4–7

Sixth-grader Liza K and her mother live in a tent in the Florida Everglades. She becomes a nature detective while searching for Dajun, a giant alligator who is marked for extinction by local officials. The book is full of detail about the local habitats and species and the forces that impact on them. She and an amateur naturalist and neighbor James James notice a drop in the numbers of mosquito fish, or gambusia. They trace the cause to runoff water polluted by pesticides. They also note the disturbing presence of the weed, hydrilla, which because of its rapid growth can suffocate the big fish and stunt the little fish. In another scene, they discover an intrusion of salt water into a canal. "James James counted the dead fish in a small area and multiplied it by the size of the entire area to get an estimate of the number of fish killed. He made a note of each species." The constant search for Dajun, the numerous other observations and the well-honed environmental consciousness make this book a particularly strong connection to the GEMS animal observation activities.

Shark Beneath the Reef

by Jean C. George
HarperCollins, New York. 1989
Grades: 5–12

Fourteen-year-old Tomas has two loves, school and fishing, and is supported by his proud fisherman grandfather and his caring high school science teacher. Tomas comes from a family of shark fishermen on the island of Coronado on the Sea of Cortez whose livelihood is threatened by governmental plans for tourism and Japanese factory fishing boats. Reflecting his Indian and Spanish heritage, Tomas calls on both Quetzalcoatl and Our Lady of Guadalupe to help his people. The oceanic environment flows through the book, as Tomas observes the activity in a tide pool or tracks a fish underwater, giving a real sense of the interrelation between marine life and its habitats. The chapter "A Warning from a Fish and a Bird" is an excellent accompaniment to the observation skills and understandings of movement and habitat that students gain from the GEMS activities.

> I throw myself to the left.
> I turn myself to the right.
> I am the fish
> Who glides in the water, who glides,
> Who twists himself, who leaps.
> Everything lives,
> Everything dances,
> Everything sings.
>
> — *African Pygmy poem*

MOONS of JUPITER

Grades 4–9

Five Activities

Students become "Galileos" as they recreate, through viewing beautiful slides (provided with the guide), historic telescopic observations of Jupiter's moons. They observe and record orbits of the moons over time and learn why these observations helped signal the birth of modern astronomy. In subsequent sessions, students model moon phases of Earth's Luna and the moons of Jupiter; experiment to learn how craters are formed; take a grand tour of the Jupiter system as viewed by the Voyager spacecraft; and make scale models to better understand size and distance. In the final session, students work in teams to create, from an assortment of common materials, a settlement on one of the moons of Jupiter. Fascinating background information is also provided.

Activity 1

"Tracking Jupiter's Moons" recreates Galileo's 1609 first sighting of four curious "stars" in a straight line in the vicinity of Jupiter. Students track the moons of Jupiter and calculate their periods of revolution, just as Galileo did.

Activity 2

The students build a model of the Moon's pulverized surface and pummel it with "meteors" to see what will happen. They can create some craters very much like those seen on Earth's Moon.

Activity 3

This activity models the scale of the Jupiter system. Using a classroom scale model of the Earth-Moon system as a basis of comparison, your students mark the distances within the Jupiter system in the school yard.

Activity 4

The students tour the Jupiter System through slides and draw their own pictures to illustrate one of these four moons: Io, Europa, Ganymede, or Callisto.

Activity 5

Students are challenged to design and build a model of a space settlement on one of the four major moons of Jupiter.

Skills

Creating/Using Models, Synthesizing, Visualizing, Observing, Explaining, Recording, Measuring, Using a Map, Evaluating Evidence, Inferring, Drawing Conclusions, Creative Designing

Concepts

History of Astronomy, Systems of Planets and Satellites, Revolution and Rotation, Comparing Surface Features, Crater Formation, Geocentric and Heliocentric Models of the Solar System, Relative Sizes and Distances of Solar System Bodies, Space Settlements

Themes

Systems and Interactions, Stability, Models and Simulations, Patterns of Change, Energy, Scale, Structure

MAKING CONNECTIONS

Science fiction provides many fine literary connections and extensions for this GEMS guide. The play by Brecht gives students dramatic insight into Galileo's life, and even contains information on his observations—the same observations the students make via slides in the classroom. Other books about Galileo or about telescopic observation also make good connections.

Several of the books involve exploring and colonizing other planets. In *The Planets* there is a highly sophisticated science fiction story about colonization of one of Jupiter's moons. Look for any good stories about exploration and settlement of the Earth's Moon, or other planets/galaxies. Ethical issues are often involved, including: nationalistic competition on Earth in the "race for space," imagined diplomatic and/or antagonistic relations with extraterrestrials, and whether or not to colonize or exploit natural resources found in space.

Venus, Mars, and Jupiter, because of their prominence in the nighttime sky, have inspired a large number of folktales, legends, and myths. Their rising and setting times are often important agricultural markers. Books such as *Star Tales: North American Indian Stories* or *They Dance in the Sky* are but two examples from many that provide great opportunities for literary connections and increased cultural understandings.

CROSS REFERENCES

LITERATURE CONNECTIONS

2010: Odyssey Two
by Arthur C. Clarke
Ballantine Books, New York. 1982
Grades: 10–Adult

This complex, mysterious, and thought-provoking sequel to Clarke's *2001: A Space Odyssey* had the benefit of being written subsequent to the Voyager mission. Chapter 13 specifically, "The Worlds of Galileo," focuses on the four main moons of Jupiter, although there are fascinating observations, accurate scientific information, and lots of interesting speculation about Jupiter and its moons throughout the book, not to mention spirits of intergalactic intelligence and Jupiter becoming a second sun. More advanced students may want to evaluate the accuracy of Clarke's descriptions of Io, Europa, Ganymede, and Callisto.

Against Infinity
by Gregory Benford
Simon & Schuster, New York. 1983
Grades: 10–Adult

This science fiction novel is an account of human settlement on Jupiter's largest moon, Ganymede. The story takes place several hundred years into the colonization process, and begins from the perspective of a 13-year-old boy whose father is one of the leaders of the settlement. The novel ties in well with the final session in which student teams undertake scientific missions to devise and build moon settlements on one of Jupiter's moons. Advanced students may want to read this novel to gather ideas about constructing biospheres, melting ice, obtaining minerals, and other ways humans might possibly survive on the moons of Jupiter. (The author is a Professor of Physics at the University of California, Irvine.)

Einstein Anderson Makes Up for Lost Time
by Seymour Simon; illustrated by Fred Winkowski
Viking Penguin, New York. 1981
Grades: 4–7

Chapter 6 poses the question "How can Einstein tell a planet from a star without using a telescope?" He explains to his friend Dennis that although stars twinkle, planets usually shine with a steady light. Looking through the telescope, he thinks the steady light he sees is Jupiter. The four faint points of steady light nearby are Jupiter's moons.

The Faces of Ceti
by Mary Caraker
Houghton Mifflin, Boston. 1991
Grades: 6–12

In this science fiction thriller, colonists from Earth form two settlements on adjoining planets of the Tau Ceti system. One colony tries to survive by dominating the natural forces that they encounter, while those who land on the planet Ceti apply sound ecological principles and strive to live harmoniously in their new environment. Nonetheless, the Cetians encounter a terrible dilemma—the only edible food on the planet appears to be a species of native animals called the Hlur. Two teen-age colonists risk their lives in a desperate effort to save their fellow colonists from starvation without killing the gentle Hlur.

Every great advance in science has issued from a new audacity of imagination.

— *John Dewey*
The Quest for Certainty

Galileo
by Leonard Everett Fisher
Macmillan Publishing, New York. 1992
Grades: 3–7

This nonfiction work provides a carefully written and well-illustrated account of Galileo's life and accomplishments. His observations of Jupiter's moons are placed in the context of a life of remarkable discoveries in many fields. The book handles the conflict with the Catholic Church in an interesting and balanced way, including modern Papal statements in a brief "More About Galileo" section at the end of the book. An excellent way to provide students with a concise assessment of Galileo's many accomplishments, triumphs, and tragedies.

Jupiter Project
by Gregory Benford
Bantam Books, New York. 1990
Grades: 7–10

A teen-ager lives with his family as part of a large scientific laboratory that orbits Jupiter, but he is ordered to return home. He has one chance to stay; if he can make an important discovery. There is a nice mix of physics and astronomy with teen-age rebellion and growing maturity, some love interest, and an exciting plot. The descriptions in Chapters 6, 7, and 8, which are part of an account of an expedition to Ganymede, could be compared by students to the information they observe and learn about this mammoth moon.

The Jupiter Theft
by Donald Moffitt
Ballantine Books, New York. 1977
Grades: 7–Adult

Strange, advanced beings from somewhere near the constellation of Cygnus encounter a Jupiter expedition from Earth. The Cygnans want to take Jupiter away to use as a power source as they migrate through the universe. There is some graphic violence as various life forms attack and/ or ally with each other, but in general the focus is on scientific speculation. In addition to interesting descriptions of Jupiter and its moons, the book has a wealth of cogent speculation on the possibilities and varieties of life on other worlds.

> **T**o command the professors of astronomy to confute their own observations is to enjoin an impossibility, for it is to command them not to see what they do see, and not to understand what they do understand, and to find out what they do not discover.
>
> — *Galileo Galilei*

When the moon
is in the seventh house
And Jupiter
aligns with Mars,
Then peace
will guide the planets,
And love
will steer the stars.

— *Jasmes Rado
and Gerome Ragni
Hair*

Life of Galileo

by Bertolt Brecht
Grove Weidenfeld, New York. 1966
Grades: 9–12

This play is a rich literary extension to this unit. During the early scenes of the play, there are several references to, explanations of, and controversies about the telescope, Galileo's observations of the Earth's moon, and the implications of his tracking of Jupiter's moons—the exact activity that the students recreate during the first activity. Much of the rest of the play focuses on science and society, church policies of the time, and other incisive social criticism characteristic of Brecht. Dramatizing several of the early scenes would make a great extension, and Galileo's long speech near the end of the play raises relevant issues. (The play is also available in the collection *Brecht: Collected Plays*, Vintage Books, Random House, 1972.)

The Planets

edited by Byron Preiss
Bantam Books, New York. 1985
Grades: 8–Adult

This extremely rich, high-quality anthology pairs a nonfiction essay with a fictional work about the earth, moon, each of the planets, and asteroids and comets. Introductory essays are by Issac Asimov, Arthur C. Clarke and others. The material is dazzlingly illustrated with color photographs from the archives of NASA and the Jet Propulsion Laboratory, and paintings by astronomical artists such as the movie production designers of *2001* and *Star Wars*. "The Future of the Jovian System" by Gregory Benford (about colonization and development of Jupiter's moon Ganymede) is a perfect match to the final activity. The vocabulary is sophisticated, so it may be more suitable for higher-level readers.

Star Tales: North American Indian Stories

retold and illustrated by Gretchen W. Mayo
Walker & Co., New York. 1987
Grades: 5–12

The nine legends in this collection explain observations of the stars, moon, and night sky. Accompanying each tale is information about the constellation or other heavenly observation and how various peoples perceived and interpreted it. Stories like these from Native American and other world cultures can be interwoven with astronomy activities.

The Three Astronauts

by Umberto Eco; illustrated by Eugenio Carmi
Harcourt, Brace, Jovanovich, San Diego. 1989
Grades: K–5

An American, a Russian, and a Chinese astronaut take off separately in their own rockets with the goal of being first on Mars. They all land at the same time, immediately distrusting each other. When they encounter a Martian, their cultural differences disappear as they unite against him. In a surprise happy ending, they recognize the humanity of the Martian after observing his charity toward a baby bird and extend this understanding to differences between all peoples. Younger children may not get the full benefit of the sophisticated illustrations and humor. The astronauts are all male, with no female characters or references.

They Dance in the Sky: Native American Star Myths

by Jean Guard Monroe and Ray A. Williamson
Houghton Mifflin, Boston. 1987
Grades: 5–12

Stories from many Native American regions and peoples, including the Southwest and Southeast, the Plains Indians, the Pawnee, the Northwest Coast, and California Indians are particularly suitable for reading aloud. Stories about the Pleiades, the Big Dipper, and the "Star Beings" are noteworthy, but all are imaginative and intriguing. Stories like these from Native American and other world cultures can be interwoven with astronomy activities, provide a sense of careful observation over time, and highlight how the stars and planets have always inspired the human imagination.

MAGNIFIERS

More than

Grades 6–9

Using the same two lenses, students find out how lenses are used in magnifiers, simple cameras, telescopes, and slide projectors. They learn that lenses have certain measurable properties that can help determine which lenses are best for specific purposes. (Class sets of inexpensive lenses for doing this activities are available by separate purchase from the GEMS project.)

Activity 1

Students experiment with a jug of water, draw their finger as it appears under different lenses, and discover that the curvature of a lens determines its magnifying power and its field of view.

Activity 2

Students make a model camera by projecting images of a light bulb onto a piece of paper, discuss how cameras work, and measure focal length.

Activity 3

A short-focus lens and a long-focus lens are used together as students make a telescope and discuss how it works. The "Going Further" activities challenge students to make a telescope that sees things right-side-up (three lenses required) and to make telescopes like those Galileo used to see the mountains of the moon.

Four Activities

Activity 4

Using a flashlight, slide, and lens, students make table-top slide projectors and discuss how they work. Students learn the meaning of long-throw and short-throw lenses.

Skills

Observing, Comparing, Measuring, Graphing

Concepts

Lenses, Images, Focal Length, Focus, Magnifiers, Cameras, Telescopes, Projectors, Field of View

Themes

Systems & Interactions, Models & Simulations, Patterns of Change, Scale, Energy, Matter

Notes: Background section includes information on lenses, optics, the human eye, and how the instruments made in class work.

MAKING CONNECTIONS

Books connecting to *More Than Magnifers* should provide ways for your students to extend and deepen their knowledge of lenses through literature. Two books listed here are from the Einstein Anderson series. A chapter from them can be read in class, leaving the students with a challenge to solve for homework. The other book is a brief overview of the life of the inventor of the microscope.

Be on the lookout for other books that feature telescopes, microscopes, or cameras, and let us know about them. Non-fiction works and biographies on scientific breakthroughs, such as the classic Paul de Kruif book *Microbe Hunters* can be very stimulating reading for some students, as can books that reveal the microscopic world or other changes in scale. Two books about Galileo and his historic telescopic observations are listed under the GEMS guide *Moons of Jupiter*.

CROSS REFERENCES

LITERATURE CONNECTIONS

Einstein Anderson Lights Up the Sky
by Seymour Simon; illustrated by Fred Winkowski
Viking Press, New York. 1982
Grades: 4–7

In Chapter 3, Einstein's water purification experiment is interrupted by Herman's questions about eyeglass lenses.

Einstein Anderson Makes Up For Lost Time
by Seymour Simon; illustrated by Fred Winkowski
Viking Press, New York. 1981
Grades: 4–7

In "The Night Sky," Einstein Anderson and his friend Dennis observe the Milky Way and Jupiter and its moons. There is an explanation of refracting and reflecting telescopes.

The Microscope
by Maxine Kumin; illustrated by Arnold Lobel
Harper & Row, New York. 1968
Grades: 4–8

A beautifully written and illustrated book, which, while poking fun at many things, portrays Anton Leeuwenhoek grinding lenses, the appearance of many common objects under a microscope, and provides accurate historical information. Although written in rhyme and with few words, its language is fairly sophisticated, and it can be read with delight by older students and adults. This light-hearted literary connection includes good illustrations of the scales of things as viewed through a microscope.

To see a world in a grain of sand
And a heaven in a wild flower.
Hold infinity in the palm of your hand
And eternity in an hour.

— *William Blake*
Auguries of Innocence

MYSTERY FESTIVAL

Mr. Bear

Grades 2–8

Ten Sessions
(five per mystery)

*T*wo imaginative and compelling mysteries, one for younger and one for older students, are featured in this exciting GEMS festival guide. Students observe the "crime scene," conduct crime lab tests on the evidence at classroom learning stations, analyze the results, and try to solve the mystery. "Who Borrowed Mr. Bear?" is for younger students and "The Mystery of Felix" is for older ones. Many key content areas are explored, and the important distinction between evidence and inference is emphasized. Crime lab tests include thread tests, powder tests, DNA, chromatography, fingerprinting, and many more. Students use careful experimentation, logical thinking, and make real-life connections to forensic science. The guide is spiral-bound for ease in copying the many station signs, data sheets, footprints, and assorted other clues. Modifications are suggested for presenting the Mr. Bear mystery to first grade.

Session 1

A make-believe "crime scene" that will launch an entire five-session festival of exciting investigations is set up beforehand. Students enter the classroom to find a roped-off area full of curious clues. Working in teams of two, students examine the evidence and record their observations very carefully, just like real forensic scientists.

Session 2

The class uses a Clue Board to organize their observations of the evidence they gathered at the crime scene. Students discuss the case and the distinction between what they know for sure ("hard evidence") and their guesses or ideas ("inferences").

Sessions 3 and 4

The classroom is transformed into a forensic laboratory. During each of the two sessions, students circulate to five different activity stations and perform tests on the various clues found at the scene of the crime in Session 1. They do thread comparison tests, chromatography, pH tests, and powder tests, among others. Signs at each learning station explain the procedures. At the end of both forensic station sessions, the class pools results and revises their evidence charts.

Session 5

After all the observations, experiments, and discussions, the class attempts to solve the mystery.

Skills

Observing, Comparing, Relating, Sorting, Classifying, Analyzing and Evaluating Evidence, Making Inferences, Distinguishing Evidence from Inference, Problem Solving, Drawing Conclusions, Communicating, Describing, Working in Teams, Logical Thinking, Organizing Data, Role Playing, Debating, Drawing and Mapping

Concepts

Forensic Science, Evidence, Fingerprints, Footprints, Chromatography, Acids, Bases, Neutrals, pH Testing, Powder Testing, Thread Comparison Testing, Crystals, Dissolving, DNA

Science Themes

Systems & Interactions, Matter, Patterns of Change, Models & Simulations, Stability, Scale, Structure, Diversity & Unity

Mathematics Strands

Number, Pattern, Logic, Measurement, Algebra

Nature of Science and Mathematics

Scientific Community, Interdisciplinary Connections, Cooperative Efforts, Real-Life Applications, Creativity and Constraints, Theory-Based and Testable, Objectivity and Ethics

MAKING CONNECTIONS

Naturally, the books listed on the next several pages are mysteries of one kind or another. While some are simple and others more complex, they all have in common the collection of evidence or clues, and all use analysis to make inferences or conclusions. The distinction between evidence and inference is emphasized in the *Mystery Festival* unit and comes alive through its application to a diverse collection of mysterious situations. Some teachers also make use of newspaper articles to further explore the evidence/inference distinction—using articles that describe a crime (usually unsolved), the evidence, and some possible inferences.

There are of course numerous books that focus on solving mysteries. Many of them can make wonderful connections to science and math activities. The process of science and mathematics is, after all, parallel to that of detection and making inferences to solve a problem.

You and your students no doubt have your own favorite mystery stories/books and authors. There are a number of well-known series, such as those involving the Boxcar Children, Encyclopedia Brown, Einstein Anderson, Two-Minute Mysteries, and even the Nancy Drew and Hardy Boys mysteries. Teachers who tested these activities also suggested, among many nominations: Nate the Great books by Marjorie Weinman Sharmat; the Piggins series by Jane Yolen; and the Miss Mallard series by Robert Quackenbush. There are also many new books that combine environmental awareness and respect for diverse cultures with a compelling mystery plot. We would especially welcome hearing about those mysteries that include details of scientific tests and evaluation of evidence similar to those in *Mystery Festival*.

Age-appropriate nonfiction books that focus on how famous cases were solved scientifically, or tell more about the work of forensic scientists, would also make good accompaniments to this unit. Such accounts would be particularly apt if they involved one or more of the scientific tests that students conduct in *Mystery Festival*. Students could be asked to write the story of a great scientific discovery in typical mystery-story style. The classic book *Microbe Hunters* by Paul de Kruif is as exciting as any fictional mystery, as is *The Double Helix* by James D. Watson, recounting how the puzzle pieces were put together in the search for the structure of DNA.

CROSS REFERENCES

LITERATURE CONNECTIONS

Cam Jensen and the Mystery of the Gold Coins
by David A. Adler; illustrated by Susanna Natti
Viking Press, New York. 1982
Dell Publishing, New York. 1984
Grades: 3–5

Cam Jensen uses her photographic memory to solve a theft of two gold coins. Cam and her friend Eric carry around their 5th grade science projects throughout the book and the final scenes take place at the school science fair. (Other titles in the series include *Cam Jensen and the Mystery at the Monkey House* and *Cam Jensen and the Mystery of the Dinosaur Bones* in which she notices that three bones are missing from a museum's mounted dinosaur.)

Chip Rogers: Computer Whiz
by Seymour Simon; illustrated by Steve Miller
William Morrow, New York. 1984
Out of print
Grades: 4–8

Two youngsters, a boy and a girl, solve a gem theft from a science museum by using a computer to classify clues. A computer is also used to weigh variables in choosing a basketball team. Although some details about programming the computer may be a little dated, this is still a good book revolving directly around sorting out evidence, deciding whether or not a crime has been committed, solving it, and demonstrating the role computers can play in human endeavors. By the author of the Einstein Anderson series.

The Eleventh Hour
by Graeme Base
Harry N. Abrams, New York. 1989
Grades: 4–9

This uniquely illustrated picture book is about an elephant's eleventh birthday party. In addition to being an illustrative and poetic *tour de force*, this book is a compelling mystery. We learn that one of the animals at the party has gobbled up the special birthday banquet (elephants being magnificent chefs, of course). All eleven animals are suspects, and the solution is contained in clues provided throughout the book. The end of the book has sealed pages to encourage you not to open the pages and find out who did it until you think you have it solved!

It is a riddle
wrapped in a mystery
inside an enigma.
— *Winston Churchill
(about Russia)*

From the Mixed-Up Files of Mrs. Basil E. Frankweiler

by E.L. Konigsburg

Atheneum, New York. 1967

Dell Publishing, New York. 1977

Grades: 5–8

> Twelve-year-old Claudia and her younger brother run away from home to live in the Metropolitan Museum of Art and stumble upon a mystery involving a statue attributed to Michelangelo. This book is a classic, and has been recommended to GEMS by many teachers. Because of the detecting techniques, which include a mention of fingerprinting, this is a good connection to *Mystery Festival*.

The Great Adventures of Sherlock Holmes

by Arthur Conan Doyle

Viking Penguin, New York. 1990

Grades: 6–Adult

> These classic short stories are masterly examples of deduction. Many of the puzzling cases are solved by Holmes in his chemistry lab as he analyzes fingerprints, inks, tobaccos, mud, etc. to solve the crime and catch the criminal. As noted in the guide's background section, inspiration for the first real crime lab is said to have come from these very stories. Various collections of these stories are available from many different publishers and in many editions and make a great connection to *Mystery Festival*.

Let's Go Dinosaur Tracking

by Miriam Schlein; illustrated by Kate Duke
HarperCollins, New York. 1991
Grades: 2–5

The many different types of tracks dinosaurs left behind and what these giant steps reveal are explored. Was the creature running … chasing a lizard … browsing on its hind legs for leaves … traveling in pairs or in a pack … walking underwater? At the end of the book, you can measure your stride and compare the difference when walking slowly, walking fast, and running. The process involved in attempting to draw conclusions about an animal's behavior or movement patterns from its tracks is similar to the way inferences are drawn from evidence in this guide's activities.

The Missing 'Gator of Gumbo Limbo: An Ecological Mystery

by Jean C. George
HarperCollins, New York. 1992
Grades: 4–7

Sixth-grader Liza K becomes a nature detective while searching for Dajun, a giant alligator who plays a part in a waterhole's oxygen-algae cycle, and is marked for extinction by local officials. She is motivated to study the delicate ecological balance by her desire to keep her outdoor environment beautiful. This "ecological mystery" combines precise scientific information and important environmental concerns with excellent characterization, a strong female role model, and an exciting, complex plot. For older students, this book connects their detective work in *Mystery Festival* to the real world needs of environmental protection.

Elementary, and older students too, my dear GEMS

That grand investigative detective Sherlock Holmes hardly needs an introduction. The short stories by Arthur Conan Doyle about the master sleuth and his companion Dr. Watson are brilliant lessons in logic, deduction, inference, and probability. The stories are not long, and in each—which is one of the reasons they are considered classics of the mystery genre—all the clues necessary to solve the riddle are presented before the solution is revealed. You may want to stop reading prior to the final paragraphs and see if your students can solve the puzzle. Especially recommended are the stories in the early collections: *The Adventures of Sherlock Holmes, The Memoirs of Sherlock Holmes,* and *The Return of Sherlock Holmes.*

Motel of the Mysteries

by David Macaulay

Houghton Mifflin, Boston. 1979

Grades: 6-Adult

Presupposing that all knowledge of our present culture has been lost, an amateur archaeologist of the future discovers clues to the lost civilization of "Usa" from a supposed tomb, Room #26 at the Motel of the Mysteries, which is protected by a sacred seal ("Do Not Disturb" sign). This book is an elaborate and logically constructed train of inferences based on partial evidence, within a pseudo-archaeological context. Reading this book, whose conclusions they know to be askew, can encourage students to maintain a healthy and irreverent skepticism about their own and other's inferences and conclusions, while providing insight into the intricacies and pitfalls of the reasoning involved. This book can help deepen the practical experiences students have gained in distinguishing evidence from inference. It also helps demonstrate, in a humorous and effective way, the connection between detective work and the science of archaeology.

Mystery Day

by Harriet Ziefert; illustrated by Richard Brown

Little, Brown & Co., Boston. 1988

Grades: 1–4

Mystery Day is a school day full of surprises for Mr. Rose's students. They have to guess the identity of five mystery powders. The students test their guesses with simple experiments as they look at the powders, touch them, taste them, and mix them with various liquids to see what happens. Once they are correctly guessed, several powders are mixed together and the investigation process starts all over again. This book works well with the Mr. Bear activities.

The Mystery of the Stranger in the Barn

by True Kelley

Dodd, Mead, & Co., New York. 1986

Grades: K–4

A discarded hat and disappearing objects seem to prove that a mysterious stranger is hiding out in the barn, but no one ever sees anyone. A good opportunity to contrast evidence and inference.

New Guys Around the Block

by Rosa Guy

Dell Publishing, New York. 1992

Grades: 7–10

Imamu is a "street smart" genuinely caring boy who lives in a very tough area of Harlem, surrounded by inner city social ills. He wants to do good, but it's often a struggle for him in his surroundings. Imamu and his friends are suspects, and have to dodge the police, as they attempt to solve a series of phantom burglaries. The author has made a laudable attempt at preserving the authenticity of street language, while cleaning it up considerably.

> To doubt everything or to believe everything are two equally convenient solutions; both dispense with the necessity of reflection.
>
> — *Jules Henri Poincaré*

What Is A Mystery?

A mystery's a magnet
Attracting eager minds
We gather clues and evidence
And ponder what we find.

A mystery's a puzzle
Of pieces all askew
Putting it together
Is what detectives do.

A mystery's a winding maze
With choices we can make
We follow lines of evidence
To find which path to take.

A mystery's a questioning
A who, where, what, when, why
We reconstruct what happened
To distinguish truth from lie.

A mystery's the great unknown
Just waiting to be found
A speck of dust, a strand of hair,
Did someone hear a sound?

A mystery's the sudden shout
Clock ticking, creaking door,
Suspense and thrill and scariness
"Quoth the Raven—nevermore."

A mystery's exciting book
Or you can write your own
Create a plot and characters
Jessica Fletcher, Sherlock Holmes.

A mystery gets curiouser
And curiouser, they say
Clues pile up inside our heads
Until, at last, one day

We see a way to solve it
A light begins to shine
It starts to fit together
In a logical design.

Then something else might happen
Another clue pops up
It's back then to the drawing board
To figure out what's up.

Solutions are elusive
It's wise to bide your time
It isn't always obvious
Who perpetrates a crime.

Elementary my dear Watson
Aha! Eureka! too
The moment of discovery
Jumps out at me and you.

A mystery's a Festival
Of stations to explore
Whose fingerprints are left behind?
Whose tracks are on the floor?

A mystery's a Festival
It's yours to sleuth and share
In the *Mystery Festival* Hall of Fame
We're sure to find you there!

— *Lincoln Bergman*

The One Hundredth Thing About Caroline

by Lois Lowry
Houghton Mifflin, Boston. 1983
Dell Publishing, New York. 1991
Grades: 5–9

Fast-moving and often humorous book about 11-year-old Caroline, an aspiring paleontologist, and her friend Stacy's attempts to conduct investigations. Caroline becomes convinced that a neighbor has ominous plans to "eliminate" the children and Stacy speculates about the private life of a famous neighbor. Due to hasty misinterpretations of real evidence, both prove to be wildly wrong in their inferences. Gathering evidence, weighing it, and deciding what makes sense are good accompanying themes. A somewhat inaccurate portrayal of "color blindness" is a minor flaw.

The Real Thief

by William Steig
Farrar, Straus & Giroux, New York. 1973
Grades: 4–8

King Basil and Gawain, devoted Chief Guard, are the only two in the kingdom who have keys to the Royal Treasury. When rubies, gold ducats, and finally the world-famous Kalikak diamond disappear, Gawain is brought to trial for the thefts. But is he the real thief? As the mystery unfolds, it becomes clear that it is important to investigate fully before making judgments or drawing conclusions.

Susannah and the Blue House Mystery

by Patricia Elmore
E.P. Dutton, New York. 1980
Scholastic, New York. 1990
Grades: 5–7

Susannah (an amateur herpetologist) and Lucy have formed a detective agency. They check into the death of a kindly old antique dealer who lived in the mysterious "Blue House." They attempt to piece together clues hoping to find the treasure they think he has left to one of them. The detectives evaluate evidence, work together to solve problems, and prevent a camouflaged theft from taking place.

Susannah and the Poison Green Halloween

by Patricia Elmore
E.P. Dutton, New York. 1982
Scholastic, New York. 1990
Grades: 5–7

Susannah and her friends try to figure out who put the poison in their Halloween candy when they trick-or-treated at the Eucalyptus Arms apartments. Tricky clues, changing main suspects, and some medical chemistry make this an excellent choice, with lots of inference and mystery. (There is also mention of a field trip to the Lawrence Hall of Science!)

> A scientist doesn't make up his mind until he's examined all the evidence.
>
> — *E.L. Konigsburg*
> *From the Mixed-up Files of*
> *Mrs. Basil E. Frankweiler*

The Tattooed Potato and Other Clues
by Ellen Raskin
E.P. Dutton, New York. 1975
Penguin Books, New York. 1989
Grades: 6–9

Answering an advertisement for a portrait painter's assistant in New York City involves a 17-year-old in several mysteries and their ultimate solution, such as the "Case of the Face on the Five Dollar Bill" where the smudged thumbprint of the counterfeiter is a clue.

The Trouble With Lemons
by Daniel Hayes
David R. Godine, Publisher, Boston. 1991
Grades: 6–9

In this suspenseful page turner with great character development, two 8th graders who think of themselves as misfits ("lemons") discover a body and attempt to solve what they think is a murder. Woven throughout the adventures are explorations of self-acceptance and understanding others, as well as accurate portrayals of the agonies of adolescence. In a great twist at the end, they find out they have jumped to conclusions about the death. Nice connection to the Felix mystery in the sense that this GEMS guide leaves open the question of how—and even whether—Felix dies. The lesson about not jumping to conclusions is one of the basic messages in *Mystery Festival*.

The Westing Game
by Ellen Raskin
E.P. Dutton, New York. 1978
Avon, New York. 1984
Grades: 6–10

The mysterious death of an eccentric millionaire brings together an unlikely assortment of 16 beneficiaries. According to instructions contained in his will, they are divided into eight pairs and given a set of clues to solve his murder and thus claim the inheritance. Newbery award winner.

Who Really Killed Cock Robin?
by Jean Craighead George
HarperCollins, New York. 1991
Grades: 3–7

This compelling ecological mystery examines the importance of keeping nature in balance, and provides an inspiring account of a young environmental hero who becomes a scientific detective.

Whose Footprints?
by Masayuki Yabuuchi
Philomel Books, New York. 1983
Grades: K–4

A good guessing game for younger students that depicts the footprints of a duck, cat, bear, horse, hippopotamus, and goat.

YOU are the JUDGE
YOU are the JURY

If your students became intensely involved in evaluating evidence and determining possible guilt as you worked your way through *Mystery Festival*, we recommend the *Be the Judge • Be the Jury* ™ book series by award-winning children's author Doreen Rappaport. They are an excellent extension to these GEMS activities. These books describe famous court cases, simplified for young people.

In the beginning of each book the different roles and procedures of a courtroom trial are clearly and simply defined. The reader becomes the jury. First the prosecution, then the defense, present their evidence and witnesses. After closing statements, the instructions of the judge to the jury are included. As the reader, you are expected to come up with your own verdict, so you can't help but be drawn in and become very actively involved in the trial. At the end of each book, the actual verdict of the trial is revealed, with historical perspective.

The following books are available: *The Lizzie Borden Trial, The Sacco-Vanzetti Trial, The Alger Hiss Trial,* and *Tinker vs. Des Moines: Student Rights on Trial.*

These books provide a double whammy—they strongly involve young readers in the important process of evaluating evidence and drawing inferences, and give them a historical and social perspective on the law. The author said she wrote the series "because I was on a jury once and found out how hard and scary and important it was deciding the fate of another human being." The series is published by HarperCollins Publishers, 10 East 53rd Street, New York, N.Y. 10022.

Of Cabbages and Chemistry

Grades 4–8

Four Sessions

This series of activities offers students a chance to explore acids and bases using the special indicator properties of red cabbages. The color-change game "Presto Change-O" helps students discover the acid-neutral-base continuum. Students discover that chemicals can be grouped by behaviors, and relate acids and bases to their own daily experience. The guide includes an "Acid and Aliens from Outer Space" extension activity. The activities in this unit are an excellent lead-in to *Acid Rain.*

Session 1

Students mix red cabbage juice with a variety of household liquids and record the color of the resulting mixture. They classify the results by the color produced and place the liquids into groups they designate.

Session 2

Students hold a "scientific convention" to analyze the results of the first session. Disagreements about a color result provide opportunity for discussion. The terms acid, base, and neutral are introduced. Students play the color-change game "Presto Change-O" and make and test predictions.

Session 3

Students experiment to determine the effect that different concentrations of acids or bases have on the color of cabbage juice. Teacher introduces the concept of a continuum. Students experiment to figure out the number of drops of various acids needed to neutralize a standard amount of base, then try to determine relative concentration of several acids.

Session 4

Students bring in household chemicals from home (beverages, cleaning products, bathroom products) to determine if they are acids, bases, or neutrals.

Skills

Observing, Recording Results, Comparing, Classifying, Experimenting, Titrating, Drawing Conclusions

Concepts

Chemistry, Acid, Base, Neutral, Indicators, Pigments, Safety, Neutralize, Concentration, Titration, Continuum

Themes

Systems & Interactions, Stability, Patterns of Change, Scale, Matter

Notes: Guide includes poem on cabbages and chemistry; print of cut cabbage; using flower petals as indicators; extensions include using more vivid chemical indicators; exploring the topic of acid rain; and doing the GEMS unit *Vitamin C Testing*; background information on acids, bases, and pH.

MAKING CONNECTIONS

This guide, though suggestive of many potential connections to literature, resulted in few good finds. A book about the nineteenth century army surgeon who studied a patient with a healed bullet hole in his stomach reveals much concrete information about **stomach acid** and other items of interest. After participating in this unit, your students will be primed to the challenge of **"neutralizing"** a solution and will enjoy a picture book about a woman who tries to make her salty coffee drinkable. Finally, a collection of humorous poems about **food and eating** is certainly relevant to this unit—all it takes is smelling a classroom where cabbage juice was made to begin the poetic flow in all of us.

Keep on the lookout for literature connections to acids and bases, indicators, neutralization, chemical reactions, cabbages and other plants that are used in science. Classification is an important part of this unit, and age-appropriate books on sorting and classification would make good connections.

CROSS REFERENCES

LITERATURE CONNECTIONS

Doctor Beaumont and the Man with the Hole in His Stomach

by Beryl and Samuel Epstein
Coward, McCann & Geoghegan, New York. 1978
Grades: 4–6

Interesting experiments about digestion are described in this army surgeon's biography. In the 1820s he had a patient with a bullet hole in his stomach. By inserting a tube, the doctor was able to directly observe and monitor the circulation of gastric juices and bile fluids. The surgeon published a book on these experiments, including findings on acidity in the stomach.

June 29, 1999

by David Wiesner
Clarion Books, Houghton Mifflin, New York. 1992
Grades: 3–6

The science project of Holly Evans takes an extraordinary turn—or does it? This highly imaginative and beautifully illustrated book not only has a central experimental component, but ranges outward into the world of extraterrestrials. Holly's careful preparations, analysis of results, and dawning awareness that the giant vegetables that are landing on Earth are not the results of her experiment have a lot to say about scientific pursuits, interspersed with great humor and lots of giant vegetables, including cabbages!

"The time has come," the Walrus said,
"To talk of many things:
Of shoes—and ships—and sealing wax—
Of cabbages—and kings—
And why the sea is boiling hot—
And whether pigs have wings."

— *Lewis Carroll*
Through the Looking Glass

The Lady Who Put Salt in Her Coffee
by Lucretia Hale
Harcourt, Brace, Jovanovich, San Diego. 1989
Grades: K–6

When Mrs. Peterkin accidentally puts salt in her coffee, the entire family embarks on an elaborate quest to find someone to make it drinkable again. A visit to a chemist, an herbalist, and a wise woman result in a solution, but not without having tried some wild experiments first.

Top Secret
by John R. Gardiner; illustrated by Marc Simont
Little, Brown & Co., Boston. 1984
Grades: 4–7

Humorous saga of a fourth grader's search for the secret of "human photosynthesis." Although his science teacher instructs him not to pursue his efforts and orders a science project on lipstick instead, he succeeds in transforming himself and then his teacher into human organisms with plant-like characteristics. The intervention of undercover government agents adds suspense. The plot is an imaginative use of material about plants, human and plant nutrition, chemistry, chemical reactions, and the process of scientific investigation.

The Secret of Top Secret

In *Top Secret* (see annotation above), the grandfather of the main character compares the process of scientific discovery to

"an avocado that has been cut into many different pieces, and then the pieces hidden in different places. Some pieces are very hard to find. Others are right in front of your nose, so close, in fact, that you cannot see them. It is your job, Allen, the job of any scientist, any thinker, to find the different pieces, one by one, and then put them in the proper order until, at last you can see..." "The avocado," I said..."Right and wrong." Grandpop held up the avocado in his hand and then turned it around to show that he had cut away the back half. I thought for a moment before I figured out what Grandpop was trying to tell me. "You don't have to find *all* the pieces to see the *whole* picture."

(Another lesson here, also very useful in science, literature, and life might be—things are not always as they appear.)

OOBLECK:

WHAT DO SCIENTISTS DO?

Grades 4–8

Four Sessions

Students investigate and analyze the properties of a strange green substance, Oobleck, said to come from another planet. The class holds a scientific convention to critically discuss experimental findings. Students design a spacecraft to land on an ocean of "Oobleck."

Session 1

Students form small lab teams to investigate the substance Oobleck that the teacher describes as coming from another planet in another solar system. They learn how scientists describe the "properties" of a substance and discuss when Oobleck acts as a solid or as a liquid.

Session 2

Students hold a "scientific convention" to discuss and analyze their findings. They challenge each other to define is properties more accurately, generate some "laws of Oobleck," and do further experiments as prompted by discussion.

Session 3

Students become engineers, using the knowledge they've gained to design (by drawing) a spacecraft that would be able to land on and take off from an ocean of Oobleck on a distant planet.

Session 4

The methods the students used to analyze Oobleck are compared to those of professional scientists, such as those on the Mars Viking Project. A large poster illustrating the Mars project is included with the guide.

Skills

Experimenting, Recording Data, Engineering, Critical Discussion, Communicating, Group Brainstorming, Decision-Making

Concepts

Scientific Methods, Solids and Liquids, Properties, Space Probes, Designing Models

Themes

Systems & Interactions, Models & Simulation, Stability, Patterns of Change, Structure, Matter

Notes: Going Further asks students to make up stories about the "Oobleckians," creatures who live on a planet with Oobleck oceans.

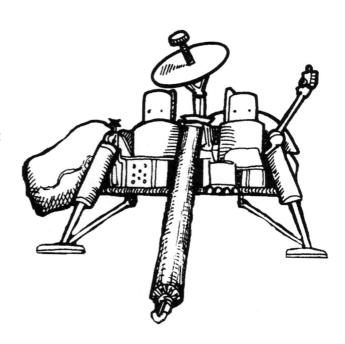

MAKING CONNECTIONS

The name of the strange substance in this unit came from none other than Dr. Seuss (with permission, of course). Even though the Dr. Seuss book, *Bartholomew and the Oobleck*, is generally geared for younger students than the fourth–eighth graders for whom these science activities are designed, the Dr. Seuss book, properly introduced, can be used to good advantage by older students as well. It's also true that the Oobleck science activities are often adapted for younger students, and in that case the literary connection is age-appropriate and highly recommended!

Be on the lookout for stories about exploration and open-ended investigation such as *The Wise Woman and Her Secret*. Such stories fit well with Oobleck activities (or any discovery-based science lesson) in which students model scientific behavior.

Many of the stories are about "goopy" substances, from slime to quicksand to ketchup, and relate to Session 1.

Stories about defining words, as in *The Search for Delicious*, could go along with Session 2. Stories involving technology, as in *Chitty Chitty Bang Bang*, can relate to Session 3. Stories you may find about exploring other worlds, as in *The Three Astronauts*, fit well with Session 4. Science fiction and nonfiction about lunar and space exploration would also connect well with this session.

CROSS REFERENCES

Dear GEMS

For younger students, the Dr. Seuss book is an excellent accompaniment to this unit. Some teachers interweave it with hands-on experience and discussion. The book is read out loud up to a point, then students are invited to do experiments to see if the substance they are investigating acts the same as the Oobleck described in the book. Then more of the book is read, students compare properties, and so on. Some of the properties are similar, but some are quite different, further enriching the activity.

Do not assume that older students might not enjoy such an activity, or in some way deepen their perspective of Oobleck's properties by using the book or excerpts from it. Barbara R. Pietrucha, an eighth grade Earth/Environmental Science teacher, wrote to the GEMS project to describe how she introduced this unit:

"I read out loud *Bartholomew and the Oobleck*. At first my 'cool' eighth graders wanted no part of it. They were honors students; they were too old for Dr. Seuss; they were too mature! So many excuses! Yet I persisted and read to them. (They finally gave in if I would close the door so they wouldn't be embarrassed.) They loved it! And the Oobleck meant so much more to them! I can't wait to see other reading ideas!"

Barbara R. Pietrucha

LITERATURE CONNECTIONS

Bartholomew and the Oobleck
by Dr. Seuss
Random House, New York. 1949
Grades: K–9

> A king orders his royal magicians to cause something new to rain down from the sky. A green rain called "Oobleck" falls onto the kingdom, in too much abundance, and its strange properties cause quite a mess until the ruler learns some humility. See the previous page for information about how this book can be used with younger and older students.

Chitty Chitty Bang Bang: The Magical Car
by Ian Fleming; illustrated by John Burningham
Alfred A. Knopf, New York. 1964
Grades: 6–Adult

> Wonderful series of whimsical adventures featuring a magical transforming car, an eccentric explorer and inventor, and his 8-year-old twins. It's a nice combination of technical and scientific information, much of it accurate, with a magical sense of how some machines seem to have a mind of their own. The car has humorous and fantastic technological adaptations: it flies when it encounters traffic jams, becomes a boat when the tide comes in, senses a trap, and helps catch some gangsters. The inventor and invention flavor is a fun extension to Session 3.

Einstein Anderson Tells a Comet's Tale
by Seymour Simon; illustrated by Fred Winkowski
Viking Press, New York. 1981
Grades: 4–7

> Chapter 5 presents some very interesting information about ketchup, a substance that sometimes acts as a solid and sometimes as a liquid.

Horrible Harry and the Green Slime
by Suzy Kline; illustrated by Frank Remkiewicz
Viking Penguin, New York. 1989
Grades: 2–4

> Four stories about Miss Mackle's second grade class. In "Demonstrations," Horrible Harry and his assistant Song Lee show how to make green slime from cornstarch, water, and food coloring. It's a big success, ending with the librarian taking it home to her husband who is interested in science. In another story, they celebrate reading *Charlotte's Web* by making cobwebs and hanging them all over the school.

The Quicksand Book

by Tomie dePaola
Holiday House, New York. 1977
Grades: 2–5

A jungle girl learns about the composition of quicksand, how different animals escape it, and how humans can use precautions to avoid getting stuck. Her "teacher," an overly confident jungle boy, turns out not to be so superior. A variety of graphics and a helpful monkey give visual interest. A recipe for making your own quicksand is included.

The Search For Delicious

by Natalie Babbitt
Farrar, Straus & Giroux, New York. 1969
Grades: 5–8

After an argument between the king and queen over the meaning of the world "delicious," the quest for its meaning begins. Everyone has a different personal definition of the word and war looms. In Session 2 of the GEMS activities, students gathered in a "scientific convention" often find the need to define a word, and refine their descriptive language, just as scientists do.

The Slimy Book

by Babette Cole
Jonathan Cape, London. 1985
Random House, New York. 1986
Out of print
Grades: Preschool–4

Here is a lighthearted look at slime of the "sticky, sludgy, slippy, sloppy, ploppy, creepy kind" and where it may be found—around the house, in invertebrate creatures, in foods, and maybe even in outer space. Excellent and fun descriptive language of the properties of an intriguing form of matter.

The Origin of "Oobleck"

As this GEMS handbook goes to press
We wanted to at least express
Brief thankfulness to Dr. Seuss
Modern-day rhymester, Father Goose,
For his stories of wild imagination
Gentle but righteous indignation
As if these accomplishments weren't enough
He gave us a word for green, gooey stuff.
We borrowed it, with all due respect,
Thanks, Dr. Seuss, for naming "Oobleck!"

— *Lincoln Bergman*

The Three Astronauts

by Umberto Eco; illustrated by Eugenio Carmi
Harcourt, Brace, Jovanovich, San Diego. 1989
Grades: K–5

An American, a Russian, and a Chinese astronaut take off separately in their own rockets with the goal of being first on Mars. They all land at the same time, immediately distrusting each other. When they encounter a Martian their cultural differences disappear as they unite against him. In a surprise happy ending, they recognize the Martian's kindness toward a baby bird and extend this understanding to differences between all peoples. Younger children may not get the full benefit of the sophisticated illustrations and humor. The astronauts are all male, with no female characters or references.

The Toothpaste Millionaire

by Jean Merrill; illustrated by Jan Palmer
Houghton Mifflin, Boston. 1972
Grades: 5–8

Incensed by the price of a tube of toothpaste, twelve-year-old Rufus tries making his own from bicarbonate of soda with peppermint or vanilla flavoring. Assisted by his friend Kate and his math class (they become known as Toothpaste 1), his company grows from a laundry room operation to a corporation with stock and bank loans. Beginning on page 47, Rufus designs a machine for filling toothpaste tubes, which is a nice tie-in to the spacecraft designing activities in Session 3 of this GEMS guide.

The Wise Woman and Her Secret

by Eve Merriam; illustrated by Linda Graves
Simon & Schuster, New York. 1991
Grades: K–3

A wise woman is sought out by many people for her wisdom. They look for the secret of her wisdom in the barn and in her house, but only little Jenny who lags and lingers and loiters and wanders finds it. The wise woman tells her, "The secret of wisdom is to be curious—to take the time to look closely, to use all your senses to see and touch and taste and smell and hear. To keep on wandering and wondering." This book captures the essence of what is meant by discovery, student-centered, use-your-senses learning, and, as such, serves as a fine accompaniment to science activities.

PAPER TOWEL TESTING

Grades 5–9   **Four Sessions**

*I*n a series of experiments, students rank the wet strength and absorbency of four brands of paper towels. Based on their findings and the cost of each brand, they determine which brand is the best buy. These activities provide a stimulating introduction to both consumer science and the concept of controlled experimentation.

Session 1
A fair test or controlled experiment is devised, using four brands labeled A, B, C, D. Each is tested for absorbency and wet strength.

Session 2
Results of Session 1 experiments are posted and discussed, which variables could be better controlled is discussed, and follow-up experiments are planned.

Session 3
Students conduct follow-up experiments, discuss results, and calculate the average class ranking for each brand of paper towel.

Session 4
Based on their results so far, and with information provided by teacher on the cost of the rolls and the number of sheets per roll, students calculate cost per sheet of paper towel. Students decide which brand is the best buy.

Skills
Designing Controlled Experiments, Measuring, Recording, Calculating, Interpreting Data

Concepts
Consumer Science, Absorbency, Wet Strength, Unit Pricing, Cost-Benefit Analysis, Decision-Making

Themes
Systems & Interactions, Models & Simulations, Stability, Patterns of Change, Structure, Matter

Notes: Going Further activities include: Students write and deliver commercial/sales pitch for the brand they consider best based on their experiments, and discuss paper towels vs. sponges.

MAKING CONNECTIONS
The books listed relate to the ideas of **controlling variables**, **definition of terms**, **advertising**, and **issues of the "best buy"** all in the context of fun and imaginative stories. We found more books than we initially expected, which is probably an indication that there are many more strong connections out there. Let us hear from you!

CROSS REFERENCE

LITERATURE CONNECTIONS

Better Mousetraps:
Product Improvements That Led to Success
by Nathan Aaseng
Lerner Publications, Minneapolis. 1990
Grades: 5–10

The book's focus is on "improvers, refiners, and polishers" and not on pioneers or trailblazers. To dramatize the results of safety testing, Elisha Otis set up an elevator at a big exposition in New York and had an assistant intentionally cut the cable with Otis aboard! The safety device brought the elevator to a halt in midfall. Getting heavy machinery to travel over muddy ground was the challenge faced by Caterpillar Tractor Company—what was learned in product development was applied to tank technology in World War I. The chapter on Eastman Kodak introduces the concept of a brand name, showing how Eastman promoted the names "Kodak" and "Brownie."

Charlie and the Chocolate Factory
by Roald Dahl; illustrated by Joseph Schindelman
Alfred A. Knopf, New York. 1964
Penguin Books, New York. 1988
Grades: 4–8

Five lucky children find golden tickets wrapped in their candy bars that allows them into the chocolate factory where the candy made is the "best in the world." Television commercials could be discussed as part of the hypnotizing potential of TV described in the song of the Oompa-Loompas. Advertising, consumer appeal, and technology are merged in wonky ideas such as lickable wallpaper for nurseries, cows that give chocolate milk, and stickjaw candy for talkative parents. Mention is also made of Vitamin C (Supervitamin Candy) and color (rainbow drop candies that allow you to spit in six different colors).

Einstein Anderson Sees Through the Invisible Man
by Seymour Simon; illustrated by Fred Winkowski
Viking Press, New York. 1983
Grades: 4–7

In "The Huck Finn Raft Race" a class competition to build a raft involves taking into account controlling variables such as the weight of the passengers. This book also contains strong connections to other subjects, such as "The Allergic Monster" (GEMS guide: *Convection: A Current Event*), "Thinking Power" (Science Themes: Matter and Energy), and "A Cold Night" about fireflies and luminescence (Science Theme: Energy).

> Knowledge must come through action; you can have no test which is not fanciful, save by trial.
>
> — *Sophocles*
> *Trachiniae*

> Science is
> organized knowledge.
>
> — *Herbert Spencer*
> *Education*

Einstein Anderson Tells a Comet's Tale

by Seymour Simon; illustrated by Fred Winkowski

Viking Press, New York. 1981

Grades: 4–7

Chapter 10 describes a soapbox derby race in which both teams have to build soapbox racing cars that weigh the same amount and are started in the same way. Our hero identifies the one test variable that allows his team to win the race. Just as students create a "fair test" of the paper towel brands, Einstein Anderson explains how the rules of the soapbox derby create a fair test of the student racers. They must be the same weight and have the same push. You could discuss with your students questions like: "How could you change the rules to be an even fairer test of the wheel size—of the streamlined shape—of the kids' driving abilities?"

June 29, 1999

by David Wiesner

Clarion Books, Houghton Mifflin, New York. 1992

Grades: 3–6

The science project of Holly Evans takes an extraordinary turn—or does it? This highly imaginative and humorous book has a central experimental component, and conveys the sense of unexpected results.

The Search for Delicious

by Natalie Babbitt

Farrar, Straus & Giroux, New York. 1969

Grades: 3–7

After an argument between the king and queen over the meaning of the word "delicious," the quest for its meaning begins. Everyone has a different personal definition of delicious and war looms. In the paper towel tests, students grapple with the meaning of such words as "absorbency" and "wet strength."

The Toothpaste Millionaire

by Jean Merrill; illustrated by Jan Palmer

Houghton Mifflin, Boston. 1972

Grades: 5–8

Incensed by the price of a tube of toothpaste, twelve-year-old Rufus tries making his own from bicarbonate of soda with peppermint or vanilla flavoring. Assisted by his friend Kate and his math class (which becomes known as Toothpaste 1), his company grows from a laundry room operation to a corporation with stocks and bank loans. Many opportunities for estimations and calculations are presented including cubic inches, a gross of toothpaste tubes bought at auction, manufacturing expenses, and profits. Your students may have trouble believing that 79 cents could be considered an outrageous price for toothpaste, and maybe that could lead to a discussion of inflation. A price war erupts between "Sparkle," "Dazzle," "Brite," and Rufus's noncommercial "Toothpaste." This ties in nicely with Session 4 of the GEMS guide in which students calculate which brand of paper towel is the best buy.

Penguins
And Their Young

Grades: Preschool–1   **Four Activities**

*P*enguins and Their Young features the emperor penguin, the tallest of the penguins. Youngsters learn about its body structure, its cold home of ice and water, what it eats, and how emperor penguin parents take care of their young.

Life science, math, and physical science are integrated with language activities. Children have fun learning through drama, role playing, and creative play as well as by observing and comparing icy shapes in water.

Activity 1

The children explore a penguin's home of ice and water by creating a play area of ice blocks, water, cork penguins, and toy fish. Children use a life-size drawing of an emperor penguin to compare their own height and body structures to those of this interesting four-foot-tall bird.

Activity 2

The youngsters learn more about penguins when they watch a drama about parent penguins caring for their eggs and young. Later, they make paper-bag penguins and create dramas on a paper ocean scene with floating icebergs. Role playing of penguin parenting continues when baby penguins hatch from plastic eggs.

Activity 3

Important math concepts and skills are introduced when children pretend to be hungry penguins and "catch" fish crackers to eat in a fun series of multi-sensory math games.

Activity 4

The children learn how penguins stay warm and how ice can have different shapes, colors, and flavors in this group of ice activities. They conduct simple ice investigations, explore ice shapes, and taste frosty ice treats.

Skills

Observing, Comparing, Communicating, Creative and Logical Thinking, Role Playing

Concepts

Penguin Habitat, Body Structure, Parenting, Feeding Strategies, Heat and Warmth, Melting, Freezing, Ice, Floating, Size, Shape

Themes

Patterns of Change, Scale, Structure, Energy, Systems & Interactions, Diversity & Unity, Models & Simulations

Math Strands

Measurement, Number, Pattern, Logic

MAKING CONNECTIONS

At the heart of this GEMS guide are inquiries into the shape, behavior, and life style of the emperor penguin. Measurement and number activities also figure prominently in the sessions. Any books that deal with the life styles of animals, how and what they eat, and how they make a home for themselves would make excellent connections. There does not seem to be much fiction about penguins available for younger students. In addition, some of the books selected have penguins doing fantastic things that no real penguin would do. For instance, in *Penguin Small* the penguin hero flies and goes all over the world. But children love to hear about these amazing adventures and invariably they comment that no real penguin could do those things. That's a great response!

CROSS REFERENCES

Penguin deluxe,
Bird in a tux,
Elegant, debonair.
Birds of a feather
Who do flock together
But never take to the air!

— Lincoln Bergman

LITERATURE CONNECTIONS

Antarctica
by Helen Cowcher
Scholastic, New York. 1990
Grades: Preschool–1

Although this is a nonfiction book, it is an exciting and dramatic tale about emperor penguins and contemporary life in Antarctica. The story neatly dovetails with all the ideas in the drama in Activity 2 and extends them by introducing other animals that live on and around the icy continent. The impact people have on the animals and the habitat of Antarctica is also presented. The text and large, colorful illustrations provide a resource for first graders who compare the Arctic and Antarctic regions as well as the nesting behaviors of the Adélie and emperor penguins.

Little Penguin
by Patrick Benson
Philomel Books, New York. 1990
Grades: Preschool–1

Little Penguin is the story of Pip, an Adélie penguin, who wants to be as big as an emperor penguin. As Pip playfully explores the snow, ice, and water of her Antarctic home, she wonders why some animals are big and some small. This story ties in beautifully with the activities on relative sizes in *Penguins And Their Young*. The children can use the illustrations to compare the appearance of Adélie penguins with emperor penguins.

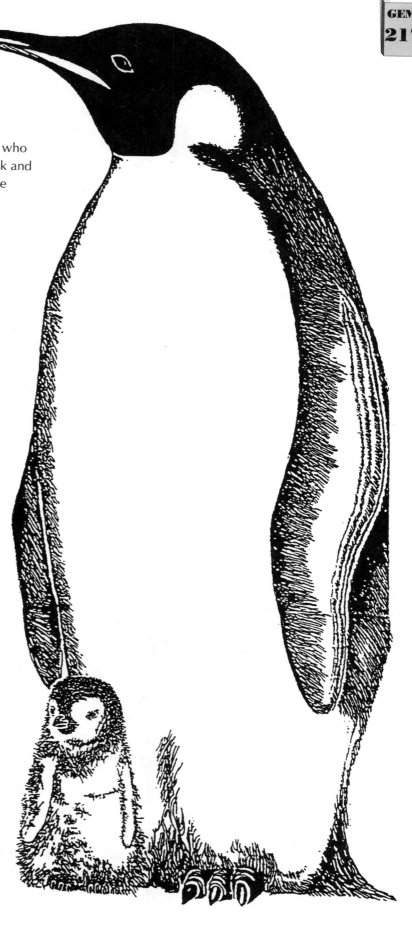

Mr. Popper's Penguins

by Richard and Florence Atwater
Little, Brown & Co., Boston. 1938
Grades: K–6

This well-loved classic about Mr. Popper's experiences with a house full of curious Adélie penguins delights children who are old enough for a book with small black and white drawings. Relevant chapters include Chapter IV: Captain Cook, in which the penguin Captain Cook eats Mr. Popper's goldfish, and Chapter VII: Captain Cook Builds A Nest, in which he collects marbles, a half-eaten lollipop, and other miscellaneous objects resembling stones to build a nest in the refrigerator. At one point Mr. Popper dresses up like a penguin and struts around the house. The nesting behavior of these Adélie penguins provides a wonderful contrast to the emperor penguin, which does not build a nest.

Out on the Ice
in the Middle of the Bay

by Peter Cumming
Annick Press, Toronto. 1993
Grades: Preschool–2

The little girl Leah wanders away from home and encounters a baby polar bear whose mother is close by. Leah's father plans to shoot the mother bear to save his child and the mother bear prepares to destroy the man to defend her cub. This suspenseful story gives children a view of Eskimo life in a contemporary Arctic setting, and the beautiful ending provides an alternative to violence in people's encroachment on nature. Although this book does not deal with penguins, first graders who compare the Arctic and Antarctic regions in "Extensions for First Graders" under Activity 2 should find this story especially relevant.

Penguin Small

by Mick Inkpen

Harcourt, Brace, Jovanovich, San Diego. 1992

Grades: Preschool–1

This beautifully illustrated story about a penguin's journey from the North Pole to the South Pole provides many opportunities for young children to contrast some of the things the fictional Penguin Small does to what real penguins do. Read the story to the children after they learn about real penguin behaviors and habitats in Activities 1 and 2. Since Penguin Small's behavior and encounters are so fantastic, children enjoy the comparison.

Pinkie Leaves Home

by Peter O'Donnell

Scholastic, New York. 1991

Grades: Preschool–1

Pinkie the Penguin, who lost all his feathers in an oil slick, does not like cold weather. One day he sets off to find a warm place by the sea where there is no ice and snow. This adventure story reinforces the concepts that feathers keep penguins warm, penguins eat fish, and penguins live by the sea. The story's surprise ending supports the idea that wearing a hat, scarf, and coat is another way of staying warm. This book provides an amusing extension to Activity 2, where the children learn about penguin feathers.

The ice was here, the ice was there,
The ice was all around:
It cracked and growled, and roared and howled,
Like noises in a swound!

— *Samuel Taylor Coleridge*
The Rime of the Ancient Mariner

QUADICE

Grades 4–8

Five Sessions

This challenging and fun mathematics game encourages students to perform mental calculations, handle fractions with greater confidence, and explore probability. Teams of three students, using a special set of four dice, play 12 rounds that involve the **arithmetic skills** of addition, subtraction, division, and multiplication. A cooperative version of the game helps students **work together** to solve problems. The guide also contains **mystery puzzles** to solve and encourages players to create mystery puzzles of their own!

Session 1

The teacher explains the rules and demonstrates how to play the game. Copies of rules are given to each student. Students play a practice round and learn how to score.

Session 2

After discussion of the first session, and review of the game, students play a full game of QUADICE and score it. Teacher evaluates score sheets to make sure students understand the game.

Session 3

The students learn and play a cooperative version of the game, working together in an effort to obtain the highest group score. The session begins with a discussion on what students have learned so far about successful game strategies.

Session 4

Students use their experience with the game to analyze the probability that particular combinations will occur. In "How Many Ways?" they analyze the probability of specific sums, differences, and quotients and discuss their results and the consequences for successful strategies.

Session 5

Students review the probability discussion, then discuss and attempt to solve two Mystery QUADICE puzzles. Later, they can make up their own puzzles.

Skills

Mental Arithmetic, Basic Operations (Addition, Subtraction, Division), Fractions

Concepts

Mathematics, Probability, Strategy Development, Problem Solving

Themes

Systems & Interactions, Models & Simulations

Notes: The guide includes some information on the benefits of cooperative learning.

MAKING CONNECTIONS

Books that involve computation and several storybooks in which numbers play an important role are cited. In addition, there is a book about a game in which the roll of two dice determines the outcome. Because dice are used to play QUADICE and the game involves probability, be sure to check the books listed in the Probability and Statistics math strand and under the GEMS guide *In All Probability*.

CROSS REFERENCES

LITERATURE CONNECTIONS

Anno's Math Games
by Mitsumasa Anno
Philomel Books/Putnam & Grosset, New York. 1982
Grades: 2–5

Picture puzzles, games, and simple activities introduce the mathematical concepts of multiplication, sequence, ordinal numbering, measurement, and direction.

Anno's Math Games II
by Mitsumasa Anno
Philomel/Putnam & Grosset, New York. 1987
Grades: 2–4

Here are more picture puzzles, games, and simple activities that introduce the mathematical concepts of multiplication, sequence, ordinal numbering, measurement, and direction.

Anno's Mysterious Multiplying Jar
by Masaichiro and Mitsumasa Anno
Philomel/Putnam & Grosset, New York. 1983
Grades: 3–8

The simple text and illustrations introduce the mathematical concept of factorials. Through an understanding of multiplication, you can learn about factorials and the way that numbers can expand. On a second reading of the book, students can follow along using calculators to verify the large number of jars at the end of the story.

Erin McEwan, Your Days Are Numbered
by Alan Ritchie
Alfred A. Knopf, New York. 1990
Grades: 4–8

Erin, a sixth grader with an intense fear of numbers, takes a job at the delicatessen and needs to learn bookkeeping to stay employed. With the encouragement of the owner, Erin surprises herself by not only improving her math skills, but she catches a bookkeeping error that saves thousands of dollars! Though stereotypical in its portrayal of women as being unskilled in math, the book is an opportunity to discuss and dispel that belief.

Jumanji

by Chris Van Allsburg
Houghton Mifflin, Boston. 1981
Grades: K–5

A bored brother and sister left on their own find a discarded board game (called Jumanji) which turns their home into an exotic jungle. A final roll of the dice for two sixes helps them escape from an erupting volcano. After reading the book, and based on their experiences with the GEMS QUADICE activities, students could investigate dice and discover more about the probability of rolling two sixes in a row.

The King's Chessboard

by David Birch; illustrated by Devis Grebu
Dial Books, New York. 1988
Grades: K–6

A proud king, too vain to admit what he does not know, learns a valuable lesson when he readily grants his wise man a special request. One grain of rice on the first square of a chessboard on the first day, two grains on the second square on the second day, four grains on the third square on the third day and so on. After several days the counting of rice grains gives way to weighing, then the weighing gives way to counting sackfuls, then to wagonfuls. The king soon realizes that there is not enough rice in the entire world to fulfill the wise man's request.

The Phantom Tollbooth

by Norton Juster; illustrated by Jules Feiffer
Random House, New York. 1989
Grades: 2–8

Milo has mysterious and magical adventures when he drives his car past the Phantom Tollbooth and discovers The Lands Beyond. On his journey, Milo encounters numbers, geometry, measurement and problem solving in amusing situations. The play on words in the text is delightful.

The Toothpaste Millionaire

by Jean Merrill; illustrated by Jan Palmer
Houghton Mifflin, Boston. 1972
Grades: 5–8

Twelve-year-old Rufus Mayflower doesn't start out to become a millionaire—just to make toothpaste. Assisted by his friend Kate and his math class (which becomes known as Toothpaste 1), his company grows from a laundry room operation to a corporation with stocks and bank loans. Many opportunities for estimations and calculations are presented including cubic inches, a gross of toothpaste tubes bought at auction, manufacturing expenses, and profits. An ideal book to illustrate the need for, and use of, mathematics in real-world problem solving.

RIVER CUTTERS

Grades 6–9   Seven Sessions

Geological time passes very quickly in *River Cutters,* as students build a **model** of a river and **simulate** the creation of a river system in minutes. Using pool filter diatomaceous earth and a simple dripping system, students create rivers, observing and recording their results. They acquire geological terminology and begin to understand rivers as dynamic, ever-changing systems. By "running" many rivers, young "river-ologists" are introduced to the **sequence of geological events** and to differences between "old" and "young" rivers. The concepts of **erosion, pollution, toxic waste**, and **human manipulation of rivers** are also introduced.

Session 1
The teacher demonstrates the river cutting model (plastic tub and diatomaceous earth, with a dripper made from a plastic cup, wire, and a coffee stirrer). Students, organized in teams, acquaint themselves with the model and materials, and make a practice river.

Session 2
Students create and chart a river, record their results on a data sheet in both pictures and their own descriptive words.

Session 3
The class, using data sheets from the previous session, discusses river features and learns geological terminology. The concept of erosion is introduced and discussed. Students make small, river-features flags for use in the next session.

Session 4
Student teams create two rivers and compare them, focusing on the differences between "young" and "old" rivers.

Sessions 5, 6, 7 (Optional)
These sessions focus on the effects of slope on rivers (5), the construction of dams (6), and toxic waste dumps and how they can pollute rivers and the water table (7).

Skills
Recording Data, Experimenting, Communicating, Decision-Making, Designing and Refining Models

Concepts
Geology, Rivers, Erosion, Sequencing of Geological Events, Pollution, Human Impact on Environment

Themes
Systems & Interactions, Models & Simulations, Patterns of Change, Evolution, Scale, Structure, Energy, Matter

Notes: Background includes brief information on diatomaceous earth, the history of geology and understanding of erosion, mathematics and statistics applied to river tributaries. Going Further ideas include letting rivers run for much longer periods, using actual soil, mapping with grid techniques, photographing changes over time, experimenting to see how plants affect erosion, and inviting spokespeople to class to discuss toxic waste, pollution, or other related issues.

MAKING CONNECTIONS

In the following collection of books, **geology**, **ecology**, and **human impact on river systems** are emphasized. Two fine books by Holling Clancy Holling chronicle journeys down rivers by a turtle and a wooden figure in a canoe. Real-life stories focusing on pollution of the Nashua River and at Love Canal are also featured. Many other fine books, from the classic fantasy of *Wind in the Willows* to the light-giving river of the Afro-American story called *The River That Gave Gifts*, can provide creative and imaginative connections to these river-related GEMS activities. Books that focus on erosion or its consequences, as well as those that relate to models and simulations could likewise make strong connections. We're sure you and your students will add your own favorites.

CROSS REFERENCES

We catched fish and talked, and we took a swim now and then to keep off sleepiness. It was kind of solemn, drifting down the big, still river, laying on our backs looking up at the stars, and we didn't ever feel like talking loud, and it warn't often that we laughed—only a little kind of a low chuckle.

— *Mark Twain*
Huckleberry Finn

LITERATURE CONNECTIONS

Abel's Island

by William Steig
Farrar, Straus & Giroux, New York. 1976
Grades: 3–5

Abel is an urban mouse who suddenly finds himself on an uninhabited island trying to survive. He discovers skills and talents in himself that help him think of ways to forage for food, cross the river, and return home. Abel's time on the island brings him a new understanding of the world from which he's separated as he re-examines the easy way of life he had previously accepted.

The Adventures of Huckleberry Finn

by Mark Twain
Viking Penguin, New York. 1953
Grades: 6–Adult

A boy and a runaway slave start down the Mississippi on a borrowed raft in this exciting and sometimes dangerous trip. The mighty Mississippi courses through much of Twain's work, even in his famous pen-name itself (from a riverboat working command). While the language and dialect reflect their time and can be discussed in class, Twain's essential humanity comes through. This classic is available from many different publishers.

He don't plant taters
He don't plant cotton
And them that plants 'em
Is soon forgotten
But ole man river
He just keeps rollin'
Along.

— *Hammerstein and Kern*
Showboat

Biography of a River: The Living Mississippi

by Edith McCall
Walker and Co., New York. 1990
Grades: 6–12

This extensive "biography" for older students details the history of human interactions with the Mississippi River. It begins with a chapter on Native American settlements, followed by European expeditions and acquisition by the United States, up through the engineering projects of the present. The first chapter in which the river "speaks" in the first person is particularly effective. Its discussion of engineering challenges relates directly to the dam building challenges of Session 6. The emphasis, throughout, on the "living" nature of the river underlines an important environmental lesson.

Danny Dunn and the Universal Glue

by Jay Williams and Raymond Abrashkin; illustrated by Paul Sagsoorian
McGraw-Hill, New York. 1977
Out of print
Grades: 4–9

Danny and his friends bring evidence to a town meeting that waste from a local factory is polluting the local stream. A discussion of watersheds, water tables, and the way the pollution moved through the system of streams (on page 85) relates well to Sessions 6 and 7 of the GEMS guide.

Drylongso

by Virginia Hamilton; illustrated by Jerry Pinkney
Harcourt, Brace, Jovanovich, San Diego. 1992
Grades: 2–6

This strikingly illustrated book takes a powerful look at what happens when rivers dry up—a contrast to the flowing rivers the children create in *River Cutters*. An unknown boy blows into a village with a severe dust storm, and tells the villagers his name is Drylongso. He tells them that he was born in a time of great drought, but that his mother told him wherever he goes "life will grow better." Drylongso has special information about drought cycles, agriculture, and ways to survive; he carries a "dowser," or divining rod. He finds water beneath the ground to help a family's planting, then disappears as mysteriously as he came. There is excellent information on climate, drought, drought cycles, and soil conditions.

Follow the Water from Brook to Ocean

by Arthur Dorros
HarperCollins, New York. 1991
Grades: K–4

This book recounts water's journey as it shapes the earth through erosion. The course of water is traced as it flows from brooks to streams to rivers, over waterfalls, through canyons and dams, to eventually reach the ocean.

Love Canal: My Story

by Lois M. Gibbs

State University of New York at Albany Press, Albany, New York. 1982

Out of print

Grades: 6–12

> Autobiography of the housewife who organized a neighborhood association that eventually resulted in a clean up of the Love Canal toxic waste site and relocation of the families living there. She went on to form the Citizen's Clearinghouse for Hazardous Waste based in Arlington, Virginia.

Minn of the Mississippi

by Holling Clancy Holling

Houghton Mifflin, Boston. 1951

Grades: 6–12

> The journey of Minn, a snapping turtle, is followed from northern Minnesota to the bayous of Louisiana. Her adventures with people, animals, and the changing seasons are vividly described. Wonderful drawings and maps of her travels accompany the engaging true-life story on the Mississippi River. Newbery honor book.

The Missing 'Gator of Gumbo Limbo: An Ecological Mystery

by Jean C. George

HarperCollins, New York. 1992

Grades: 4–7

> Sixth-grader Liza K and her mother live in a tent in the Florida Everglades. She becomes a nature detective while searching for Dajun, a giant alligator who plays a part in a waterhole's oxygen-algae cycle, and is marked for extinction by local officials. The book is full of detail about the local habitats and species and the forces that impact on them. "Look how Mother Nature's plan for the Everglades has been tortured and diverted...the Everglades, which is really a slow river, is so rich with soil and nutrients that the Army Corps of Engineers was engaged to drain it for farmland..." Her neighbor explains how canals were built, fish and birds died, and the river changed. "You change one thing and you change the whole ecosystem."

An Oak Tree Dies and a Journey Begins

by Louanne Norris and Howard E. Smith, Jr.; illustrated by Allen Davis

Crown, New York. 1979

Out of print

Grades: 3–5

> A storm uproots an old oak tree on the bank of a river and its journey to the sea begins. Animals seek shelter in the log, children fish from it, mussels attach to its side. The tree, even after it dies, contributes to the environment. Older students can appreciate the fine pen and ink drawings.

Our Endangered Planet: Rivers and Lakes

by Mary Hoff and Mary M. Rogers

Lerner Publications, Minneapolis. 1991

Grades: 4–9

> An attractive and user-friendly reference book covering the dangers of surface water pollution with many illustrations and photographs. Other relevant titles in this series (all published in 1991) include: *Groundwater, Population Growth*, and *Tropical Rain Forests*.

Paddle-to-the-Sea

by Holling Clancy Holling

Houghton Mifflin, Boston. 1941

Grades: 6–9

> A Native American boy carves a wooden figure in a canoe and sets it afloat near the headwaters of a river north of the Great Lakes. This book chronicles the canoe's four-year journey to the sea. Caldecott honor book. (A detailed review of this book is on page 228.)

Rain of Troubles:
The Science and Politics of Acid Rain

by Lawrence Pringle

Macmillan, New York. 1988

Grades: 5–12

> Acid rain's discovery, formation, transportation, its effects on plant and animal life, and how economic and political forces have delayed action are discussed. The negative impact of acid rain on lakes and rivers can also be related to the toxic waste modeling activities in Session 7 of the GEMS activities.

A River Ran Wild:
An Environmental History

by Lynne Cherry

Harcourt, Brace, Jovanovich, San Diego. 1992

Grades: 1–5

> True story of the Nashua River Valley in North-Central Massachusetts from the time that the Native Americans settled there, naming it River With the Pebbled Bottom. The book traces the impact of the industrial revolution on the river and the eventual clean-up campaign mounted by a local watershed association. The graphic borders are packed with historical information, showing the original wildlife, tools and utensils used by Native Americans and early settlers, and continuing on to modern artifacts such as a plastic water jug.

The River That Gave Gifts: An Afro-American Story

by Margo Humphrey
Children's Book Press, San Francisco. 1987
Grades: K–5

Four children in an African village make gifts for wise old Neema while she still has partial vision. Yanava, who is not good at making things, does not know what to give, and seeks inspiration from the river. As she washes her hands in the river, rays of light fly off her fingers, changing into colors and forming a rainbow. After all the other gifts are presented, she rubs her hands in the jar of river water she has brought and thus gives a rainbow of light and the gift of sight to Neema. In addition to the themes of respect for elders and the validity of different kinds of achievement, the river is portrayed as a primeval source of power.

Sierra

by Diane Siebert; illustrated by Wendell Minor
HarperCollins, New York. 1991
Grades: 4–8

Long narrative poem in the voice of a mountain in the Sierra Nevada. It begins and ends with the lines:

I am the mountain
Tall and grand
And like a sentinel I stand.

Dynamic verse and glorious mural-like colored illustrations depict the forces shaping the earth as well as the plant, animal, and human roles in this ecosystem.

Three Days on a River in a Red Canoe

by Vera B. Williams
Greenwillow/William Morrow, New York. 1981
Grades: 3–6

Mom, Aunt Rosie, and two children on a three-day camping trip by canoe, encounter currents, wild winds, a rainbow, a moose, and more. In Session 4 of the GEMS guide, your students, like the children in the canoe, take careful note of the characteristics of the river they create.

The Wind in the Willows

by Kenneth Grahame; illustrated by Ernest Shepard
Aerie Books, New York. 1988
Grades: 4–Adult

This wonderful, humorous classic, filled with the bustling lives of eccentric animal characters, takes place along a river. The scenic descriptions accurately reflect the habitats of each animal. While the book is often read out loud to younger children, the pace and comic timing of the conversations makes it highly entertaining for adults.

Everything I love about America is a gift of the rivers: steamboats, pioneers, Huckleberry Finn, blue herons and snowy egrets, the Grand Canyon and the Blue Ridge hollows, jazz and catfish and ferryboats and covered bridges. None of them would be there, in memory or in fact, without the rivers. America is a great story, and there is a river on every page of it.

— *Charles Kuralt*
Down by the River

PADDLING TO THE SEA

BY JACQUELINE BARBER

River Cutters is a GEMS unit that enables students to model the passage of geological time, so they can simulate the effects of thousands of years of erosion in minutes. Students learn to identify geological formations, compare the features of old and young rivers, and investigate the effects of toxic waste dumps and dams on rivers, as they develop an understanding of river systems.

Paddle-to-the-Sea by Holling Clancy Holling (see annotation on page 226) is the story of a small canoe, carved by a Native American boy, that makes a journey from a snow bank near a river to the north of Lake Superior all the way to the Atlantic Ocean. The boy sets out to test a theory:

> "I have learned in school that when this snow in our Nipigon country melts, the water flows to that river. The river flows into the Great Lakes, the biggest lakes in the world. They are set like bowls on a gentle slope. The water from our river flows into the top one, drops into the next, and on to the others. Then it makes a river again, a river that flows to the Big Salt Water."

This enchanting book proceeds to chronicle the four-year journey of the carved boat, through rushing brooks, rapids, rivers, bays, canals, marshes, lakes and over waterfalls; through locks, a beaver dam, a sawmill, under docks, on boats, in storms, past factories. Over 50 pages long, this book is unusual in that every other page has a full-page color drawing. Margins are filled with little diagrams containing details of the Great Lakes, the path of the carved canoe, a diagram of a canal lock, some of the history of trade waterways, how a sawmill works, a diagram of a lake freighter, and more.

The story is full of picturesque language and lush detail of the wildlife surrounding the waters:

> "A muskrat swam past the drifting canoe and disappeared in the dead rushes...A buck deer waded in the shallows. He had only one antler and the weight of it made him walk with his head turned aside....The cubs caught crayfish and frogs in the mud, while the mother bear squatted on a rock beside a deep pool in the lagoon and smacked a black bass to the bank with her paw."

The reader is filled with vibrant images of these inland waterways and all the life that relies on them. Both the ingenuity and the audacity of humans is clear as the carved canoe witnesses the positive and negative effects that food gathering, transport, and industry have had on our waters.

Reading *Paddle-to-the-Sea* can lead you and your students to an immensely satisfying and rich conclusion to the GEMS *River Cutters* unit.

Jacqueline Barber is the director of the GEMS project.

Shapes, Loops, & Images

All Ages   **Four Exhibits**

*T*hese exhibits offer an opportunity to explore logic and spatial relationships through challenges involving topology, problem solving, repeated patterns, and multiple images. Visitors untangle rope loops, create their own tessellations, and venture into a backwards world of mirror challenges. This guide also delves into many different aspects of geometry through its interactive approach to exhibits about shapes, loops, and images. Now available in booklet form, rather than the more expensive folder format of the first edition. Exhibits are easy to construct, and use common materials and equipment.

Skills

Observing, Comparing, Matching, Finding Patterns, Analyzing, Relating, Visualizing, Predicting, Experimenting, Inferring

Concepts

Reflection, Mirror Images, Symmetry, Line of Symmetry, Fractions of a Circle, Perception, Kaleidoscopes, Geometry of Patterns, Spatial Relationships, Tessellations, Topology

Themes

Systems & Interactions, Patterns of Change, Models & Simulations, Structure

MAKING CONNECTIONS

This guide delves into many aspects of geometry and literature connections listed here include books about **shape** and **mirror images** as well as one non-fiction book that gives the reader a concrete understanding of **topology** through hands-on activities. The famous "Alice" works provide numerous wonderful connections, as might be expected given their author's mathematical profession.

CROSS REFERENCES

Anno's Math Games III

by Mitsumasa Anno

Philomel Books/Putnam & Grosset, New York. 1991

Grades: 4–10

> Picture puzzles, games, and simple activities introduce the mathematical concepts of abstract thinking, circuitry, geometry, and topology. The book invites active participation.

Jim Jimmy James

by Jack Kent

Greenwillow Books/William Morrow, New York. 1984

Out of print

Grades: K–2

> Jim Jimmy James makes friends and plays with his shadow. A very elementary look at the concept of reflection. As a follow-up, children can partner with a friend and play shadow games with each other. *Note: Some of the illustrations are not accurate reflections.*

Rubber Bands, Baseballs and Doughnuts: A Book about Topology

by Robert Froman;

illustrated by Harvey Weiss

Thomas Y. Crowell, New York. 1972

Out of print

Grades: 4–8

> This introduction to the world of topology requires active reader participation. The activities provide concrete examples and insights into abstract concepts.

Shadows and Reflections

by Tana Hoban

Greenwillow, New York. 1990

Grades: Preschool–5

> Color photographs without text feature shadows and reflections of various objects, animals, and people.

My Shadow

I have a little shadow that goes in and out with me,
And what can be the use of him is more than I can see,
He is very, very like me from the heels up to the head;
And I see him jump before me, when I jump into my bed.

The funniest thing about him is the way he likes to grow—
Not at all like proper children, which is always very slow;
For he sometimes shoots up taller like an India-rubber ball,
And sometimes he gets so little that there's none of him at all.

He hasn't got a notion of how children ought to play,
And can only make a fool of me in every sort of way,
He stops so close beside me, he's a coward you can see;
I'd think shame to stick to nursie as that shadow sticks to me!

One morning, very early, before the sun was up,
I rose and found the shining dew on every buttercup;
But my lazy little shadow, like an arrant sleepyhead,
Had stayed at home behind me and was fast asleep in bed!

— *Robert Louis Stevenson*

Shadows Here, There, Everywhere

by Ron and Nancy Goor
Thomas Y. Crowell, New York. 1981
Grades: K–5

Presents information about shadows, including how they are formed, why they can be of various lengths, and how they reveal the shape and texture of things. The book is user-friendly and the photographs are interesting and appealing.

The Shapes Game

by Paul Rogers; illustrated by Stan Tucker
Henry Holt, New York. 1989
Grades: Preschool–2

Fun-to-say riddles and pictures that are kaleidoscopes of brilliant colors take young children from simple squares and circles through triangles, ovals, crescents, rectangles, diamonds, spirals, and stars.

Shapes, Shapes, Shapes

by Tana Hoban
Greenwillow Books, New York. 1986
Grades: Preschool–5

Color photographs of familiar objects, such as a chair, barrettes, and manhole cover, present a study of rounded and angular shapes.

Through the Looking Glass

by Lewis Carroll
Viking Penguin, New York. 1984
Grades: All Ages

In Chapter One, when Alice first begins her further adventures in Wonderland, she enters through a mirror. She finds a world completely opposite of the one she left behind. Students may enjoy testing their perceptions by tracing simple geometric shapes while looking in a mirror or trying simple visual tasks while they are looking through the looking glass.

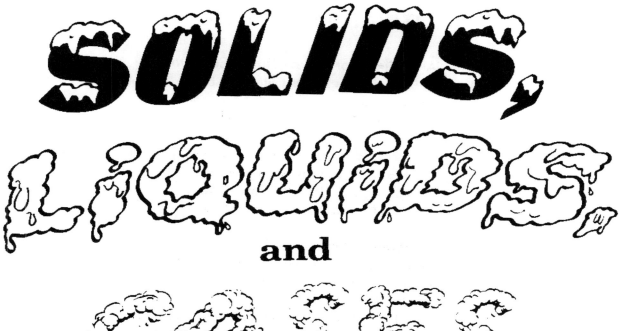

SOLIDS, LIQUIDS, and GASES

Grades 3–6

This GEMS Assembly Presenter's guide provides a script and detailed instructions and materials lists for the presentation of a program to a school assembly audience of several hundred people. In this program, the presenter invites student participation as she helps the audience discover the properties of solids, liquids, and gases; the quirks of phase change; and the idea that all matter is made up of moving atoms. The interactive nature of the presentation enhances exciting experiments.

Skills

Observing, Comparing, Predicting, Visualizing, Relating, Modeling

Concepts

Solid, Liquid, Gas, Matter, Energy, Phase, Phase Change, Evaporation, Sublimation, Condensation, Atoms

Themes

Systems & Interactions, Models & Simulations, Stability, Patterns of Change, Scale, Structure, Energy, Matter

MAKING CONNECTIONS

The following listing includes books that pertain to **solids**, **liquids**, **gases**, **and their properties**. One book considers the effects of taking these forms of matter into **extreme temperature conditions**. Several books show **phase change**, **evaporation**, **heat transfer** and **filtration** to explain how refrigeration and air conditioning work and how water is purified.

CROSS REFERENCES

LITERATURE CONNECTIONS

A Chilling Story: How Things Cool Down
by Eve and Albert Stwertka; illustrated by Mena Dolobowsky
Julian Messner/Simon & Schuster, New York. 1991
Grades: 4–8

How refrigeration and air conditioning work are explained simply, with sections on heat transfer, evaporation, and expansion. Humorous black and white drawings show a family and its cat testing out the principles in their home. This book connects well to the experiments involving dry ice and liquid nitrogen in the GEMS guide.

Everybody Needs a Rock
by Byrd Baylor; illustrated by Peter Parnall
Aladdin Books, New York. 1974
Grades: K–5

This book describes the qualities to consider when selecting the perfect rock for play and pleasure. In so doing, the properties of color, size, shape, texture, and smell are discussed in such a way that you'll want to rush out and find a rock of your own. Nice introduction or follow-up to a discussion of the properties of solids.

Hot-Air Henry
by Mary Calhoun; illustrated by Erick Ingraham
William Morrow, New York. 1981
Grades: K–3

Henry, a spunky Siamese cat, stows away on a hot air balloon and accidentally gets a solo flight. He learns that there is more to ballooning than just watching as he deals with air currents, power lines, and manipulating the gas burner. Though the format and style of the book are aimed at primary grades, information on ballooning and the more complex concept that hot air becomes less dense are also presented.

The Magic School Bus at the Waterworks
by Joanna Cole; illustrated by Bruce Degen
Scholastic, New York. 1986
Grades: K–6

When Ms. Frizzle takes her class on a field trip to the waterworks, everyone ends up experiencing the water purification system from the inside. Evaporation, the water cycle, and filtration are just a few of the concepts explored in this whimsical field trip. The phase changes of water, from solid to liquid to gas, provide a familiar example for all ages of some of the concepts explored in this assembly presenter's guide.

Thus the theory of description matters most.
It is the theory of the word for those

For whom the world is the making of the world,
The buzzing world and its lisping firmament.

It is a world of words to the end of it,
In which nothing solid is its solid self.
— *Wallace Stevens*
Description Without Place

Splash! All About Baths

by Susan K. Buxbaum and Rita G. Gelman; illustrated by Maryann Cocca-Leffler
Little, Brown & Co., Boston. 1987
Grades: K–6

Penguin answers his animal friends' questions about baths such as "What shape is water?" "Why do soap and water make you clean?" "What is a bubble?" "Why does the water go up when you get in?" "Why do some things float and others sink?" and other questions. Answers to questions are both clear and simple. Received the American Institute of Physics Science Writing Award.

Supersuits

by Vicki Cobb; illustrated by Peter Lippman
J.B. Lippincott, Philadelphia. 1975
Grades: 4–7

This book describes severe environmental conditions that require special clothing for survival such as freezing cold, fire, underwater work, and thin or non-existent air. "Going Where It's Cold" talks about solids, liquids, and gases at cold temperatures.

Very Last First Time

by Jan Andrews; illustrated by Ian Wallace
Atheneum, New York. 1986
Grades: 2–4

An Inuit girl, Eva, walks by herself (for the "very first last time") in a sea-floor cavern under the frozen ocean ice when the tide goes out, gathering mussels and making discoveries. Later, her candle goes out, and the tide starts to come in, roaring louder, while the ice shrieks and creaks. Terrified at first, Eva recovers, and eventually finds her way to the surface and her waiting mother. Although the book does not scientifically explain the freezing of the top of the sea or the action of the tides, you and your class may want to discuss these questions: "Why does only the top part of the water freeze?" "Why does the ice stay intact even when the water underneath it goes out with the tide?" The images of Eva on the sea floor beneath the ice are unique and fascinating. The descriptive language and Eva's intense interest in nature exemplify excellent scientific observation skills.

Water's Way

by Lisa W. Peters; illustrated by Ted Rand
Arcade Publishing/Little, Brown & Co., New York. 1991
Grades: K–3

"Water has a way of changing" inside and outside Tony's house, from clouds to steam to fog and other forms. Innovative illustrations show the changes in the weather outside while highlighting water changes inside the house.

GEMS 236

TERRARIUM HABITATS

Grades K–6

Five Activities

Bring the natural world into your classroom and deepen understanding of and connections to all living things. The guide begins with an exploration of soil. Teams of students design and construct terrariums for the classroom. Sowbugs, earthworms, and crickets are placed in the terrarium habitat and students observe and record changes over time. There are detailed instructions on setting up and maintaining the terrariums, along with concise and interesting biological information on a number of possible small organisms that can become terrarium inhabitants. Special features include an optional soil profile test for older students and making "decomposition bags" to learn more about the natural life cycle.

Activity 1

The students explore soil. An optional test using water and alum separates the soil into distinct layers.

Activity 2

The students work together to build a mini-forest habitat in a container made from a milk carton, using soil and a variety of natural items like leaves, bark, bird seed, and strawberry plants.

Activities 3 and 4

Pillbugs and earthworms are observed carefully and then placed into the terrariums. Continuing to observe the terrarium is the most exciting part of the activity. Did the isopods have babies? What happened to the dried leaves? Where did these white eggs come from? Look what sprouted from the seeds! Student journals of observations and drawings record the daily changes.

Activity 5

Students explore interactions within their terrarium habitats over time and as they choose, adding other animals, plants, food items, and objects of their choice.

Skills

Observing, Comparing, Describing, Measuring, Communicating, Organizing, Experimenting, Recording, Drawing Conclusions, Building Models

Concepts

Soil and Ground Habitats, Ecology, Life Cycle, Food Webs, Nutrient Cycle, Decomposition, Recycling, Adaptation, Animal Structures and Behavior

Science Themes

Systems & Interactions, Patterns of Change, Structure, Energy, Matter, Evolution

Mathematics Strands

Pattern, Number, Measurement

Nature of Science and Mathematics

Cooperative Efforts, Creativity and Constraints, Interdisciplinary Connections, Real-Life Applications

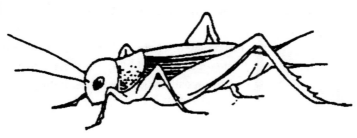

MAKING CONNECTIONS

Some of the books focus on snails, crickets, ants, or salamanders that could live in a terrarium or on a forest floor. **Interdependence**, **life cycle**, and **scale** are emphasized through fantasy stories that ask the reader to view life from an animal's perspective. Other books look at the **system** of plants and animals that is nurtured by a decomposing tree. Four books in particular are written as inviting resource books or almanacs that give ideas for other fun, hands-on **investigations of nature**.

CROSS REFERENCES

> Science, like life, feeds on its own decay. New facts burst old rules; then newly divined conceptions bind old and new together into a reconciling law.
>
> — *William James*
> *The Will To Believe*

LITERATURE CONNECTIONS

Chipmunk Song

by Joanne Ryder; illustrated by Lynne Cherry
E.P. Dutton, New York. 1987
Grades: K–5

A chipmunk goes about its activities in late summer, prepares for winter, and settles in until spring. The reader is put in the place of a chipmunk and participates in food gathering, hiding from predators, hibernating, and more. Roots, tunnels, stashes of acorns and other facets of the imagined environment loom large and lifelike.

Deep Down Underground

by Olivier Dunrea
Macmillan, New York. 1989
Grades: Preschool–3

Cumulative counting book led off by "one wee moudiewort" (Scottish word for type of mole) and a creepy cadre of earthworms, caterpillars, beetles, toads and spiders, sowbugs, garter snakes, and red ants. The dynamic language makes it great for reading aloud individually or in a group wriggling, wrangling, scooching, and scraping.

Earthworm

by Adrienne Souter-Perrot; illustrated by Etienne Delessent
Creative Editions, Mankato, Minnesota. 1993
Grades: Preschool–2

Simply and accurately written, and elegantly illustrated, this is an excellent early childhood introduction to earthworms and their revitalization of the soil.

Earthworms, Dirt, and Rotten Leaves

by Molly McLaughlin

Avon, New York. 1990

Grades: 4–7

> The earthworm and its environment are explored and suggestions made for experiments to examine the survival of the earthworm in its habitat. Answers the question, "Why would anyone want to have anything to do with earthworms?" Recipient of awards for writing from Library of Congress, American Library Association, New York Academy of Sciences, etc.

The Empty Lot

by Dale H. Fife; illustrated by Jim Arnosky

Sierra Club Books/Little, Brown & Co., Boston. 1991

Grades: 2–4

> What good is a vacant lot? City-dweller Harry Hale owns one, but when he goes to take a good look before selling it, he is amazed to find that the lot is far from empty. It's pulsing with life: birds and their nests; ants, beetles, fungi, and molds in the soil; and frogs and dragonflies near the stream. He is so impressed by the utilization of the different habitat areas that he changes his for sale sign to read "occupied lot—every square inch in use."

The Fall of Freddie the Leaf

by Leo Buscaglia

Charles B. Slack, Inc./Holt, Rinehart and Winston, New York. 1982

Grades: All ages

> This simply told story, with beautiful color photographs and a real leaf on the inside back cover, describes the growth, maturity, decay, and death of a leaf named Freddie and his friends. It is "dedicated to all children who have ever suffered a permanent loss, and to the grownups who could not find a way to explain it." Because the story is about leaves, it is also a good connection to the GEMS activities that relate to decomposition.

The Frog Alphabet Book

by Jerry Pallotta; illustrated by Ralph Masiello

Charlesbridge Publishing, Watertown, Massachusetts. 1990

Grades: K–3

> A beautifully illustrated book that shows the diversity of frogs and other "awesome amphibians" from around the world.

Frogs, Toads, Lizards, and Salamanders

by Nancy W. Parker and Joan R. Wright; illustrated by Nancy W. Parker

Greenwillow Books, New York. 1990

Grades: 3–6

> Physical characteristics, habits, and environment of 16 creatures are encapsulated in rhyming couplets, text, and anatomical drawings, plus glossaries, range maps, and a scientific classification chart. A great deal of information is presented, the rhymes are engaging and humorous, and the visual presentation terrific. "A slimy Two-toed Amphiuma terrified Grant's aunt from Yuma" (she was picking flowers from a drainage ditch).

The Girl Who Loved Caterpillars

adapted by Jean Merrill; illustrated by Floyd Cooper

Philomel Books/Putnam & Grosset, New York. 1992

Grades: 2–6

> Based on a twelfth century Japanese story, this book is a wonderful and early portrait of a highly independent and free-spirited girl, Izumi, who loves caterpillars. Izumi wonders "Why do people make such a fuss about butterflies and pay no attention to the creatures from which butterflies come? It is caterpillars that are really interesting!" Izumi is interested in the "original nature of things," and in doing things naturally. Clever poetry is interspersed as part of the plot. Great connection to the observation activities in *Terrarium Habitats*. An excellent and relevant portrayal of an independent-thinking female role model.

BOOK LICE

I was born in a
fine old edition of Schiller

           While I started life
           in a private eye thriller

We're book lice
who dwell
in these dusty bookshelves.
Later I lodged in
Scott's works—volume 50

           We're book lice
           who dwell
           in these dusty bookshelves.

           While I passed my youth
           in an Agatha Christie

We're book lice
attached
despite contrasting pasts.
One day, while in search of
a new place to eat

           We're book lice
           attached
           despite contrasting pasts.

           He fell down seven shelves,
           where we happened to meet

We're book lice
who chew
on the bookbinding glue.
We honeymooned in an
old guide book on Greece

           We're book lice
           who chew
           on the bookbinding glue.

           I missed Conan Doyle
           he pined for his Keats

We're book lice
fine mates
despite different tastes.
So we set up our home
inside Roget's Thesaurus

           We're book lice
           fine mates
           despite different tastes.

           Not far from my mysteries
           close to his Horace

We're book lice
adoring
despite her loud snoring
And there we've resided,
and there we'll remain,

           We're book lice
           adoring
           despite his loud snoring

           He nearby his Shakespeare
           I near my Spillane

We're book-loving
book lice

           We're book-loving
           book lice
           plain proof of the fact

which I'm certain I read
in a book some months back
that opposites
often are known
to attract.

           that opposites
           often are known
           to attract.

           — *Paul Fleishman*
           Joyful Noise: Poems for Two Voices

Joyful Noise: Poems for Two Voices

by Paul Fleischman; illustrated by Eric Beddows
Harper & Row, New York. 1988
Grades: K–Adult

Fantastic series of poems celebrating insects that are meant to be read
aloud by two readers at once, sometimes merging into a duet. It includes
beetles and crickets, and many others such as grasshoppers and cicadas.
The combination of rich scientific detail with poetry, humor, and a sense of
the ironic contrasts and division of labor in the lives and life changes of
insects is powerful and very involving. Students in upper level classes
might love performing these for the class. (Two of his poems are included
in this handbook: "Honeybees" is on page 56, and "Book Lice" is on page
240.) Newbery award winner.

Linnea's Almanac

by Christina Bjork; illustrated by Lena Anderson
R&S Books/Farrar, Straus & Giroux, New York. 1989
Grades: 3–6

Linnea keeps an almanac tracking her indoor and outdoor investigations of
nature over a year's time. She opens a bird restaurant in January and goes
beachcombing in July. The almanac is written in journal form with simple
monthly activities for young readers to do at home.

Linnea's Windowsill Garden

by Christina Bjork; illustrated by Lena Anderson
R&S Books/Farrar, Straus & Giroux, New York. 1988
Grades: 3–6

Linnea tells about her indoor garden. From seeds to cuttings to potted
plants, Linnea describes the care of her plants throughout their life cycle.
The friendly narration and simple information invites readers to try the
activities and games at home.

The Magic School Bus Inside the Earth

by Joanna Cole; illustrated by Bruce Degen
Scholastic, New York. 1987
Grades: 3–6

Another in this highly educational and amusing series,
Ms. Frizzle takes her class on a field trip to the center
of the Earth and back again. Much geological information is interspersed throughout. This book could serve
as a good extension to the soil activities.

Nicky the Nature Detective

by Ulf Svedberg
R&S Books/Farrar, Straus & Giroux, New York. 1988
Grades: 3–8

Nicky loves to explore changes in nature. She watches a red maple tree and all the creatures and plants that live on or near the tree throughout the seasons. Her discoveries lead her to look carefully at the structure of the nesting place, why birds migrate, who left tracks in the snow, where butterflies go in the winter, and many more things. This book is packed with information and inviting graphic elements.

An Oak Tree Dies and a Journey Begins

by Louanne Norris and Howard E. Smith, Jr.; illustrated by Allen Davis
Crown, New York. 1979
Out of print
Grades: 3–5

An old oak tree on the bank of a river is uprooted by a storm and its journey to the sea begins. Animals seek shelter in the log, children fish from it, mussels attach to its side. A story of how a tree, even after it dies, contributes to the environment. Students will appreciate the fine pen and ink drawings.

Once There Was a Tree

by Natalia Romanova; illustrations by Gennady Spirin
Dial Books, Penguin, New York. 1983
Grades: K–6

A tree is struck by lightning, cut down, and left as a stump. A bark beetle lays her eggs under its bark, its larvae gnaw tunnels. Ants make their home there. A bear uses the stump to sharpen her claws. The stump is visited by birds, frogs, earwigs, and humans. It endures the weather. As a new tree grows from the stump, a question remains: "Whose tree is it?"

One Day in the Woods

by Jean C. George; illustrated by Gary Allen
Thomas Y. Crowell, New York. 1988
Grades: 4–7

On a day-long outing in a woodland forest, Rebecca, a "ponytailed explorer," searches for the elusive ovenbird. Her observation of, and interaction with, the plant and animal life are enhanced by realistic black and white drawings.

The Salamander Room

by Anne Mazer; illustrated by Steve Johnson
Alfred A. Knopf, New York. 1991
Grades: K–3

A little boy finds an orange salamander in the woods and thinks of the many things he can do to turn his room into a perfect salamander home. In the process, the habitat requirements of a forest floor dweller are nicely described.

The Snail's Spell
by Joanne Ryder; illustrated by Lynne Cherry
Puffin Books, New York. 1988
Grades: K–5

Imagine how it feels to be a snail and in the process learn something about the anatomy and locomotion of a snail. Though the picture-book format gives a primary-level feel to the book, imagining you are a snail is interesting for older students as well. Outstanding Science Book for Young Children Award from the New York Academy of Sciences.

The Song in the Walnut Grove
by David Kherdian; illustrated by Paul O. Zelinsky
Alfred A. Knopf, New York. 1982
Grades: 4–6

A curious cricket named Ben meets Charley the grasshopper. Together they learn of each other's daytime and nighttime habits while living in an herb garden. The friendship between Ben and Charley grows when Ben rescues Charley from being buried in a pail of grain and they learn to appreciate each other's differences. This story weaves very accurate accounts of insect behavior with their contributions to the ecology of Walnut Grove.

Two Bad Ants
by Chris Van Allsburg
Houghton Mifflin, Boston. 1988
Grades: Preschool–4

When two curious ants set off in search of beautiful sparkling crystals (sugar), it becomes a dangerous adventure that convinces them to return to the former safety of their ant colony. Illustrations are drawn from an ant's perspective, showing them lugging individual sugar crystals and other views from "the small."

When the Woods Hum
by Joanne Ryder; illustrated by Catherine Stock
William Morrow, New York. 1991
Grades: 1–4

Young Jenny, who has heard her father reminisce about the wonder of hearing the woods hum, investigates periodical cicadas—"hummers." They observe the wingless creepers emerging from underground, the adult cicadas shedding their old skins, and the female laying eggs. A page of sketches at the end of the book gives some detail, differentiating between the annual and periodical cicada. The cycle motif is reinforced when the grown-up Jenny and her father take her son to the woods so he can hear the humming.

Whisper from the Woods

by Victoria Wirth; illustrated by A. Scott Banfill
Green Tiger Press (Simon & Schuster), New York. 1991
Grades: Preschool–4

> This stunningly designed book brings the woods alive with gorgeous full-page color illustrations and a poetic text that depicts in detail trees, vegetation, the seasons, and the animal life of the forest. Subtly drawn, within a tree seed is a tiny human face that grows, as the tree grows, and becomes mature. Eventually all the trees have human faces with arms and hands for roots that intertwine under the ground. In the summer "these trees, young and old, would lean together and whisper; sharing thoughts and passing the wisdom of many years." The visual effects are powerful and beautiful, conveying in a unique way the connection between all living things and illustrating the life cycle. The emotional impact makes this book a true treasure.

Wonderful Worms

by Linda Glaser; illustrated by Loretta Krupinski
The Millbrooke Press, Brookfield, Connecticut. 1992
Grades: Preschool-3

> This is an excellent and unique, cleverly illustrated introduction to earthworms and their work. The writing is particularly noteworthy for its use of a child's perspective: "Worms feel sounds with their whole bodies. They feel thunder when I walk."

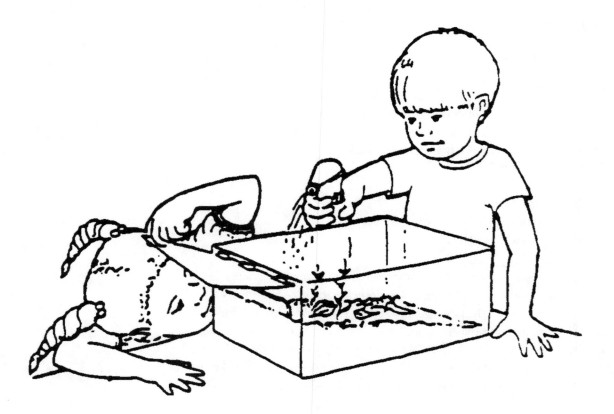

TREE HOMES

Preschool–1

Six Activities

Appreciation for trees and the animals that live in tree homes is encouraged in this guide with an emphasis on the biological need for warmth and shelter. The children become familiar with a living tree, build a child-size tree from cardboard boxes, paper, and cardboard tubes; role play dramas about a mother bear and her cubs, raccoons, and a family of owls and their tree homes. The children make paper models of raccoons and owls. Sorting, classifying, and measuring are also emphasized.

Activity 1

Children visit a living tree and examine its branches, leaves, and look for holes where animals might make a home. In the classroom they make a cardboard tree with branches and leaves. They compare the parts of the cardboard tree with a living tree.

Activity 2

A bear drama is presented with toy bears. The children sort the toy bears by color and then by size. They observe the bears, count them, and make comparisons between different bears.

Activity 3

The children test ideas of how bears, and they, stay warm. They put on clothing and blankets, snuggle together, jump up and down, get into the sunlight, and rub hands together (friction) to get warmer.

Activity 4

A toy raccoon is shown and discussed; a raccoon drama is presented. Using a poster of a raccoon, the children examine the parts of a raccoon and count legs, toes, stripes, etc. Children pretend to be raccoons and use sweaters or towels as "tails" to keep warm. At snack time, with the room partially darkened, the children reach into a bag and guess what food it contains, much like a raccoon finding food at night. The children make paper-bag raccoons.

Activity 5

An owl poster is examined and the parts of its body are counted. Real feathers are examined. An owl drama is presented and the children pretend to be owls and "fly" around the room. The children make paper-bag owls. They find a hole in the cardboard tree as a home for the owl. A drama about owls making nests is presented; the children make a nest using sticks and put a paper egg into it.

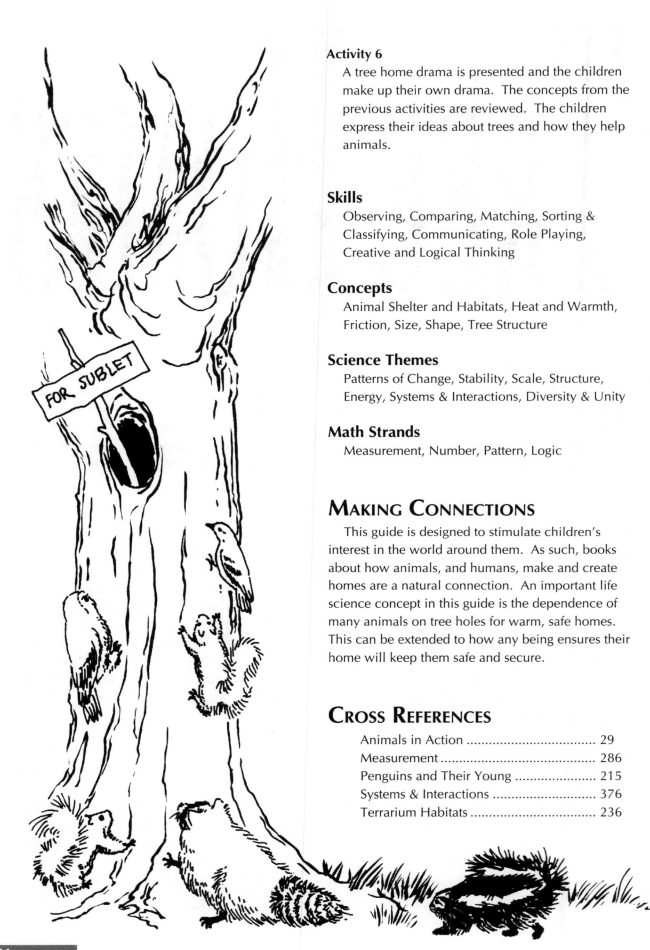

Activity 6

A tree home drama is presented and the children make up their own drama. The concepts from the previous activities are reviewed. The children express their ideas about trees and how they help animals.

Skills

Observing, Comparing, Matching, Sorting & Classifying, Communicating, Role Playing, Creative and Logical Thinking

Concepts

Animal Shelter and Habitats, Heat and Warmth, Friction, Size, Shape, Tree Structure

Science Themes

Patterns of Change, Stability, Scale, Structure, Energy, Systems & Interactions, Diversity & Unity

Math Strands

Measurement, Number, Pattern, Logic

MAKING CONNECTIONS

This guide is designed to stimulate children's interest in the world around them. As such, books about how animals, and humans, make and create homes are a natural connection. An important life science concept in this guide is the dependence of many animals on tree holes for warm, safe homes. This can be extended to how any being ensures their home will keep them safe and secure.

CROSS REFERENCES

Literature Connections

Baby Raccoon

by Beth Spanjian

Longmeadow Press, Stamford, Connecticut. 1988

Grades: Preschool–1

Baby Raccoon, along with his mother, sisters, and brothers, all leave their tree home nest for an adventurous search for food on a beautiful, moonlit summer night. This book works well with Activity 4.

Black Bear Baby

by Bernice Freschet

G.P. Putnam's Sons, New York. 1981

Out of print

Grades: Preschool–2

This story about Black Bear Baby and his sister realistically describes the early life of black bear cubs. This book works well with Activity 2.

The Day the Sun Danced

by Edith T. Hurd

Harper & Row, New York. 1965

Grades: K–3

During the darkness and cold of winter, the rabbit goes to the bear, the fox, and the deer to tell them that something is going to happen. Beautiful woodcuts contrast the bleakness of winter and the brilliant colors of spring. This book works well with Activity 3.

Good-Night, Owl!

by Pat Hutchins

Macmillan, New York. 1972

Grades: Preschool–2

Owl is kept awake during the day by all the noisy animals that live in the tree near him. Night comes and the situation changes. This book works well with Activity 5.

The Great Kapok Tree

by Lynne Cherry

Harcourt, Brace, Jovanovich, San Diego. 1990

Grades: Preschool–3

The animals of the rain forest try to convince a sleeping man not to cut down a very large Kapok tree. This book works well with Activity 1.

I think I shall never see
A poem as lovely as a tree . . .

A tree that looks to God all day
And lifts her leafy arms to pray;

A tree that may in summer wear
A nest of robins in her hair . . .

Poems are made by fools like me,
But only God can make a tree.

— *Joyce Kilmer*
Trees

> They cannot see the
> forest for the trees.
>
> — *Christoph Wieland*
> *Musarion*

Hello, Tree

by Joanne Ryder
Lodestar Books, New York. 1991
Grades: K–3

> This story encourages an appreciation for the beauty, growth, shade, sounds, smells, and textures of a tree and for the animals seen in the tree. This book works well with Activity 1.

Night Tree

by Eve Bunting
Harcourt, Brace, Jovanovich, San Diego. 1991
Grades: Preschool–3

> Each Christmas a family goes to a nearby forest to decorate a tree with food for the animals. This book works well with Activity 1.

Once There Was a Tree

by Natalia Romanova
Dial Books, New York. 1985
Grades: Preschool–Adult

> This beautifully illustrated story about the death of an old tree and the growth of a new one explores the question, "Who owns a tree—the animals that live in the tree or the bear, the birds, and the man who visit it?" This book works well with Activity 1.

Our Very Own Tree

by Lawrence F. Lowery
Encyclopaedia Britannica Educational Corp., Chicago. 1993
(To order, call 1-800-554-9862)
Grades: K–3

> Two friends have a special tree they often visit. Through the discoveries of these two little girls, children have a wonderful opportunity to learn about trees and the animals that live in them. This 12" x 18" book is beautifully illustrated and has a built-in stand. This book works well with Activity 1.

Owl Lake

by Keizaburo Tejima
Philomel Books, New York. 1982
Out of print
Grades: Preschool–2

> Bold woodcuts illustrate the story of a father owl who flies across a lake in the moonlight searching for silver fish to feed his hungry family. This book works well with Activity 5.

Owl Moon

by Jane Yolen
Philomel Books, New York. 1987
Grades: Preschool–1

Large watercolor illustrations enhance this story about a little girl and her father who go into the woods on a moonlit night in search of a Great Horned Owl. The girl learns from her father that you need to be quiet, brave, and hopeful when you go looking for owls. This book works well with Activity 5.

Raccoons and Ripe Corn

by Jim Arnosky
Lothrop, Lee and Shepard Books, New York. 1987
Grades: Preschool–3

Mother raccoon and her kits feast on corn all night long under a full moon. The large color illustrations show the raccoons in delightfully realistic poses. This book works well with Activity 4.

Stellaluna

by Janell Cannon
Harcourt, Brace & Co., New York. 1993
Grades: Preschool–4

A baby fruit bat named Stellaluna is separated from her mother when an owl attacks. Not yet able to fly, Stellaluna falls and clutches a tiny twig. She is adopted and raised by birds. Many of her "bat ways" disappear as she, for example, learns to eat bugs, but, although she tries, she never stops her habit of hanging by her feet. In the end she is reunited with her mother but stays connected to her bird friends. Two detailed pages of fascinating background material about bats are included.

A Tree is Nice

by Janice Udry
Harper & Row, New York. 1957
Grades: Preschool–1

Simple text and colorful illustrations express the many joys children find in trees. This book works well with Activity 1.

Trees

by Harry Behn

Henry Holt and Company, New York. 1992

Grades: Preschool–2

> This beautifully illustrated poem celebrates the importance of trees. This book works well with Activity 1.

Two Little Bears

by Ylla

Harper & Row, New York. 1954

Out of print

Grades: Preschool–2

> The adventures of two lost bear cubs come to an end when their mother finds them. Beautiful large photographs capture the cubs in delightful poses. This book works well with Activity 2.

Whoo-oo Is It?

by Megan McDonald

Orchard Books, Franklin Watts, New York. 1992

Grades: Preschool–1

> Mother Owl hears a noise in the night and finally finds out what is making the mysterious sound. This book works well with Activity 5.

The Wind in the Willows

by Kenneth Grahame; illustrated by Ernest H. Shepard

Aerie Books, New York. 1988

Grades: Preschool–Adult

> This wonderful, humorous classic, filled with the bustling lives of eccentric animal characters, takes place along a river and in the homes of the various animals including Badger, Toad, Mole, and Rat. The scenic descriptions accurately reflect the habitat of each animal. While the book is often read out loud to younger children, the pace and comic timing of the conversations makes it highly entertaining for adults.

Winnie-The-Pooh

by A.A. Milne; illustrations by Ernest H. Shepard

E.P. Dutton, New York. 1926

Dell Publishing, New York. 1954

Grades: Preschool–Adult

> This well-loved classic contains chapter after delightful chapter of the adventures of Christopher Robin, Pooh, and their animal friends—most of whom live in tree homes. Relevant chapters include Chapter 3 in which Pooh visits Piglet who lives in the bottom of a beech tree and Chapter 2 in which Pooh becomes stuck trying to leave rabbit's home after eating too much honey. This book works well read aloud to younger children especially with Activity 2.

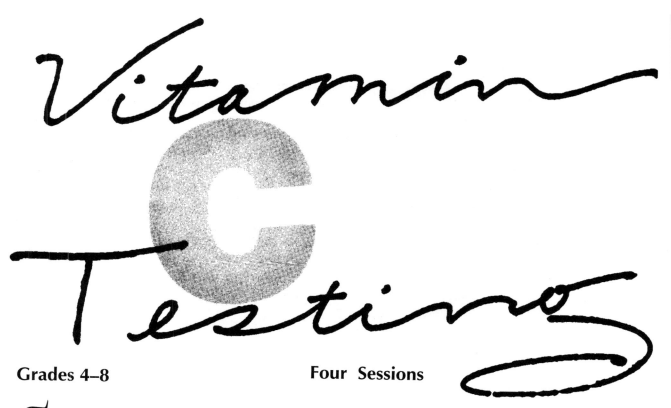

Vitamin C Testing

Grades 4–8   **Four Sessions**

This guide is a stimulating introduction to chemistry and nutrition. The students perform a simple chemical test using a vitamin C indicator to compare the vitamin C content of different juices, and graph results. Older students can examine the effects of heat and freezing on vitamin C content.

Session 1

Students learn the lab technique titration and use it to test various fruit drinks for vitamin C content.

Session 2

Using the data gathered in the first session, students pool data, calculate averages, construct a bar graph, and draw conclusions about relative amounts of vitamin C and which food processes are better for preserving it. Teacher provides scientific and nutritional information about vitamin C.

Session 3

Students use the techniques they've learned to test the vitamin C content of beverages they drink at home. This helps connect what they've learned to daily diet, and introduces what they're learning to their families.

Session 4

Students conduct experiments comparing two different treatments of the same juice (for example, one left out uncovered compared to one covered and put in a refrigerator) to see the effect on vitamin C content. The activity can be extended to comparing the effects of freezing and boiling. Students pool data and analyze results.

Skills

Chemistry Laboratory Techniques, Experimenting, Analyzing Data, Graphing, Drawing Conclusions

Concepts

Chemistry, Nutrition, Vitamin C Content, Titration, Indicator, End Point, Conditions Causing Vitamin Loss

Themes

Systems & Interactions, Stability, Patterns of Change, Scale, Matter

Notes: Background section on how vitamin C was discovered, its role in human health, and differences on its recommended nutritional amounts. Numerous Going Further ideas, including re-creation of original experiments that led to discovery of vitamin C, and ways to find the absolute, rather than the relative, amounts of vitamin C present in substances.

MAKING CONNECTIONS

An obvious connection is to books in which **vitamin deficiencies** play a role. We found one such strong connection in a book about a boy living alone in the mountains that even links animal behaviors to the need for vitamins. There certainly are other books that would provide this same kind of connection, perhaps relating to long ocean voyages or other situations of hardship where the results of vitamin deficiencies are evident.

A book about oranges tells about the lengths to which humans go to ensure constant **sources of vitamin C rich foods** in all seasons and all regions. Another book explores **food** and **nutrition** through fun activities, features, and historical anecdotes. While not related specifically to vitamin C, this book and others like it can provide a larger context for the unit.

Finally there are several books that focus on **experimentation**. In one, the scientist is young, and his experiment pertains to **vitamins and plant growth**. In the other, an army surgeon learns from a somewhat grizzly but fascinating situation—a patient with a healed bullet hole in his stomach. This leads to an improved knowledge of the digestive system and its processes.

CROSS REFERENCES

LITERATURE CONNECTIONS

Doctor Beaumont and the Man with the Hole in His Stomach
by Beryl and Samuel Epstein
Coward, McCann & Geoghegan, New York. 1978
Grades: 4–6

Interesting experiments about digestion are described in this biography of an army surgeon who in the 1820s had a patient with a bullet hole in his stomach. By inserting a tube, the doctor was able to directly observe and monitor the circulation of gastric juices and bile fluids. The surgeon eventually published a book on these experiments, including findings on acidity in the stomach.

Elliot's Extraordinary Cookbook
by Christina Bjork; illustrated by Lena Anderson
Farrar, Straus & Giroux, New York. 1990
Grades: 3–6

With the help of his upstairs neighbor, Elliot cooks wonderful recipes (including cinnamon buns and rye bread made with live yeast) and investigates what's healthy and what's not so healthy. He finds out about proteins, carbohydrates, and the workings of the small intestine. He learns about the history of chickens and how cows produce milk. His friend shows him how to grow bean sprouts, and he sews an apron. A nice context of food and nutrition.

June 29, 1999
by David Wiesner
Clarion Books, Houghton Mifflin, New York. 1992
Grades: 3–6

> The science project of Holly Evans takes an extraordinary turn—or does it? This highly imaginative and humorous book has a central experimental component, and conveys the sense of unexpected results.

My Side of the Mountain
by Jean C. George
E.P. Dutton, New York. 1959
Penguin Books, New York. 1991
Grades: 5–12

> Classic story of a boy who runs away and spends a year living alone in the Catskill mountains, recording his experiences in a diary. He struggles for survival and is supported by animal friends. Ultimately he realizes he needs human companionship. In the winter when his food is running low, he suffers nose bleeds and other symptoms of scurvy or vitamin deprivation. Page 134 describes how he finds sources of vitamins, such as liver, and what he notices about other animals' nutrition, such as a squirrel seeking the bark of a sapling and birds sitting in the sunlight as if they were trying to replenish Vitamin D. "Hunger is a funny thing. It has a kind of intelligence all its own." Winner: Newbery Honor Book, ALA Notable Book, Hans Christian Andersen International Award.

Oranges
by Zack Rogow; illustrated by Mary Szilagyi
Franklin Watts, New York. 1988
Grades: K–5

> Describes the long journey and the combined labor of the many people it takes to bring a single orange from the tree to the table. Reveals the multicultural patchwork of our nation.

Russell Sprouts
by Johanna Hurwitz; illustrated by Lillian Hoban
William Morrow, New York. 1987
Viking/Penguin, New York. 1989
Grades: 1–4

> In the "The Science Project" Russell's class is studying vitamins and plant growth. Since potatoes contain vitamin C, he chooses to sprout a potato for his project. His positive attitude that science is fun involves his whole family in his project. Poses the delightful question "Do kisses contain vitamins?"

> Reason and experiment have been indulged, and error has fled before them.
>
> — *Thomas Jefferson*
> *Notes on the State of Virginia*

THE WIZARD'S LAB

All Ages

*T*he Wizard's Lab experiments challenge students with captivating phenomena in the physical sciences including light and optics, mechanics, and electricity.

Ten of the most popular interactive table-top exhibits from the LHS Wizard's Lab provide stimulating activities in the physical sciences. Now available in booklet form, rather than the more expensive folder format of the first edition. Exhibits are easy to construct, and use common materials and equipment. Cards with cartoons instruct visitors in hands-on challenges and provide scientific background.

Exhibits include the spinning platform, solar cells and light polarizers, resonant pendula, magnets, lenses, electricity makers and the "human battery," the oscilloscope and sound, and the harmonograph. Although originally designed for use by science centers and museums, many teachers use these exhibits to create a discovery room at a school, or as table-top learning stations in a classroom.

Skills

Observing, Analyzing, Finding Patterns

Concepts

Swing Rate of a Pendulum, Resonance, Oscillating Motion, Superposition of Motion, Speed of Rotation, Angular Momentum, Magnetism, Magnetic Poles, Electricity, Batteries, Generators, Electrodes, Electrolytes, Solar Energy, Solar Cells, Series and Parallel Circuits, Light, Polarization of Light, Sound, Wave Motion, Superposition of Waves, Images, Lens Focal Length, Focal Point

Themes

Systems & Interactions, Models & Simulations, Patterns of Change, Stability, Diversity & Unity, Scale, Energy, Matter

MAKING CONNECTIONS

The inventive spirit of the Wizard's Lab is captured by some of the stories of scientists and inventors. *Chitty Chitty Bang Bang* depicts a zany professor with a magical invention that has a life of its own! Stories about Danny Dunn and Einstein Anderson, and *Dear Mr. Henshaw* introduce important concepts along with accounts of clever inventions. *Ruby Mae Has Something To Say* provides a playful rendering of a technological invention, as would any other books that involve a "Rube Goldberg-like" inventive spirit.

Stories of real inventors can be found in the biographies of Archimedes and Michael Faraday. For a real challenge to your students, you might have them read one of these biographies (or others of your choice) and invent (at least on paper) a new Wizard's Lab exhibit that might have been dreamed up by a particular inventor.

CROSS REFERENCES

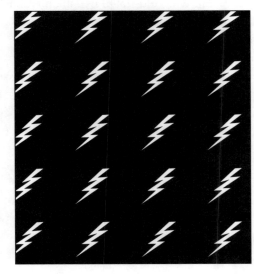

LITERATURE CONNECTIONS

Archimedes and His Wonderful Discoveries
by Arthur Jonas; illustrated by Aliki
Prentice Hall, Englewood Cliffs, New Jersey. 1963
Out of print
Grades: 3–6

This biography, after a brief biographical chapter, focuses on the problems that Archimedes solved and how his discoveries and theories are still used today. Chapter 5 describes force, fulcrum, and resistance with specific examples of levers, pulleys, a wheel and axle, inclined planes, wedges, and screws, and how these tools are based on Archimedes' discoveries. Chapter 9 describes Archimedes' contributions to the war against the Romans in which he developed use of levers, pulleys, and mirrors for offense and defense, and where he was killed. The book is full of the spirit of invention and inquiry, offering a nice example of the interrelationship between science and history.

Chitty Chitty Bang Bang: The Magical Car
by Ian Fleming; illustrated by John Burningham
Alfred A. Knopf, New York. 1964
Grades: 5–12

Wonderful series of adventures featuring a magical transforming car, an eccentric explorer and inventor, and his 8-year-old twins. Nice combination of technical and scientific information, much of it accurate, with a magical sense of how some machines seem to have a mind of their own. This one definitely does, as it flies when it encounters traffic jams, becomes a boat when the tide comes in, senses a trap, and helps catch some gangsters. A humorous and fantastic literature accompaniment to activities involving technology and inventions.

Coils, Magnets, and Rings: Michael Faraday's World
by Nancy Veglahn; illustrated by Christopher Spollen
Coward, McCann & Geoghegan, New York. 1976
Out of print
Grades: 5–8

This delightful biography of Michael Faraday, best known for his discovery of the electric generator, captures the spirit of the questioning scientist and the atmosphere of the Royal Institution, a nineteenth century institution dedicated to scientific research in all fields. The book portrays a series of fascinating experiments performed by Faraday with electricity and magnets that your students can replicate with several of the Wizard's exhibits. An excellent assignment for your students would be to have them try the "Wizard's Lab" experiments on electricity and magnetism first; then have them read the book and see if they can determine which ones Michael Faraday would have experimented with or invented.

> **In Terms of Physics**
>
> We live in a bubble chamber
> on a microsecond zigzag
> not without encounters
> and exciting crashes
> until we disappear
> or change into another
> having left remarkable
> or unremarkable designs
> our "world lines."
>
> —*Lillian Morrison*
> *Overheard in a Bubble Chamber*

> It is common sense to take a method and try it. If it fails, admit it frankly and try another. But above all, try something.
>
> — *Franklin Delano Roosevelt*

Danny Dunn and the Swamp Monster

by Jay Williams and Raymond Abrashkin; illustrated by Paul Sagsoorian

McGraw-Hill, New York. 1971

Out of print

Grades: 4–6

Danny and friends discover a superconductor, which they use in an adventure. Pages 24–26 have good descriptions of current, resistance and magnetic fields.

Dear Mr. Henshaw

by Beverly Cleary

William Morrow, New York. 1983

Grades: 4–6

A 10-year-old boy writes to his favorite author and, over time, with a growing sense of identity and self-esteem, and enhanced writing ability. A portion of the book describes how the boy sets out to catch a thief by rigging a battery-powered burglar alarm to his lunch box, and how this invention gains him respect.

Einstein Anderson Lights Up the Sky

by Seymour Simon; illustrated by Fred Winkowski

Viking Press, New York. 1982

Grades: 4–7

In Chapter 10, "The Spring Festival," the decorating committee gets help from static electricity when placing balloons for the festival.

Einstein Anderson Sees Through the Invisible Man

by Seymour Simon; illustrated by Fred Winkowski

Viking Press, New York. 1983

Grades: 4–7

In "The Huck Finn Raft Race," a class competition to design a raft involves controlling variables, even the weight of the students chosen to ride on it.

The Pet of Frankenstein

by Mel Gilden; illustrated by John Pierard

Avon Books, New York. 1988

Grades: 5–7

This book is one in an entertaining series of adventures involving Danny Keegan and his monstrous classmates. In this episode, Frankie's particular expertise is electronics, so there is a lot of it with some nicely interwoven technical language. From video games and computers to Tesla coils and robotic animals, the bizarre narrative jumps amusingly along. In the end, Frankie succeeds in creating a robotic dog from an old design of his namesake, Baron Frankenstein. Definitely not heavy reading, the book will tickle some students' imaginations and funny bones.

Ruby Mae Has Something To Say
by David Small
Crown Publishers, New York. 1992
Grades: 2–6

This zany saga traces Ruby Mae Foote's path from Nada, Texas, to the United Nations. Her message of world peace cannot be given until Billy Bob, Ruby Mae's nephew, transforms her tongue-tied and sometimes incomprehensible speech into earthshaking eloquence with a Rube Goldberg-type invention, a hat she wears called the Bobatron. The invention, though apparently non-electrical, nevertheless makes a comic connection to the technological experience students gain in *Wizard's Lab*.

LITERATURE CONNECTIONS

All new GEMS guides now contain a section on literature connections, based on the listings in this handbook. These sections list our recommendations of books that can be read in connection with the science and math activities in that guide. The listings are meant as jumping off places from which you can come up with your own favorites. There are many ways to creatively combine science and math with literature, and to weave lessons across the curriculum. Here are a few general approaches we've found useful, offered in the hope that they catalyze your own imaginative interweavings.

Reading appropriate literature before a science/math unit is a good way to motivate and excite students about the unit they are beginning. If you're presenting a unit that explores a major content area or theme—structure, for example—one way to begin is by reading several books that lay the groundwork for understanding how broad the theme is, or how it manifests itself in many different areas. Some of these areas may be more familiar and accessible to the students through literature and stories than through the scientific or mathematical territory they are about to explore.

After you do the hands-on science and math activities and any "Going Furthers" you may pursue, you can conclude the unit with several other books or stories that deepen the ideas. You could have older students write a book report highlighting how the theme of structure is explored through the science and math activities. With younger students you could introduce the theme, and discuss how a particular book illuminates it. This can help students see the relevance of structure to their lives and experiences. They also gain a sense of how structure (or any other theme or concept you choose to emphasize) in science and mathematics is echoed and amplified in art and literature.

With some units, reading several stories beforehand is a great way to help students understand what they will do in the unit. The cooperative problem-solving activities of *Group Solutions*, for example, can be complemented and strengthened by reading several books that emphasize cooperation and working together.

In the same way, kindergarten teachers who have taught *Investigating Artifacts* highly recommend reading four or five Native American myths or stories to the class before the unit. This helps younger students understand the task of making up their own stories to explain a natural event. It also greatly enriches their cultural backgrounds.

. . . AND GEMS

Many activities that connect science and mathematics to literature can be enriched by constructive interaction between students of different ages and abilities. Older students can come into a kindergarten classroom to read out loud the "myths" they have created. This is positive reinforcement for them, provides practice in oral presentation skills, and models the task wonderfully for the younger students. It can work the other way as well—groups of kindergarten students could go to a sixth grade class to present a series of bee dramas as part of their activities in *Buzzing A Hive*.

It can also be effective to read different literature books or certain sections of a specific book at appropriate times throughout a GEMS unit—interspersing the story with the science or math activity. This works especially well when the story parallels stages of the activity or relates directly to what the students actually do. The GEMS teacher's guide *Frog Math* gives explicit instructions for ways to combine Arnold Lobel's story "The Lost Button" with button sorting and classifying activities. Some of the annotations in this GEMS handbook highlight certain sections in recommended books that lend themselves to this approach. In some cases, particular chapters in longer books are especially pertinent to a particular activity. This, of course, should be done in a way that does not have a negative effect on the artistic and literary integrity of the book, and is often most effectively done with a story already familiar to the students. On page 130 a teacher details how she interweaves books with the GEMS unit *Hide A Butterfly*.

The GEMS teacher's guides that focus on mysteries are particularly "likely suspects" for literature connections. In *Fingerprinting* or *Crime Lab Chemistry*, related portions of a mystery story could be read as students advance through the activity. In the first portion the characters, like the students, might be gathering data (evidence). In subsequent portions, they might be sorting and classifying, analyzing, evaluating, narrowing down the evidence, making inferences about suspects and events, and then attempting to solve the mystery.

The GEMS crime lab and fingerprinting activities clearly lend themselves to dramatization, and several scenarios with intriguing characters appear in the guides. Other GEMS guides include drama, such as the lake animals play in *Acid Rain*, and the role playing and drama activities interwoven in *Animal Defenses*, *Buzzing A Hive*, and *Hide A Butterfly*. Other guides

Literature Connections . . . and GEMS (continued)

contain poems and legends that lend themselves to dramatization. The limericks from *Global Warming and the Greenhouse Effect* (page 112) have been dramatized in a humorous debate format. Role play and drama greatly enrich the creative writing and imaginative powers of students. Many teachers say that after role playing the writing of the students becomes richer, with more varied language and detailed description, thus building student writing confidence.

At home, sharing activities and literature can foster a strong and positive parental involvement that leads naturally to discussion and reinforcement of what students learn in the classroom. "Yeah," a youngster might say to her father after reading Dr. Seuss, "we did this with the 'Oobleck' when we experimented in the classroom but it wasn't like what happens in the book at all. We found out that sometimes it acts like a liquid and sometimes like a solid. Can we make some here, so I can show you, *please*?" Bringing home the activity and the story enriches both. The GEMS handbook *A Parent's Guide to GEMS* contains suggestions for ways parents can help foster their children's joy in learning.

The many ways science and mathematics combine with young people's literature can be very effectively integrated with writing and other impor-

tant aspects of the language arts—not to mention all the other arts! Science and mathematics activities can be intertwined with language arts in many positive ways: creative writing, journal writing, oral presentations, group reports, role playing, and dramatizations. Students can create their own books, publish reports on a scientific convention or town meeting they held during a GEMS activity, devise their own computer-generated newsletters, or make contributions to schoolwide or community publications. Many reading and dramatic activities can be enhanced by video and tape recording, computer networking, and other modern tools and technologies.

The GEMS project welcomes your suggestions and ideas and wants to add them to future editions of this handbook. Please write to us and share what you have done. Please see the suggestions form on page 412.

GEMS

Thanks!

On the Cutting Edge

Jurassic Park

For older, more mature students, many books designed for adult audiences can make excellent literary connections to scientific and mathematical themes. This is definitely true of a great deal of science fiction, of which only a few examples are included in these listings. Other novels explore controversial themes in medicine and scientific research.

One particularly striking example that is sure to get students pondering "cutting-edge" questions in science while learning more about the important mathematical strand of statistics and probability is the best-seller *Jurassic Park* by Michael Crichton (Ballantine, New York. 1993).

The plot derives from imagining the possibility that humans could re-create dinosaurs and other prehistoric animals from dinosaur and other genetic material that has been trapped in ancient amber deposits. Using this process, a secret amusement park has been set up by an iconoclastic billionaire in a remote location in Costa Rica, where scientists and math-ematicians have been flown under high security to assure that all is well before the park is opened to the public. Several youngsters are also very directly involved, adding to the book's appeal to younger readers.

Enormous disaster ensues, and be forewarned that some of the resulting scenes are quite graphic. As the tension and danger build, we learn much about certain prehistoric species, as well as a great deal about genetics, statistics and probability, computer systems, chaos, and evolution. The broader issues arising from human attempts to control biological evolution and general interference with the environment are extremely relevant today and are certain to become even more so in the future.

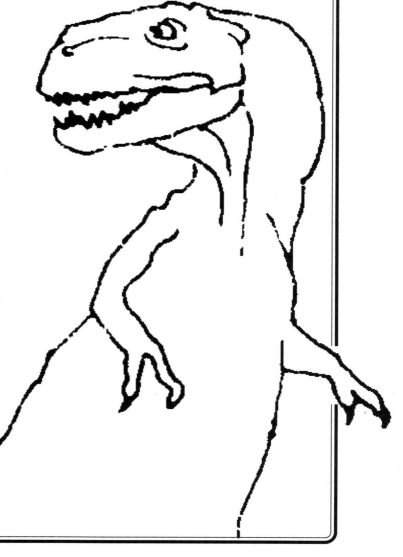

The mathematics strands that follow are listed in alphabetical order, to assist you in finding the literature connections for a particular strand. It should be noted that this is not the order in which students explore these strands, as should be evident by the listing of Algebra as the first strand. A more logical educational order would be: Geometry; Number; Measurement; Pattern; Logic; Probability and Statistics; Functions; Algebra.

It is also worth pointing out that these strands are not really separate and distinct, but overlap and intertwine with each other. Listings of math strands in other documents may differ. For example, many see "pattern" as interwoven in all of the other strands, rather than as a separate category. These strands can be helpful in analyzing the subject area, but they should be viewed flexibly and as just one possible way to comprehend mathematics as a whole.

Note: Some of the books listed are currently out of print; however, they should be readily available at your local school or public library.

Algebra → Geometry → Number → Measurement → Pattern → Logic → Probability & Statistics → Functions →

Philosophy is written in this grand book—
I mean the universe—which stands continually
open to our gaze, but it cannot be understood
unless one first learns to comprehend the
language and interpret its characters in which it
is written.

It is written in the language of mathematics,
and its characters are triangles, circles, and other
geometrical figures without which
it is humanly impossible to understand a single
word of it; without these, one is wandering
about in a dark labyrinth.

— Galileo Galilei
Il Saggiatrore

Algebra

Algebra uses **symbols to represent unknown quantities or variables**. It is often referred to as the language of mathematics and is the basis of higher mathematics.

Young students begin to use algebra when they solve equations such as 3 + □ = 7. In this case, the unknown or variable is represented by a box. Symbols or variables can be used to express any arithmetic relationship.

MAKING CONNECTIONS

There are relatively few books listed in this section, and we welcome your suggestions for future editions. Other than one very simple book that uses **symbols to represent information**, most of the books involve more sophisticated algebraic concepts, such as **exponential growth** and **sequences and series.** From these books, which allow students to explore the ideas presented by experimenting with manipulatives in the classroom, rather than writing

down or solving actual algebraic equations. Students can gain concrete experience and an intuitive understanding of algebra before moving on to more abstract processes and notation.

Many mystery thrillers are at least superficially connected to the mathematics strands of both logic and algebra, in that they require the arrangement or substitution of clues in order to solve a problem or find out something not previously known. We welcome your suggestions for literature connections that echo the search for and representation of an unknown.

CROSS REFERENCES

LITERATURE CONNECTIONS

A Grain of Rice
by Helena C. Pittman
Hastings House, New York. 1986
Bantam Books, New York. 1992
Grades: 2–5

A clever, cheerful, hard-working farmer's son wins the hand of the Emperor's daughter by outwitting the father who treasures her more than all the rice in China. Pong Lo's winning strategy is to use a mathematical ruse, asking simply for a grain of rice that is to be doubled every day for one hundred days. The book clearly illustrates exponential growth.

> Mathematics is
> the science which
> draws necessary
> conclusions.
>
> — *Benjamin Peirce*
> *Linear Associative Algebra*

The Great Adventures of Sherlock Holmes

by Arthur Conan Doyle
Viking Penguin, New York. 1990
Grades: 6–Adult

The search for, and discovery of, the unknown by putting together a number of clues to find the answer is Sherlock Holmes and algebra at their best. These, and nearly the entire canon of Holmes adventures, are classics because there is never a false step as Sherlock Holmes unravels each case logically, clearly, and cleverly.

The King's Chessboard

by David Birch; illustrated by Devis Grebu
Dial Books, New York. 1988
Grades: K–6

A too proud king learns a valuable lesson when he readily grants his wise man a special request: one grain of rice on the first square of a chessboard on the first day, two grains on the second square on the second day, four grains on the third square on the third day and so on. After several days the counting of rice grains gives way to weighing, then the weighing gives way to counting sackfuls, then to wagonfuls. The king soon realizes that there is not enough rice in the entire world to fulfill the wise man's request. This tale involves exponential growth. Students can use manipulatives in the classroom to see how quickly the rice amasses.

Melisande

by Edith Nesbit; illustrated by P.J. Lynch
Harcourt, Brace, Jovanovich, San Diego. 1989
Grades: 1–8

Princess Melisande will grow up to be bald because of a curse by an evil fairy. Upon being granted one wish, she asks for golden hair a yard long that will grow an inch every day and twice as fast when cut. Soon the princess realizes the implications of her wish. With the help of a determined godmother and a prince, order is restored. Though traditional fairy tale roles prevail, this story lends itself to an exploration of geometric progression (binomial sequence). Students can use yarn as a hands-on tool to understand how Melisande's hair grows.

The Secret Birthday Message

by Eric Carle
HarperCollins, New York. 1986
Grades: Preschool–2

Instead of a birthday package, Tim gets a mysterious letter written in code. Full-color pages, designed with cut-out shapes, allow children to fully participate in this enticing adventure, which gives an elementary look at the use of symbols to represent information. An exciting way to open a shape unit that can include a project to make shape books.

For Your Bookshelf

A New Look at a Classic Fairy Tale

BY JAINE KOPP

Though the classic fairy tale, Melisande, was written by E. Nesbit nearly one hundred years ago, it has elements that are still fresh today. Make no mistake; Melisande is a genuine fairy tale—complete with magic spells and marvels and the prerequisite marriage between a princess and a prince. But it doesn't stop there! Ms. Nesbit weaves mathematics into her tale of Melisande, the young princess, who is cursed by an evil fairy to grow up to be bald. Upon being granted one wish, Melisande says, "I wish I had golden hair a yard long, and

that it would grow an inch every day, and grow twice as fast every time it was cut, and . . ." Luckily her father, a skilled mathematician, interrupts her wish before she finishes as he sees its consequences.

This wish alone can be used to begin mathematical investigations by students across the grade levels. Students in grades K–5 can work on the unifying idea of quantifying through a unit on the process of measurement. For example, primary students can cut string into a piece that they estimate to be one yard long—the original length of hair Melisande wished for. They then can compare that to a piece of string that is one yard long. Next, using the one-yard long piece of string, they can compare that to their hair length and then to objects in the classroom. They could also use nonstandard units (unifix cubes, toothpicks, hand spans, etc.) to measure the one-yard length of string. Upper elementary students can use the standard length of Melisande's hair to apply meaning to a yard as they measure the classroom and outdoor areas in yards. In middle school, students can go further to explore how long Melisande's hair would be in a week, two weeks, a month, etc. IF she did not cut it. Now enter the variable of having it grow twice as fast every time it was cut. This exploration involves students in the unifying idea of proportional relationships that integrates the strands of algebra, functions, discrete math, measurement, and number.

However, the tale does not end here. Enter Prince Florizel. In his attempt to solve Melisande's dilemma by cutting off her hair in a clever manner, he creates yet another. Now instead of her hair growing in length, Melisande grows in height! She is soon too large to fit indoors. She becomes sad since she is separated from those she loves by her enormous height. Nonetheless, Melisande is able to use her size to save the kingdom from an attack by an invading army. Eventually, with the help of Florizel, she returns to her proper height though her hair begins to grow again. Luckily, through the suggestion of the King's godmother, a unique solution— employing a scale and the sound judgment of Florizel—is used to return Melisande to her proper size with a head of hair five feet and a quarter inches long. The Princess bestows a hundred kisses upon Florizel and not surprisingly they are married the very next day!

Melisande's growth in height and the ultimate solution to the evil spell are also worthy of further investigations. In addition, the logic used to solve problems throughout the story illustrates divergent thinking and encourages students to make predictions. In fact, you may choose to stop the story at various points and have your students come up with possible solutions. Though I wouldn't advocate fairy tales for social and political lessons, I appreciate their ability to transport us into the world of fantasy and fairy lore as well as to provide a magic carpet to fly into mathematical investigations.

Jaine Kopp teaches for the Mathematics Education Project at the Lawrence Hall of Science.

Functions

Functions are **relationships** where each member of one set is paired with just one member from a second set.

Often these **paired numbers are organized in tables or graphed on a coordinate plane**.

An example of a simple function is the relationship between a person/persons and the number of eyes they have (one person has two eyes, two people have four eyes, three people have six eyes …). Younger students begin to explore these types of functions when they begin multiplication. An example of a more complex function is the algebraic relationship between two variables, x and y, in the equation $y = x^2$. In this case, if the number 1 were substituted in for x, then $y = 1$. If $x = 2$, then $y = 4$, and if $x = 3$, then $y = 9$, and so on. In this equation, any value can be substituted for "x", and then "y" can be determined. Businesses uses similar functions to determine things such as maximum profits, efficiency of personnel, time and production parameters. Other real-world examples that relate to exponential functions include compound interest and radioactive decay.

MAKING CONNECTIONS

The books listed here reflect a number of aspects of functions. Two books focus on common objects that occur in multiples, such as wheels on a tricycle or legs on a cat, and can be used to illustrate relationships like: one tricyle/three wheels; two tricyles/six wheels, etc. A basic example of a cause and effect function is seen in *Why Mosquitoes Buzz in People's Ears*. Several books relate to another type of function, the maximum point (similar to the idea of threshold limit), as depicted by a boat sinking as it takes on too much weight, a bed collapsing, and a beet finally being pulled out of the ground. A *Grain of Rice* and *A King's Chessboard* explore functions of exponential growth, while *Melisande* (see article on page 265) relates to series and sequences functions.

CROSS REFERENCES

Literature Connections

Each Orange Had 8 Slices
by Paul Giganti, Jr.; illustrated by Donald Crews
Greenwillow Books, New York. 1992
Grades: Preschool–3

> Through a series of questions, you count things such as balloons, seeds, animals, and wheels in several different ways by multiples. Can be used as a very basic introduction to functions: Students can explore the relationship between the number of tricycles and the wheels on the tricycles.

A Grain of Rice
by Helena C. Pittman
Hastings House, New York. 1986
Bantam Books, New York. 1992
Grades: 2–5

> A clever, cheerful, hard-working farmer's son wins the hand of the Emperor's daughter by outwitting her father who treasures her more than all the rice in China. Pong Lo's winning strategy is to use a mathematical ruse, asking simply for a grain of rice that is to be doubled every day for one hundred days. The book clearly illustrates an exponential function.

The King's Chessboard
by David Birch; illustrated by Devis Grebu
Dial Books, New York. 1988
Grades: K–6

> A too proud king learns a valuable lesson when he readily grants his wise man a special request: one grain of rice on the first square of a chessboard on the first day, two grains on the second square on the second day, four grains on the third square on the third day and so on. After several days the counting of rice grains gives way to weighing, then the weighing gives way to counting sackfuls, then to wagonfuls. The king soon realizes that there is not enough rice in the entire world to fulfill the wise man's request. This tale involves a binomial sequence. Students can use real objects to see how quickly the rice amasses.

Melisande
by Edith Nesbit; illustrated by P.J. Lynch
Harcourt, Brace, Jovanovich, San Diego. 1989
Grades: 1–8

> Because of a curse, Princess Melisande will grow up to be bald. Upon being granted a wish, she asks for golden hair a yard long that will grow an inch every day and twice as fast when cut. Soon the princess realizes the implications of her wish. With the help of a determined godmother and a prince, order is restored. Though traditional fairy tale roles prevail, this story lends itself to an exploration of geometric progression (binomial sequence). Students can use yarn as a hands-on tool to understand how Melisande's hair grows.

The Napping House

by Audrey Wood; illustrated by Don Wood

Harcourt, Brace, Jovanovich, San Diego. 1984

Grades: K–3

A flea atop a number of sleeping creatures causes a commotion, with just one bite. The simple, repetitive pattern is contagious. Reading this book is a fun and easy way to illustrate the idea of the maximum point (or threshold limit), more popularly known as "the straw the broke the camel's back."

The Turnip: An Old Russian Folktale

by Katherine Milhous and Alice Dalgiesh; illustrated by Pierr Morgan

Philomel Books, New York. 1990

Grades: K–3

One of Dedoushka's turnips grows to such an enormous size that the whole family, including the dog, cat, and mouse, is needed to pull it up. Shows that even small beings can be important in relation to larger patterns, and illustrates a relation between the two.

Who Sank the Boat?

by Pamela Allen

Sandcastle Books/Putnam & Grosset, New York. 1990

Grades: K–2

Who causes the boat to sink when five animal friends of varying sizes go for a row? Reading this book is a fun and easy way to illustrate the idea of the maximum point (or threshold limit), more popularly known as "the straw the broke the camel's back."

Why Mosquitoes Buzz in People's Ears

by Verna Aardema;

illustrated by Diane and Leo Dillon

Dial Books, New York. 1975

Grades: K–5

A mosquito teases an iguana. The chain reaction causes a sense of panic among the animal community that ends with the killing of one of mother's owls. This story is a simple example of a cause and effect function.

On the Threshold

The concept of threshold is important throughout mathematics and science and it is beautifully illustrated in literature. A number of stories and folktales involve a plot where many people must all pull to dislodge an enormous turnip, or a whole menagerie of animals must climb in a boat before it capsizes, or it takes one hundred monkeys to cause something to happen. It is not the last tiny field mouse who pulls the turnip; nor is it the 100th monkey that does the work, rather it is the mouse or the last monkey that causes a threshold to be attained.

Choose your favorite stories that demonstrate the idea of a threshold, and seize the opportunity to help your students understand and articulate it.

Geometry

eometry is the study of lines, angles, shapes and relationships.

As such, it is ever-present in our everyday lives. For example, food items come in a variety of shapes and the relationship between their shapes has an impact on how they can best be stored in a limited space, such as a cabinet or shelf. Street signs are made in different shapes, railroad tracks illustrate a set of parallel lines, both housing construction and interior decoration make use of repeated shapes and interrelated shapes and structures.

Young children's first experience with geometry involves playing with objects in their physical world. Through that experience they develop a vocabulary for common geometric shapes and learn to identify each one.

MAKING CONNECTIONS

There are many engaging books listed here that invite the reader to find and identify **shapes**. There are also two books that focus on shapes in the context of quilt-making that can be used as a springboard to creating quilt squares out of a variety of paper or cloth shapes. In addition to two-dimensional shapes, (circles, polygons) students can explore three-dimensional ones (cones, rectangular prisms). Books about the **structure** of three-dimensional things include topics such as bridges and buildings. Other books, such as *Sadako and the Thousand Paper Cranes*, that contain origami figures are included because origami transforms a two-dimensional piece of paper into a three-dimensional object.

There are also books on many other topics in geometry, such as **symmetry** (fold a shape in half and both sides match exactly), **reflection** (mirror image), and even a user-friendly book on **topology** (the twisting and bending of shapes in space). This is by no means an exhaustive list; these were the topic areas in which we found quality material. Any additions would be greatly appreciated.

CROSS REFERENCES

LITERATURE CONNECTIONS

Anno's Math Games III

by Mitsumasa Anno

Philomel Books/Putnam & Grosset, New York. 1991

Grades: 4–10

> Picture puzzles, games, and simple activities introduce the mathematical concepts of abstract thinking, circuitry, geometry, and topology. The book invites active participation. An exploration of triangles includes origami shapes, while a section on ever-popular mazes encourages logical thinking.

Bridges

by Ken Robbins

Dial Books, New York. 1991

Grades: K–5

> From delicate webs of steel spanning a vast river to stone arches reaching over a highway, bridges expand our world by joining one place with another. This book of hand-tinted photographs of bridges includes many types with descriptions of their design and use.

Flatland: A Romance of Many Dimensions

by Edwin A. Abbott

Viking/Penguin, New York. 1984

Princeton University Press, Princeton, New Jersey. 1991

Grades: 8-Adult

> Geometry serves as a vehicle for social satire in this famous and complex Victorian classic. The hero, A. Square, describes the consistent two-dimensional world of Flatland to his spherical visitor from Spaceland, and in so doing manages to combine many geometric concepts with incisive wit and social commentary that supports equal educational opportunity for women and the "lower classes," as well as tolerance for social non-conformity. The book explores the use of analogy in visualizing higher dimensions, now important in theoretical physics. This emphasis on visualization also relates to the emergence of modern computer graphics.

Grandfather Tang's Story

by Ann Tompert; illustrated by Robert A. Parker

Crown, New York. 1990

Grades: K–5

> Grandfather tells Little Soo a story about shape-changing fox fairies who try to outdo each other until a hunter brings danger to both of them. The seven shapes that grandfather uses to tell the story are the pieces of an ancient Chinese puzzle, a tangram. Students can make their own tangrams, replicating the animals in the story or creating their own. This book is a wonderful and powerful way to connect mathematics to literature because it embodies the connection and because creating and solving tangrams is an involving activity for all ages.

The geometrical mind is not so closely bound to geometry that it cannot be drawn aside and transferred to other departments of knowledge. A work of morality, politics, criticism, perhaps even eloquence will be more eloquent, other things being equal, if it is shaped by the hand of geometry.

— *Bernard Le Bovier de Fontenelle*
Preface sur l'Utilite des Mathematiques et las Physique

Jim Jimmy James

by Jack Kent
Greenwillow Books/William Morrow, New York. 1984
Out of print
Grades: K–2

One boring rainy day, Jim Jimmy James makes friends and plays with his shadow. A very elementary look at the concept of reflection. As a follow-up, children can partner with a friend and play shadow games with each other. Shadows and reflections are among the earliest phenomena related to shape and geometry that children experience. (Note: Some of the illustrations in the book are not accurate reflections.)

The Keeping Quilt

by Patricia Polacco
Simon & Schuster, New York. 1988
Grades: K–5

A homemade quilt ties together the lives of four generations of an immigrant Jewish family, remaining a symbol of their enduring love and faith. Strongly moving text and pictures. A resource to begin a quilt project. Quilts are creative real-life examples of fitting shapes into a defined space, including tessellations, the intriguing mathematical and creative art of *exactly* fitting similar shapes into a defined space. Sidney Taylor Award winner.

My Cat Likes to Hide in Boxes

by Eve Sutton; illustrated by Lynley Dodd
Scholastic, New York. 1973
Grades: K–2

Delightful book with rhymes about cats all over the world and "my cat" who likes to hide in boxes! The predictable pattern encourages reading participation. The idea of boxes and using shapes as homes is an early connection to structure and geometry.

Opt: An Illusionary Tale

by Arline and Joseph Baum
Viking Penguin, New York. 1987
Grades: 2–6

A magical tale of optical illusions in which objects seem to shift color and size while images appear and disappear. You are an active participant in this book as you are guided through the land of Opt. Included are explanations of the illusions and information to assist readers in making their own illusions!

The Paper Airplane Book

by Seymour Simon; illustrated by Byron Barton

Viking Press, New York. 1971

Grades: 3–6

> A user-friendly book on the aerodynamics of airplanes, complete with instructions on how to construct paper airplanes. Emphasis on the structure of airplanes and how changes in structure/shape impact the forces in flight. Additional experiments are included.

The Patchwork Quilt

by Valerie Flournoy; illustrated by Jerry Pinkney

Dial Books, New York. 1985

Grades: K–5

> Using scraps cut from the family's old clothing, Tanya helps her grandmother piece together a quilt of memories. When Grandma becomes ill, Tanya's family also gets involved in the project and they work together to complete the quilt. Quilts, like geometry, are fascinating explorations of shapes and how they fit together.

The Phantom Tollbooth

by Norton Juster; illustrated by Jules Feiffer

Random House, New York. 1989

Grades: 2–8

> Milo has mysterious and magical adventures when he drives his car past The Phantom Tollbooth and discovers The Lands Beyond. On his journey, Milo encounters amusing situations that involve numbers, geometry, measurement, and problem solving. The play on words in the text is delightful.

Round Trip

by Ann Jonas

Greenwillow Books/William Morrow, New York. 1983

Grades: K–3

> Illustrated solely in black and white, this story of a trip between the city and the country is read at first in the standard way, then, on reaching the end, the book is flipped over as the story continues. Lo and behold, the illustrations turned upside down are transformed to depict the new scenes of the story. The strong black/white contrast helps provide a startling demonstration of the ways shapes and images fit into each other and can change, depending on one's perspective.

Rubber Bands, Baseballs and Doughnuts: A Book about Topology

by Robert Froman; illustrated by Harvey Weiss

Thomas Y. Crowell, Minneapolis. 1972

Out of print

Grades: 4–8

> An introduction into the world of topology through active reader participation. The activities provide concrete examples and insights into abstract concepts.

The perfection of mathematical beauty is such . . . that whatsoever is most beautiful and regular is also found to be most useful.

> — *D'Arcy Wentworth Thompson*
> *On Growth and Form*

— Geometry —

Sadako and the Thousand Paper Cranes
by Eleanor Coerr; illustrated by Ronald Himler
Dell Books, New York. 1977
Grades: 3–6

In this true story a young Japanese girl is dying of leukemia as a result of radiation from the bombing of Hiroshima. According to a Japanese tradition, if she could fold 1,000 paper cranes, the gods would grant her a wish and make her well. But, she had folded only 644 paper cranes before she died. In her honor, a Folded Crane Club was organized and each year on August 6, members place thousands of cranes beneath her statue to celebrate Peace Day. The moving story can introduce a class origami project to make 1,000 cranes or other origami figures, and of course connects strongly to social studies and current events issues.

Block City

What are you able to build with your blocks?
Castles and palaces, temples and docks.
Rain may keep raining, and others go roam,
But I can be happy and building at home.

Let the sofa be mountains, the carpet be sea,
There I'll establish a city for me;
A kirk and a mill and a palace beside,
And a harbor as well where the vessels may ride.

Great is the palace with pillar and wall,
A sort of a tower on top of it all,
And steps coming down in an orderly way
To where my toy vessels lie safe in the bay.

This one is sailing and that one is moored:
Hark to the song and the sailors on board!
And see on the steps of the palace, the kings
Coming and going with presents and things!

Now I have done with it, down let it go!
And all in a moment the town is laid low.
Block upon block lying scattered and free,
What is there left of my town by the sea?

Yet as I saw it, I see it again,
The kirk and the palace, the ships and the men,
And as long as I live and where'er I may be,
I'll always remember my town by the sea.

—*Robert Louis Stevenson*

The Secret Birthday Message
by Eric Carle
Harper & Row, New York. 1986
Grades: Preschool–2

Instead of a birthday package, Tim gets a mysterious letter written in code. Full-color pages, designed with cut-out shapes, allow children to fully participate in this enticing adventure. This book could serve as an exciting way to launch a series of lessons on shapes, which could also include student projects of making "shape books."

The Shapes Game
by Paul Rogers; illustrated Stan Tucker
Henry Holt & Co., New York. 1989
Grades: Preschool–2

Fun-to-say riddles and pictures that are kaleidoscopes of brilliant colors take young children from simple squares and circles through triangles, ovals, crescents, rectangles, diamonds, spirals, and stars.

Shapes, Shapes, Shapes
by Tana Hoban
Greenwillow/William Morrow, New York. 1986
Grades: Preschool–5

Color photographs of familiar objects, such as a chair, barrettes, and a manhole cover, are a way to study round and angular shapes.

Spaces, Shapes and Sizes

by Jane J. Srivastava; illustrated by Loretta Lustig

Thomas Y. Crowell, Minneapolis. 1980

Grades: 1–6

> This inviting and well-presented nonfiction book about volume shows the changing forms and shapes a constant amount of sand can take. The book includes estimation activities, an investigation of volume of boxes using popcorn, and a displacement activity. The reader will want to try the activities listed.

The Tipi: A Center of Native American Life

by David and Charlotte Yue

Alfred A. Knopf, New York. 1984

Grades: 5–8

> This excellent book describes not only the structure and uses of tipis, but Plains Indian social and cultural life as well. Some of the cultural language and oversimplification are less vital than they might be, but it is written in an accessible style. There are good charts, exact measurements, and information on the advantages of the cone shape. The central role played by women in constructing the tipi and in owning it are discussed. While this book includes some mention of the negative consequences of European conquest, noting that in some places tipis were outlawed, it is weak in this important area, and should be supplemented with other books.

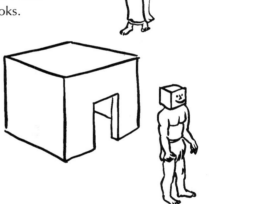

The Village of Round and Square Houses

by Ann Grifalconi

Little, Brown & Co., Boston. 1986

Grades: K–4

> A grandmother explains to her listeners why the men live in square houses and the women live in round ones in their African village on the side of a volcano. The village of Tos really exists in the remote hills of the Cameroons. This book can begin an exploration of shape and structure. Caldecott honor book.

Wrinkle in Time

by Madeleine L'Engle

Fararr, Strauss & Giroux, New York. 1962

Dell Books, New York. 1976

Grades: 7–12

> In this popular and imaginative novel, Meg accidentally transports herself to a two-dimensional planet and finds out what it must be like to be two-dimensional. On pages 73–76, there are explanations of dimensions going from two-dimensional to four-dimensional, including discussion of a **tesseract**, which is the mathematical term used to describe the generalization of a cube to four dimensions.

Logic

Logic is making good sense out of something in an organized fashion.

In mathematics, logic begins with **sorting** and **classifying**. It includes all types of **problem solving**. The methods used to solve problems illustrate different types of **logical thinking**, from applying a known principle to a new situation to making **inferences** from everyday experiences. Logic is used by everyone on a daily basis as they encounter real world problems. At times the logic used is very informal such as estimating the number of baskets of strawberries needed for eight adults. On the other hand, the logic can be very formal and follow a clear strategy or plan, such as developing a marketing plan for a new product. Logical thinking takes place on many levels, from elementary to highly complex.

MAKING CONNECTIONS

The book listings under Logic are divided into two main headings: **Sorting and Classifying**; **Problem Solving**; and **Logical Thinking**.

Books involving **Sorting and Classifying** lend themselves to investigations of attributes (characteristics). Young children begin sorting and classifying with real objects and sort them by concrete attributes such as color, shape, and size. *Alligator Shoes* is one good example and can be used as a springboard for students to sort and classify a real-life object—their own shoes. As students progress in their classification skills, they can sort by more than one attribute and/or use more sophisticated categories. Books by Ruth Heller, such as *Chickens Aren't The Only Ones*, lend themselves to more complex sorts, such as by the number of legs an animal has, its environment, and so on.

Books involving **Problem Solving and Logical Thinking** encourage students to use a variety of logic skills. Students must analyze what is known and think ahead about what might be. These books range from simple problem solving (*Miss Nelson Is Missing*) to more complex problems that require the use of divergent thinking skills, with less apparent, sometimes hidden or surprising solutions that require the problem solver to adopt perspectives different from the most evident. The solutions to *Stories to Solve Folktales From Around the World*, for example, can be found only by looking beneath the surface, perhaps considering mutiple meanings of a key word. Other books, such as *The Westing Game*, illustrate the process of gathering information, making inferences, and applying knowledge to a real world situation, thus representing a more complex process of logic. The solution of many mysteries also is revealed through a process of logical thinking.

CROSS REFERENCES

LITERATURE CONNECTIONS

SORTING AND CLASSIFYING

Alice's Adventures in Wonderland
by Lewis Carroll
Viking Penguin, New York. 1984
Grades: 4–Adult

Logic is turned on its head during Alice's adventures in Wonderland. Nearly every page contains some twist on what we usually perceive and expect to happen. Logical thinking in all its paradoxes abounds in this classic as the mathematican-author has a field day caricaturing the processes of human thought. Especially hilarious, and a test of logical thinking, is Alice's meeting with the caterpillar in Chapter 3.

"I know what you are thinking about," said Tweedledum; "but it isn't so, nohow."

"Contrariwise," said Tweedledee, "if it was so, it might be; and if it were so, it would be; but as it isn't, it ain't. That's logic."

—*Lewis Carroll*
Through the Looking Glass

Alligator Shoes

by Arthur Dorros
E.P. Dutton, New York. 1982
Grades: K–2

Being locked in a shoe store by mistake is Alvin Alligator's dream come true. He tries on an endless variety of footwear, but his decision about what footgear he prefers will surprise and delight your students. Great to read before having students use their own shoes to provide the data for a real-life graph.

Amos and Boris

by William Steig
Fararr, Strauss & Giroux, New York. 1971
Penguin Books, New York. 1977
Grades: K–4

The mouse Amos and the whale Boris discover their common links when Boris returns Amos to land after he almost drowns at sea. After many years of mousing about, Amos is able to return the favor and rescue Boris when he is stranded by a tidal wave. A tender and comical story of friendship enriched by sophisticated vocabulary. The unusual pairing of animals can lead to a mammal classification activity.

Aunt Ippy's Museum of Junk

by Rodney A. Greenblat
HarperCollins, New York. 1991
Grades: K–2

A brother and sister visit their ecology-minded Aunt Ippy and her world-famous Museum of Junk, which includes treasures such as a barrel of one-of-a-kind shoes and a sack of clocks. This story can inspire a class to begin a collection of "junk" boxes, an activity that can raise awareness of recycling and reusing materials in new and different ways. Such a collection also provides new math manipulatives for teaching many concepts, including sorting and classifying.

The Big Orange Splot

by Daniel Manus Pinkwater
Scholastic, New York. 1977
Grades: K–3

After a seagull drops a can of orange paint on Mr. Plumbean's house, he transforms it into his dream house, much to the horror of his neighbors who prefer a street where all the houses are the same. Slowly, all the other houses take on unique qualities reflecting their owner's dreams. Students can create their own dream houses and sort a variety of ways.

Chickens Aren't the Only Ones

by Ruth Heller
Grosset & Dunlap, New York. 1981
Grades: Preschool–5

> Many animals lay eggs and their eggs and egg-laying behaviors differ. The brightly colored, almost 3-D, illustrations depict the diversity of all egg-laying (oviparous) animals. Students could bring in a variety of eggs or make their own paper eggs to sort and classify.

A House Is a House for Me

by Mary Ann Hoberman; illustrated by Betty Fraser
Viking Penguin, New York. 1978
Grades: K–3

> The dwellings of various animals, peoples, and things such as a shell for a lobster or a glove for a hand are listed in rhyme. The concept of diversity is cleverly illustrated as the notion of a "house" is explored. Students can make replicas of their own houses or collect pictures of dwellings to classify. Hoberman's poem "Cricket" is on page 30 of this handbook.

People

by Peter Spier
Doubleday, New York. 1980
Grades: Preschool–6

> The differences between the billions of people that populate the earth are celebrated: different noses, different clothes, different customs, different religions, different pets, and so on. You are left with a feeling of awe and respect for the differences, and enhanced appreciation for the very basic similarities that all humans share.

The Reason for a Flower

by Ruth Heller
Grosset & Dunlap, New York. 1983
Sandcastle/Putnam & Grosset, New York. 1992
Grades: Preschool–5

> This beautifully illustrated book, with rhyming text, shows the diverse forms flowering plants take. It focuses on the function of seeds and pollen, different methods of seed dispersal, and some examples of atypical flowers. Using real seeds, plants and flowers, students can sort and classify by various characteristics.

Shoes

by Elizabeth Winthrop; illustrated by William Joyce
Harper & Row, New York. 1986
Grades: Preschool–2

> A survey of the many kinds of shoes in the world concludes that the best of all are the perfect natural shoes that are your feet. Great to read before doing a survey of shoes or sorting and classifying a group of real shoes. Can be used to generate data that can be graphed about the size of feet or left- or right-handedness.

Logical consequences are the scarecrows of fools and the beacons of wise men.

— *Thomas Huxley*
Animal Automatism

I see a certain order in the universe, and math is one way to make it visible.

— May Sarton
As We Are Now

Sylvester and the Magic Pebble

by William Steig

Simon & Schuster, New York. 1969

Grades: Preschool–3

One day Sylvester finds a magic pebble that has the power to grant all wishes. Sylvester uses the pebble to make a wish that keeps him separated from his parents until the story's end. Extension activities can include making a collection of pebbles to sort, classify, and graph. Students can also write about a wish they might make if they found a magic pebble. Caldecott award winner.

Through the Looking Glass

by Lewis Carroll

Viking Penguin, New York. 1984

Grades: All Ages

Logic is turned on its head during Alice's further adventures in Wonderland. She enters Wonderland through a mirror and finds a world completely opposite of the one she left behind. Nearly every page contains some twist on what we usually perceive and expect to happen.

Whose Hat Is That?

by Ron Roy; photographs by Rosemarie Hausherr

Clarion Books/Ticknor and Fields, New York. 1987

Grades: Preschool–3

The appearance and function of eighteen types of hats, including a top hat, a jockey's cap, and a football helmet, are examined. At the same time, the rainbow diversity of the people under the hat is celebrated.

PROBLEM SOLVING AND LOGICAL THINKING

Abel's Island

by William Steig

Farrar, Straus & Giroux, New York. 1976

Grades: 3–5

Abel is an urban mouse who suddenly finds himself stranded on a river island. As time goes by, he discovers skills and talents that help him think of ways to cross the river and return home. Abel solves the problems of daily survival and the ultimate river crossing. Newbery honor book.

Doctor DeSoto

by William Steig

Farrar, Straus & Giroux, New York. 1982

Grades: K–3

Doctor DeSoto, a dentist mouse, does very good work even on very large animals. When a fox comes in with a horrid toothache, he and his wife have to do some fast thinking to figure out a way to fix the tooth and to stay alive! Delightful illustrations accompany the text. This book offers a good example of divergent thinking and creative problem solving for younger students.

The Eleventh Hour: A Curious Mystery

by Graeme Base
Harry N. Abrams, New York. 1989
Grades: 3–8

> An elephant's eleventh birthday party is marked by eleven games preceding the banquet to be eaten at the eleventh hour, but when the time to eat arrives the birthday feast has disappeared. Rhyming text and gloriously detailed illustrations contain cryptic clues and hidden messages to keep sleuths searching for the thief. Great book for developing visual-discrimination and logical-thinking skills.

The Great Adventures of Sherlock Holmes

by Arthur Conan Doyle
Viking Penguin, New York. 1990
Grades: 6–Adult

> The search for, and discovery of, the unknown by putting together a number of clues to find the answer is Sherlock Holmes and algebra at their best. These, and nearly the entire canon of Holmes adventures, are classics because there is never a false step as Sherlock Holmes unravels each case logically, clearly, and cleverly.

If You Give a Moose a Muffin

by Laura J. Numeroff; illustrated by Felicia Bond
HarperCollins, New York. 1991
Grades: Preschool–3

> Chaos can ensue if you give a moose a muffin and start him on a cycle of urgent requests. A great way to introduce young children to the concept of a cycle and logical progression.

If You Give a Mouse a Cookie

by Laura J. Numeroff; illustrated by Felicia Bond
HarperCollins, New York. 1985
Scholastic, New York. 1989
Grades: Preschool–3

> A mouse is likely to make a cycle of requests after you give him a cookie. "He's going to ask for a glass of milk … and that's only the beginning!" Great introduction to the logic of cause and effect. The "Big Book" edition includes a teaching guide.

Kimako's Story

by June Jordan; illustrated by Kay Burford
Houghton Mifflin, Boston. 1981
Grades: 2–3

> Kimako works poetry puzzles indoors and has outdoor adventures with the dog she is taking care of for a friend. The story includes poems for the reader to complete and a map of Kimako's city paths. The book promotes logical-thinking skills. The text reflects the cultural awareness of the author, a leading African-American studies professor and poet.

The Man Whose Name Was Not Thomas

by M. Jean Craig; illustrated by Diane Stanley
Doubleday, New York. 1981
Out of print
Grades: K–3

A country lad is neither a carpenter nor a farmer nor a bricklayer nor a fisherman, but "something else" instead. The text tells what does *not* happen and sparks you to guess what does happen. The full-color illustrations provide the answers! A clever story and guessing game that encourages logical thinking. Keep in mind that the book contains traditional gender roles.

Many Moons

by James Thurber; illustrated by Louis Slobodkin
Harcourt, Brace & World, New York. 1943
Grades: K–5

A little princess wants the Moon. Neither the King, the Lord High Chamberlain, the Royal Wizard, the Royal Mathematician, nor the Court Jester are able to solve the problem—it takes the 10-year old princess to figure out how to get the Moon.

Ming Lo Moves the Mountain

by Arnold Lobel
Greenwillow Books/William Morrow, New York. 1982
Grades: K–2

After offering many unhelpful suggestions, a wise man finally helps Ming Lo move the mountain that drops stones on his house and always casts a dark shadow. The special "dance" he and his wife must do to move the mountain actually moves their house away from the mountain. The solution is usually obvious for young children. This is a good first problem-solving story.

Miss Nelson Is Missing!

by Harry Allard and James Marshall
Houghton Mifflin, Boston. 1977
Scholastic, New York. 1987
Grades: K–2

When the behavior of Miss Nelson's class reaches an unacceptable level, action is taken. Enter the substitute, Miss Viola Swamp, who means business. The class actually begins to miss Miss Nelson and to speculate on all the likely or least likely scenarios to explain her absence. As for who Viola Swamp really is . . .

> **A**s I was going up the stair
> I met a man who wasn't there.
> He wasn't there again today
> I wish, I wish he'd stay away.
> — *Hugh Mearns*
> *The Psychoed*

More Stories to Solve:
Fifteen Folktales from Around the World
by George Shannon; illustrated by Peter Sis
Greenwillow Books/William Morrow, New York. 1989
Grades: 3–8

> This further collection of brief folktales from a variety of cultures invites
> you to solve a mystery or problem before the resolution is presented.
> Great book for developing divergent-thinking and problem-solving skills.

The Phantom Tollbooth
by Norton Juster; illustrated by Jules Feiffer
Random House, New York. 1989
Grades: 2–8

> Milo has mysterious and magical adventures when he drives his car past
> the Phantom Tollbooth and discovers The Lands Beyond. On his journey,
> Milo encounters numbers, geometry, measurement, and problem solving in
> amusing situations. The play on words in the text is delightful.

The Pied Piper of Hamelin
by Mercer Mayer
Macmillan, New York. 1987
Grades: Preschool–6

> After the pied piper pipes the village free of rats, the villagers refuse to pay
> him for the service. So he pipes away their children as well. A classic
> illustration of an ecosystem gone out of whack, as well as a challenge
> requiring creative problem-solving skills.

Sideways Arithmetic from Wayside School
by Louis Sachar
Scholastic, New York. 1989
Grades: 3–5

> This series of problems and puzzles uses "sideways arithmetic" to stimulate
> divergent thinking skills and the funny bone. Sideways arithmetic
> approaches arithmetic as you have never seen before.

Socrates and the Three Little Pigs
by Tuyosi Mori; illustrated by Mitsumasa Anno
G.P. Putnam, New York. 1986
Grades: 4–8

> Socrates, a wolf, attempts to catch one of three pigs for his wife's dinner.
> These three pigs collectively own five cottages. With the help of his frog-
> friend, the mathematician Pythagoras, Socrates tries to determine the
> possible cottages the pigs might be in. As the story unfolds, the illustrations
> show the many possible locations of the pigs, and in doing so, visually and
> clearly show the difference between permutations and combinations. This
> type of math, known as combinatorial analysis, forms the basis for com-
> puter programming and problem-solving and this connection is explained
> on a more advanced level in the back of the book. This book is also
> published under the title: *Anno's Three Little Pigs.*

Good, too, Logic,
of course; in itself,
but not in fine
weather.

— *Arthur Hugh Clough*
Dipsychus

Stories to Solve:
Folktales from Around the World
by George Shannon: illustrated by Peter Sis
Greenwillow Books/William Morrow, New York. 1985
Grades: 3–8

These brief folktales from a variety of cultures invite you to solve a mystery or problem before the resolution is presented. Great book for developing divergent-thinking and problem-solving skills.

The Stupids Have a Ball
by Harry Allard; illustrated by James Marshall
Houghton Mifflin, Boston. 1978

The Stupids Step Out
by Harry Allard; illustrated by James Marshall
Houghton Mifflin, Boston. 1974

The Stupids Die
by Harry Allard; illustrated by James Marshall
Houghton Mifflin, Boston. 1981
Grades: K–3

These three books about the Stupid family have children laughing with delight as they see the Stupid children do such silly things as slide up a banister or take a bath fully clothed in an empty bath tub. These books promote logical-thinking skills as kids find all the outrageous things and suggest ways to correct them.

The Westing Game
by Ellen Raskin
Avon, New York. 1984
Grades: 6–10

The mysterious death of an eccentric millionaire brings together an unlikely assortment of 16 beneficiaries. According to instructions contained in his will, they are divided into 8 pairs, and given a set of clues to solve his murder and thus claim the inheritance. Mysteries like this one are an involving way to promote logical thinking skills. Newbery award winner.

Whose Footprints?
by Masayuki Yabuuchi
Philomel Books, New York. 1983
Grades: K–4

A good guessing game that depicts the footprints of a duck, cat, bear, horse, hippopotamus, and goat. Drawing inferences as to which animal made which prints provides very basic experience in the logical process of making inferences from evidence.

Winnie-The-Pooh
by A.A. Milne; illustrations by Ernest H. Shepard
E.P Dutton, New York. 1926
Grades: Preschool–4

This classic children's book describes the adventures of Pooh, Christopher Robin, and his animal friends. Of particular interest for logical thinking is Chapter 1, in which bees are encountered and Pooh uses a "logical" process to conclude that a buzzing noise is made by bees, and bees mean honey, and honey is for Pooh!

Everybody Needs A Rock
Go Out and Find Yours!

Byrd Baylor's book, *Everybody Needs A Rock*, is bound to get you in a rock-collecting frame of mind. As you go out to find one, you're likely to be selective, reflecting the ten rules in the book, which include: the special rock should be able to fit in your hand; bounce in your pocket as you run; be just the right color; and you need not ask anyone to help you choose a rock because your special rock will make itself known to you alone.

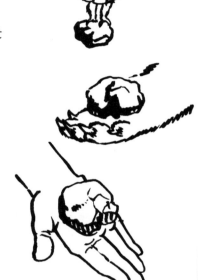

Whether you go out rock hunting as a class of students, or as a family, the possibilities for further investigation open up like a sparkling crystal geode. Start with observation of rocks and their surroundings, including close attention to detail, to acquire a rock-solid and very useful scientific and mathematical skill.

Having a class collection of rocks is a natural pathway to all sorts of sorting and classification activities. Observe the variety of colors. Put the rocks in water and watch for any color changes. Study the shapes of your rocks to see whether or not they will roll, then try to rock and roll! Put the rocks in a bag and try to find yours just by touch alone! Become geologists and find out what kind of rocks you have, and the role they may have played in the Earth's crusty development. Learn what the layers in some rocks mean—explore the fascinating world of fossils.

Some cultures speak of spirits in the rocks and other natural objects. What do you think? If your rock observed its surroundings over the last 100 years, what would it make of all the changes? Write a story about your rock. Most of all enjoy your rock, and remember, "everybody needs a rock."

Measurement

Measurement is a process used to find out the size of something. It often involves using a measuring tool that delineates uniform units.

There are various kinds of sizes including **length**, which measures how long something is or the distance between two things; **area**, which measures the flat surface enclosed in a shape; **weight**, which measures how light or heavy something is; **volume**, which measures how much space a three-dimensional shape takes up; and **capacity** which measures how much of something a container (be it a pool, a jar or a movie theater) can hold.

The two systems we use for measurement are the **English system**, which measures in inch/foot/yard/mile (length), ounce/pound/ton (weight), cup/pint/quart/gallons (liquid volume); and the **metric system** which measures in centimeter/meter/kilometer (length), gram/kilogram (weight), milliliter/liter (liquid volume).

Before children begin to use standard tools to measure, they measure things with nonstandard items such as their hand spans, blocks, scoops, and plastic teddy bears. In each case, the item chosen must be of uniform size. After students gain experience with the process, they can begin to use standard tools to measure. Two other measurement systems that students study are currency and time. The money system in the United States uses base ten as the metric system does and time is measured using clocks and calendars.

MAKING CONNECTIONS

The literature connections for measurement fall into several main categories: **size** (length, area, weight, volume and/or capacity), **money**, and **time**. Size books include those (such as *How Big Is A Foot?*) that graphically and humorously demonstrate the need for a standard unit of measure, as well as cookbooks, for their recipes. *If You Made A Million* explores all aspects of the money system, while several other storybooks listed have a key monetary component. *How Old Is Old?* portrays how age is relative, and several other books explore concepts of time.

CROSS REFERENCES

LITERATURE CONNECTIONS

Alexander Who Used to Be Rich Last Sunday

by Judith Viorst; illustrated by Ray Cruz
Atheneum, New York. 1978
Grades: K–3

A humorous look at how Alexander spends the dollar that his grandparents give him on a Sunday visit. Although Alexander would like to save the money for a walkie-talkie, saving money is hard! He and his money are quickly parted on such items as bubble gum, bets, a snake rental and a garage sale.

Anno's Sundial

by Mitsumasa Anno
Philomel, New York. 1985
Grades: K–3

The history of using shadows to tell time is revealed. As a follow-up, students can trace their shadows outdoors over the course of a day. Reinforces the concept that with the passage of time, the position of the sun in the sky changes the shadows that are formed.

As the Crow Flies: A First Book of Maps

by Gail Hartman; illustrated by Harvey Stevenson
Bradbury Press, New York. 1991
Grades: Preschool–2

Maps chart the different worlds and favorite places of an eagle, rabbit, crow, horse, and gull, each from its own perspective. The last two pages form a large map incorporating all the geographical areas seen earlier—the mountains, meadow, lighthouse and harbor, skyscrapers, hot dog stand, etc. A wonderful book with which to introduce or reinforce mapping skills.

A Chair for My Mother

by Vera B. Williams
Greenwillow Books/William Morrow, New York. 1982
Grades: K–3

A child, her waitress mother and her grandmother save coins to buy a comfortable armchair after all their furniture is lost in a fire. The accumulation of coins of various denominations are then exchanged for dollars at the bank. As a follow-up, students can sort and count small jarfuls of coins. For more on using this book in a money unit, see page 121.

> We especially need imagination in science. It is not all mathematics, nor all logic, but it is somewhat beauty and poetry.
>
> — *Maria Mitchell*
> *Diary*

Clocks and More Clocks
by Pat Hutchins
Puffin Books, Middlesex, England. 1970
Out of print
Grades: K–3

Confused because his hall clock reads one time while the attic clock reads a different time, a man solves this mysterious puzzle of changing times when he purchases a watch from the clockmaker. A delightful story to begin a unit on telling time with young students.

Counting on Frank
by Rod Clement
Gareth Stevens Children's Books, Milwaukee, Wisconsin. 1991
Grades: K–5

A young boy and his dog compare and count various things. For example, knowing that his gum tree grows six and one-half feet in a year, the boy calculates that he would be 53 feet tall if he grew at that rate. Not bad, except that he could never find clothes that fit! Each situation can be used to pose an open-ended problem for students to investigate.

Elliot's Extraordinary Cookbook
by Christina Bjork; illustrated by Lena Anderson
R&S Books/Farrar, Straus & Giroux, New York. 1990
Grades: 3–6

With the help of his upstairs neighbor, Elliot cooks wonderful foods and investigates what's healthy and what's not so healthy. Measurement is intrinsic to the recipes. One section relates population to the amount of food necessary to feed the world.

The Grouchy Ladybug
by Eric Carle
Harper & Row, New York. 1986
Grades: Preschool–2

A grouchy ladybug challenges everyone she meets regardless of their size or strength. This book is a wonderful springboard to measurement activities involving size for young children. There is a clock on each page to chronicle the day in hours for older children.

Gulliver's Travels
by Jonathan Swift
Penguin Books, New York. 1983
Grades: 4–Adult

On his first two voyages, an Englishman becomes shipwrecked in a land where people are six inches high and then stranded in a land of giants. This classic tale can inspire an exploration of ratio, proportion, and scale. This book is available from a number of different publishers.

How Big Is a Foot?

by Rolf Myller

Dell/Bantam Doubleday Dell Group, New York. 1990

Grades: K–5

When the king asks the apprentice carpenter to build the queen a bed for her birthday, he readily agrees and asks for the measurements. The king obliges and measures her bed using his feet. Somehow the bed that gets made is much smaller. This delightful story clearly shows the need for a standard unit of measure.

How Old Is Old?

by Ann Combs; illustrated by J.J. Smith-Moore

Price Stern Sloan, Los Angeles. 1987

Grades: K–3

With examples from nature, including chickadees, trees, stars, and snakes, Alistair's grandfather explains to him that being old is a relative concept. Raises interesting questions to investigate using varying units of measurement.

The Hundred Penny Box

by Sharon B. Mathis; illustrated by Leo and Diane Dillon

Puffin/Viking Penguin, New York. 1975

Grades: 3–6

Michael's love for his 100-year-old great-great-aunt leads him to intercede with his mother who wants to throw out an old penny box and buy a new one. Each of the hundred pennies in Aunt Dew's box marks memories of one year of her life, and Michael listens attentively as she tells her stories. This book can lead to a discussion of aging and the passage of time. Newbery honor book.

If You Made A Million

by David M. Schwartz; illustrated by Steven Kellogg

Lothrop, Lee & Shepard, New York. 1989

Grades: K–5

Describes the various forms that money can take, including coins, paper money, and personal checks; and how it can be used to make purchases, pay off loans, or build interest in the bank. Ends with the question of how would you would spend a million dollars. Includes an adult-level section on the history of money and its use.

Inch By Inch

by Leo Lionni

Astor-Honor, New York. 1960

Grades: Preschool–2

An inchworm cleverly avoids being eaten by demonstrating how useful he is as a measuring tool. The inchworm proceeds to measure an assortment of birds' body parts until he inches himself out of sight on the final measurement. Children can make their own model inchworms and measure familiar objects with them.

> Lack of money
> is the root of all evil.
> — *George Bernard Shaw*
> *Man and Superman*

James and the Giant Peach

by Roald Dahl; illustrated by Nancy E. Burkett
Alfred A. Knopf, New York. 1961
Penguin Books, New York. 1988
Grades: 3–8

A fantasy adventure about young James who escapes from his horrific Aunt Sponge and Aunt Spiker via a magically created giant peach. Inside the peach are insects of huge proportions who become his friends. Students can make large models of insects to get an appreciation of James' comrades inside the giant peach. There is a strong element of interesting scale comparison and measurement in many passages and episodes. It is worth noting that some reviewers have strongly criticized this author for negative and stereotypical female characters such as the cruel and greedy aunts in this book.

Let's Go Dinosaur Tracking

by Miriam Schlein; illustrated by Kate Duke
HarperCollins, New York. 1991
Grades: 2–5

Explores the many different types of tracks dinosaurs left behind and what these giant steps reveal. Was the creature running? chasing a lizard? browsing on its hind legs for leaves? traveling in pairs or in a pack? walking underwater? At the end of the book, the reader is invited to measure their stride and compare the difference when walking slowly, walking fast, and running.

Linnea's Windowsill Garden

by Christina Bjork; illustrated by Lena Anderson
Farrar, Straus & Giroux, New York. 1988
Grades: 3–6

From seeds to cuttings to potted plants, Linnea describes the care of her plants in her indoor garden throughout their life cycle. The friendly narration and simple information invites readers to grow an avocado pit, or try other activities and games at home. The planning and recording of her gardening efforts often involves careful measurement and related pursuits.

Momo

by Michael Ende
Doubleday, New York. 1985
Grades: 5–Adult

Momo has a special gift, an extraordinary ability to take the time to really listen. Her presence enables people to find the truth within themselves and to open their imagination. But the Men in Gray are only interested in the value of an hour or a minute and want time to be used efficiently at all costs. As Momo foils the gray invaders, the book explores time on many levels, from calculating the number of seconds spent doing an activity to understanding the importance of having time for creativity, relaxation, and pleasure.

Mr. Archimedes' Bath

by Pamela Allen

Lothrop, Lee & Shepard, New York. 1980

Grades: K–2

> Upset by the overflow of his bath and puzzled by the changing water level, Mr. Archimedes tries to blame one of his three bath companions—a kangaroo, a goat and a wombat—for missing bath water. Finally, Mr. Archimedes resorts to some scientific testing and measuring to find out who the culprit is. Humorously done, this book will make the reader want to jump in and out of the bath and see the results.

Music, Music for Everyone

by Vera B. Williams

Mulberry Books, New York. 1984

Grades: K–3

> Rosa plays her accordion with her friends in the Oak Street Band and earns money to help her mother with expenses while her grandmother is sick. Children can brainstorm ways to earn money and what wages they can charge for the jobs. A class could choose an item to purchase and figure out how to fund raise for it! (An extended article on this book is on page 121.)

My Cat Likes To Hide In Boxes

by Eve Sutton; illustrated by Lynley Dodd

Scholastic, New York. 1973

Grades: K–2

> Delightful book with rhymes about cats all over the world and "my cat" who likes to hide in boxes! The predictable pattern encourages reader participation. Children can bring in boxes from home and can investigate the size of the boxes and devise ways they might be used for measurement.

On the Day You Were Born

by Debra Frasier

Harcourt, Brace, Jovanovich, San Diego. 1991

Grades: K–6

> This beautiful book explores the relativity of time by portraying other phenomena that take place in nature at the same time when "you" (or any child) are born. The age range of the book is extended by a special section "More About the World Around You," which provides concise scientific information about the natural world.

The Phantom Tollbooth

by Norton Juster; illustrated by Jules Feiffer

Random House, New York. 1989

Grades: 2–8

> Milo has mysterious and magical adventures when he drives his car past The Phantom Tollbooth and discovers The Lands Beyond. Milo encounters amusing situations that involve numbers, geometry, measurement, and problem solving. The play on words in the text is delightful.

There is no money in poetry, but there is no poetry in money either.

— *Robert Graves*

The Purse
by Kathy Caple
Houghton Mifflin, Boston. 1986
Grades: K–2

Katie keeps her money in a band-aid box until her older sister convinces her to buy a purse. Because she uses all her money for the purse, she has nothing left to put in it! Katie does earn more money and the way she spends it provides a novel twist to the end of the story. A fun introduction to the monetary system.

The Topic Is Measurement

It's February in Rosa Franco's second grade classroom—the topic for the month is **measurement.**

Rosa's students have measured bubbles at one of the learning station activities in the GEMS guide, *Bubble Festival*. They have measured and graphed the outside temperature each day—first thing in the morning, and again right before lunch. Also, a parent came in and did a cooking project with small groups of students in which they measured liquid and solid volume using teaspoons and cups. On another day, Rosa had her students measure how many hands tall they are.

For story time each week, Rosa selected books that portrayed the concept of measurement.

The first week they read *How Big Is a Foot?* by Rolf Myller in which a carpenter builds the queen a bed using his own foot as the foot measure. Needless to say, the bed is much too small, illustrating the need for standard units of measure.

The second week they read *Thunder Cake* by Patricia Polacco in which a girl and her grandmother measure the distance of an approaching storm by counting the seconds between when they saw lightning and heard thunder.

In week three, they read *Anno's Sundial* by Mitsumada Anno which shows the history of how shadows have been used to measure time.

In the final week, they read *How Old is Old?* by Ann Combs in which a boy's grandfather helps him learn that old is a relative concept, as together they measure the ages of chickadees, trees, stars, and snakes.

As the month progresses, Rosa and her students keep a list of all the ways that people measure things.

Something Special For Me
by Vera B. Williams
Mulberry Books, New York. 1983
Grades: K–3

Rosa has difficulty choosing a special birthday present to buy with the coins her mother and grandmother have saved in a jar, until she hears a man playing beautiful music on his accordion. Since Rosa has a limited budget, she has to choose her gift carefully. Students can be given a fictitious budget and spend it as they choose! (An extended article on this book is on page 121.)

Spaces, Shapes and Sizes
by Jane J. Srivastava; illustrated by Loretta Lustig
Thomas Y. Crowell, Minneapolis. 1980
Grades: 1–6

This inviting nonfiction book about volume shows the changing form a constant amount of sand can take. It includes estimation activities, an investigation of volume of boxes using popcorn, and a displacement activity.

Thunder Cake

by Patricia Polacco

Philomel/Putnam & Grosset, New York. 1990

Grades: K–6

> Grandma finds a way to dispel her granddaughter's fear of thunderstorms. The two race to gather ingredients and bake "thunder cake" before the storm arrives. By counting the seconds between when they see the lightning and hear the thunder, they are able to measure the distance of the approaching storm. Both counting and cake-making provide an engaging and empowering distraction from the sounds and sights of the thunder-storm. The recipe for thunder cake, with its secret ingredients, is included!

The Toothpaste Millionaire

by Jean Merrill; illustrated by Jan Palmer

Houghton Mifflin, Boston. 1972

Grades: 5–8

> Twelve-year-old Rufus doesn't start out to become a millionaire—just to make toothpaste. Incensed by the price of a tube of toothpaste, Rufus tries making his own from bicarbonate of soda with peppermint or vanilla flavoring. Assisted by his friend Kate and his math class (which becomes known as Toothpaste 1), his company grows from a laundry room opera- tion to a corporation with stocks and bank loans. An ideal book to illus- trate the need for and use of mathematics in real-world problem solving.

Tuck Everlasting

by Natalie Babbitt

Farrar, Straus & Giroux, New York. 1975

Grades: 5–7

> A 10-year-old girl stumbles upon the Tuck family, who have been blessed with eternal life after drinking from a magic spring. The book suggests a somewhat sobering view of what it might be like to live forever. Discus- sions of time and age are likely to be prompted by this story.

Who Sank the Boat?

by Pamela Allen

Sandcastle Books/Putnam & Grosset, New York. 1983

Grades: K–2

> Who causes the boat to sink when five animal friends of varying sizes go for a row? This book explores the concept of threshold, sometimes popularly expressed as "the straw that broke the camel's back." The book also includes an easy-to-replicate experiment using simple rafts in a tub of water and standard units of measurement.

Number

Numbers are the aspect of mathematics with which people are the most familiar.

The system of numbers starts with the counting numbers (1,2,3 …) which are often learned through song and rote practice. After developing an understanding of the counting numbers, students learn to compute using addition, subtraction, division, and multiplication. As they solve arithmetic problems, the solutions include a wider set of numbers: the integers (negative and positive whole numbers and zero) and rational numbers (the integers and fractions).

MAKING CONNECTIONS

Many children's books relate to number in some central way. The books listed here are in several major categories.

First, to provide young children with experience with numbers, there are **counting** books. We have listed some that have a unique creative strength of their own, in addition to clear mathematical depiction of the numbers, such as one that shows two different arrays of 20 animals in their natural habitat. These counting books give children an opportunity to connect written numerals with a corresponding number of items. Other counting books, such as *How Many Snails?* ask open-ended questions that allow the reader to count a variety of items.

Other books listed here build upon the manipulation of numbers and increased student skills to involve **computation** as well as **estimation** and **mental math** skills. **Number storybooks** include selections like *The Phantom Toll Booth*, in which numeric concepts play an important role in the story.

Another aspect of numbers is **large numbers**. One book introduces young children to the number 100, while *How Much is A Million?* gives a sense of how large the number one million is. At an even more sophisticated level, *A Grain of Rice* is a clever story that gets into large numbers through **exponential growth**.

CROSS REFERENCES

LITERATURE CONNECTIONS

The 329th Friend

by Marjorie Weinman Sharmat; illustrated by Cyndy Szekeres
Scholastic, New York. 1979
Four Winds Press, New York. 1979
Grades: K–3

Emery Raccoon wakes up one morning feeling discontent with his own company. His solution is to invite somebody to his home. The one somebody quickly evolves into a total of 328 somebodies—all of whom he invites to lunch! Emery discovers to his delight that not only does he have 328 friends, he is a good friend to himself—which makes him the 329th friend! Numbers are creatively used and illustrated throughout the story.

Animal Numbers

by Bert Kitchen
Dial Books, New York. 1987
Grades: K–3

A counting book in which exotic and familiar animals are shown. This large format book received a number of awards for its stunning illustrations. Counts from 1–10 and continues with 15 pigs, 25 snakes, 50 sea horses, 75 turtles, 100 tadpoles and tadpoles-to-be!

Anno's Math Games

by Mitsumasa Anno
Philomel Books/Putnam & Grosset, New York. 1987
Grades: 2–5

Picture puzzles, games, and simple activities introduce the mathematical concepts of multiplication, sequence, ordinal numbering, measurement, and direction.

Anno's Math Games II

by Mitsumasa Anno
Philomel/Putnam & Grosset, New York. 1982
Grades: 2–4

More picture puzzles, games, and simple activities that introduce the mathematical concepts of multiplication, sequence, ordinal numbering, measurement, and direction.

Anno's Mysterious Multiplying Jar

by Masaichiro and Mitsumasa Anno
Philomel/Putnam & Grosset, New York. 1983
Grades: 3–8

Through an understanding of multiplication, the reader can learn about factorials and the way that numbers can expand. On a second reading of the book, students can follow along using calculators to verify the large number of jars at the end of the story.

For deeds do die, however nobly done,
And thoughts of men do as themselves decay,
But wise worlds taught in numbers for to run,
Recorded by the Muses, live for ay.

— *Edmund Spenser*
The Ruines of Time

A Chair For My Mother
by Vera B. Williams
Greenwillow Books/William Morrow, New York. 1982
Grades: K–3

A child, her waitress mother, and her grandmother save coins to buy a comfortable armchair after all their furniture is lost in a fire. The accumulation of coins of various denominations is exchanged for dollars at the bank. The money jar in the story can provide a model for classroom estimation jar activities. (A further look at this book is on page 121.)

Deep Down Underground
by Olivier Dunrea
Macmillan, New York. 1989
Grades: K–3

Animals present the numbers from one to ten, as earthworms, toads, ants, and others march and burrow, scurry and "scooch" deep down underground.

The Doorbell Rang
by Pat Hutchins
Greenwillow Books, New York. 1989
Grades: K–3

Each time the doorbell rings, there are still more people who have come to share Ma's wonderful cookies. A great book to introduce the concept of division with a high interest material—cookies! The book also contains a repeating word pattern.

Each Orange Had 8 Slices
by Paul Giganti, Jr.; illustrated by Donald Crews
Greenwillow Books/William Morrow, New York. 1992
Grades: Preschool–3

Through a series of questions, the reader is invited to count and add balloons, seeds, animals, and wheels, in several different ways such as in multiples of 2, 3, 4, and 5. Can lead to an investigation of items that come in multiples to begin a unit on multiplication.

If I could write the beauty of your eyes
And in fresh numbers number all your graces,
The age to come would say, "This poet lies;
Such heavenly touches ne'er touch'd earthly faces."

— *William Shakespeare*
Sonnets

Eating Fractions

by Bruce McMillan
Scholastic, New York. 1991
Grades: 1–3

> From bananas to pizza to corn on the cob, food is cut into halves, thirds, and fourths to illustrate how parts make a whole. Through these colorful photographs, you will enjoy a feast of fractions. Recipes are included.

Erin McEwan, Your Days Are Numbered

by Alan Ritchie
Alfred A. Knopf, New York. 1990
Grades: 4–8

> Erin, a sixth grader with an intense fear of numbers, takes a job at the delicatessen where she needs to learn bookkeeping to stay employed. With the encouragement of the owner, Erin surprises herself by not only improving her math skills—she even catches a bookkeeping error that saves thousands of dollars! Though stereotypical in its portrayal of women and girls as being unskilled in math, the book presents an opportunity to discuss and dispel that belief.

From One to One Hundred

by Teri Sloat
E.P. Dutton, New York. 1991
Grades: Preschool–3

> Detailed illustrations of people and animals introduce the numbers 1 through 10 and then continue to 100, counting by tens. Counting to 100 by tens gives children a sense of large numbers.

A Grain of Rice

by Helena C. Pittman
Hastings House, New York. 1986
Bantam Books, New York. 1992
Grades: 2–5

> A clever, cheerful, hard-working farmer's son wins the hand of the Emperor's daughter by outwitting her father, who treasures the daughter more than all the rice in China. Pong Lo's winning strategy is to use a mathematical ruse, asking simply for a grain of rice that is to be doubled every day for one hundred days. Illustrates exponential growth.

How Many Snails?

by Paul Giganti, Jr.; illustrated by Donald Crews
Greenwillow Books/William Morrow, New York. 1988
Grades: Preschool–3

> A young child takes walks to different places and wonders about the amount and variety of things seen on the way, from fish to fire trucks to cupcakes. This book invites the reader to actively participate and count by attributes.

All the mathematical sciences are founded on relations between physical laws and laws of numbers . . .

— *James Maxwell*
On Faraday's Lines of Force

> When angry, count to ten before you speak; if very angry, an hundred.
>
> *— Thomas Jefferson*
> *A Decalogue of Canons*
> *for Observation in*
> *Practical Life*

How Much Is A Million?

by David M. Schwartz; illustrated by Steven Kellogg

Lothrop, Lee & Shepard, New York. 1985

Grades: K–5

> With detailed, whimsical illustrations that include children, goldfish, and stars, this book leads the reader to conceptualize what at first seems inconceivable—a million, a billion, and a trillion. An adult-level note explains the calculations used.

Jumanji

by Chris Van Allsburg

Houghton Mifflin, Boston. 1981

Grades: K–5

> A bored brother and sister left on their own find a discarded board game (called Jumanji) which turns their home into an exotic jungle. A final roll of the dice for two sixes helps them escape from an erupting volcano. In connection with probability activities in several GEMS guides, students can investigate the probability of rolling two sixes in a row.

The King's Chessboard

by David Birch; illustrated by Devis Grebu

Dial Books, New York. 1988

Grades: K–6

> A proud king learns a valuable (and exponential) lesson when he readily grants his wise man a special request for rice. One grain of rice on the first square of a chessboard on the first day, two grains on the second square on the second day, four grains on the third square on the third day and so on. After several days the counting of rice grains gives way to weighing, then the weighing gives way to counting sackfuls, then to wagonfuls. The king soon realizes that there is not enough rice in the entire world to fulfill the wise man's request.

Moja Means One: Swahili Counting Book

by Muriel Feelings; illustrated by Tom Feelings

Dial Books, New York. 1971

Grades: K–5

> As you count from one through ten in Swahili, the charcoal-toned drawings show market stalls, Mount Kilimanjaro, musical instruments, and other facets of life in East Africa. Caldecott honor book.

The Phantom Tollbooth

by Norton Juster; illustrated by Jules Feiffer

Random House, New York. 1989

Grades: 2–8

> Milo has mysterious and magical adventures as he drives his car past the Phantom Tollbooth and discovers The Lands Beyond. On his journey, Milo encounters numbers, geometry, measurement and problem-solving in amusing situations. The play on words in the text is delightful.

The Right Number of Elephants
by Jeff Sheppard; illustrated by Felicia Bond
Harper & Row, New York. 1990
Grades: Preschool–3

A little girl imagines situations in which different numbers of elephants from 10 down to 1 might be helpful. Questions such as "How many elephants would you need to escape a summer storm?" are posed. Elephants skateboard, cast shadows on the beach, and paint the ceiling with their trunks in the hilarious illustrations.

Sadako and the Thousand Paper Cranes
by Eleanor Coerr; illustrated by Ronald Himler
Dell Books, New York. 1977
Grades: 3–6

In this true story, a young Japanese girl is dying of leukemia as a result of radiation from the bombing of Hiroshima. According to Japanese tradition, if she could fold 1,000 paper cranes, the gods would grant her wish and make her well. But she folded only 644 paper cranes before she died. In her honor, a Folded Crane Club was organized, and each year on August 6 members place thousands of cranes beneath her statue to celebrate Peace Day. The moving story can introduce a class origami project to make 1,000 cranes, and has strong social studies and current events connections.

Seventeen Kings and Forty-Two Elephants
by Margaret Mahy; illustrated by Patricia MacCarthy
Doubleday, New York. 1990
Grades: K–3

Seventeen kings and forty-two elephants romp with a variety of jungle animals during their journey through a wild, wet night. The vibrant illustrations, originally done as batik paintings on silk, celebrate the joyful noise of the playful, rhyming tongue-twisting poem. Young people will also enjoy verifying that there are indeed 17 kings and 42 elephants.

Ten, Nine, Eight
by Molly Bang
Greenwillow Books/William Morrow, New York. 1983
Grades: K–2

Numbers from 10 to 1 are part of this lullaby that details a little girl's room ending with "one big girl all ready for bed."
Awarded the Caldecott honor prize.

This is the third time;
I hope good luck lies in
odd numbers . . . There
is divinity in odd num-
bers, either in nativity,
chance, or death.

— *William Shakespeare*
Merry Wives of Windsor

The Toothpaste Millionaire
by Jean Merrill; illustrated by Jan Palmer
Houghton Mifflin, Boston. 1972
Grades: 5–8.

Twelve-year-old Rufus doesn't start out to become a millionaire—just to make toothpaste. Assisted by his friend Kate and his math class (which becomes known as Toothpaste 1), his company grows from a laundry room operation to a corporation with stock and bank loans. Many opportunities for estimations and calculations are presented including cubic inches, a gross of toothpaste tubes bought at auction, manufacturing expenses, and profits. An ideal book to illustrate the need for, and use of, mathematics in real-world problem solving.

Two Ways to Count to Ten: A Liberian Folktale
retold by Ruby Dee; illustrated by Susan Meddaugh
Henry Holt & Co., New York. 1988
Grades: Preschool–3

In this traditional African folk tale, King Leopard invites all the animals to a spear-throwing contest whose winner will marry his daughter and inherit the throne. The contest provides a framework for counting to 10 by ones until one clever animal counts a new way and wins the contest.

What Comes In 2's, 3's and 4's?
by Suzanne Aker; illustrated by Bernie Karlin
Simon & Schuster, New York. 1990
Grades: K–3

A delightful counting book that describes in various ways how the numbers 2, 3, and 4 occur in daily life. Easily recognizable examples are used (tricycle, wings on a bird, legs on a cat) and concepts such as symmetry and time are incorporated. The book invites the reader to count and make comparisons. As a follow-up activity, students can create their own books using magazines, pictures, photos, and drawings.

The Wildlife 1-2-3: A Nature Counting Book
by Jan Thornhill
Simon & Schuster, New York. 1989
Grades: K–3

Beautifully illustrated counting book that uses the diversity of wildlife in natural habitats to accompany the numbers 1 to 20 as well as the larger numbers 25, 50, 100 and 1,000. Each set of animals is displayed twice on each page (once inside a frame, and again within the border that surrounds the frame, giving the reader two representations of the same number). Included at the back of the book are helpful notes that describe the animals' environment, diet, offspring, and other interesting facts.

The Wolf's Chicken Stew

by Keiko Kasza

G.P. Putnam's Sons, New York. 1987

Grades: K–3

A hungry wolf's attempts to fatten a chicken for his stew pot has unexpected results. A delightful book to begin an investigation of 100 as the delectable items that the wolf brings to the chicken's doorstep come in quantities related to 100. When he arrives at the chicken's house to capture her for his stew, he has quite a surprise, and the unexpected ending is very touching.

SMART

My dad gave me one dollar bill
'Cause I'm his smartest son,
And I swapped it for two shiny quarters
'Cause two is more than one!

And then I took the quarters
And traded them to Lou
For three dimes—I guess he don't know
That three is more than two!

Just then, along came old blind Bates
And just 'cause he can't see
He gave me four nickels for my three dimes,
And four is more than three!

And I took the nickels to Hiram Coombs
Down at the seed-feed store,
And the fool gave me five pennies for them,
And five is more than four!

And then I went and showed my dad,
And he got red in the cheeks
And closed his eyes and shook his head—
Too proud of me to speak!

— *Shel Silverstein*
Where the Sidewalk Ends

Pattern

attern is an underlying concept in all of mathematics, and is used as a powerful problem-solving tool throughout all the math strands. A pattern can be found in anything that **repeats itself over and over**.

Looking for patterns develops logical thinking and the ability to better predict future events. As the mind searches for patterns, sense can be made from what may at first appear to be **discrete, separate**, and **unrelated** things or events. Patterns are found throughout **nature**, as part of **life cycles**, and in many other aspects of our lives, including of course, literature and art.

MAKING CONNECTIONS

One of the books listed involves a **repeating pattern** as a young boy creates patterns with his feet and a stick. Other books, such as *Brown Bear, Brown Bear, What Do You See?* exemplify the patterns of **repeating verse**. Many books explore some aspect of **life cycles** including the seasons, days in a week, months in a year, evolution, and death.

CROSS REFERENCES

"Night and Day" by M.C. Escher

LITERATURE CONNECTIONS

Badger's Parting Gifts

by Susan Varley

Lothrop, Lee & Shepard, New York. 1984

Grades: K–3

> Badger's friends are sad when he dies, but they treasure the legacies he left them. Mole, in particular, has difficulty adjusting to the loss of his dear friend and finds solace in the memories of Badger. A sensitive look at death as part of the life cycle.

Brown Bear, Brown Bear, What Do You See?

by Bill Martin, Jr.; illustrated by Eric Carle

Henry Holt & Co., New York. 1983

Grades: Preschool–2

> Popular classic asks the question "What do you see?" The answers all involve color and animals. The repetitive pattern in the story is infectious and children can expand beyond the story to add more animals.

Chicken Soup with Rice

by Maurice Sendak

Harper & Row, New York. 1962

Scholastic, New York. 1986

Grades: Preschool–3

> This delightful book of rhyming verse about eating chicken soup through the year lends itself to both exploring word patterns and marking the cycle of months throughout a year.

Fortunately

by Remy Charlip

Four Winds Press/Macmillan, New York. 1987

Grades: K–2

> Ned has a series of fortunate and unfortunate events as he travels from New York to Florida for a surprise party. Students will catch on to the pattern and use their logical-thinking skills to predict the next event that will unfold.

Linnea's Almanac

by Christina Bjork; illustrated by Lena Anderson

R&S Books/Farrar, Straus & Giroux, New York. 1989

Grades: 3–6

> Linnea uses an almanac that she writes in to keep track of her indoor and outdoor investigations of nature over the cycle of one year's time. She opens a bird restaurant in January and goes beachcombing in July. The almanac is written in journal form with simple monthly activities for young readers to do at home.

I have heard of a man who had in mind to sell his house, and therefore carried a piece of brick in his pocket, which he showed as a pattern to encourage purchasers.

— *Jonathan Swift*
The Drapier's Letters

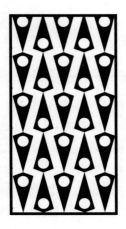

Nicky the Nature Detective
by Ulf Svedberg; illustrated by Lena Anderson
R&S Books/Farrar, Straus & Giroux, New York. 1988
Grades: 3–8

Nicky loves to explore the changes in nature. She watches a red maple tree and all the creatures and plants that live on or near the tree through the seasons of the year. Her discoveries lead her to look carefully at the structure of a nesting place, why birds migrate, who left tracks in the snow, where butterflies go in the winter, and many more things. This book is packed with information.

Once There Was A Tree
by Natalia Romanova; illustrated by Gennady Spirin
Dial Books, New York. 1985
Grades: K–3

A tree is split in two by lightning and left a stump. A variety of animals claim ownership of the tree stump for food and shelter while others, including a man, make the claim for reasons other than survival. The question of ownership is resolved when a new tree regenerates and all can cooperatively share the tree.

Ox-Cart Man
by Donald Hall; illustrated by Barbara Cooney
Viking Press, New York. 1979
Grades: K–3

Describes the day-to-day life of an early nineteenth century rural New England family throughout the changing seasons. They grow crops, make food, clothing, and other items to sell and for their own consumption. Caldecott award winner.

Polar Bear, Polar Bear, What Do You Hear?
by Bill Martin, Jr.; illustrated by Eric Carle
Henry Holt & Co., New York. 1991
Grades: Preschool–2

Zoo animals from a polar bear to a walrus make their distinctive sounds for each other. Children imitate the sounds for the zookeeper. This pattern book lends itself to group participation.

The Snowy Day
by Ezra Jack Keats
Viking Press, New York. 1962
Grades: Preschool–2

Peter goes for a walk on a snowy day. He makes different patterns in the snow with his feet, a stick, and then his whole body. He tries to save a snowball but is disappointed when it melts. That night Peter dreams that the sun melted all the snow outside, but when he wakes up, it's snowing again!

The Tenth Good Thing about Barney
by Judith Viorst; illustrated by Erik Blegvad
Atheneum, New York. 1971
Grades: K–4

A young boy's beloved cat, Barney, dies. At the funeral, he remembers nine good things about Barney and later comes up with the tenth—Barney is helping the flowers grow. Patterns of the life cycle are explained simply and sensitively.

Tessellations

Tessellations in their simplest form are tilings, like those your students are familiar with in their homes. The word tessellation comes from the Latin **"tessella,"** which means small square stone or tile. Look for examples of tessellations in bathroom and kitchen tilings, architectural designs, quilting, art forms and nature after building tessellations with pattern blocks.

For a tiling to be a tessellation, three elements are necessary:

◆ There are no gaps between shapes,

◆ There are no overlays of tiles on top of one another, and

◆ There is a pattern of one or more shapes that repeats itself and can be extended in the plane.

A LITTLE BACKGROUND

Tessellations connect mathematics, history and art. Starting as early as 4000 B.C., Sumerians created geometric mosaics for decorations. Archimedes (287-212 B.C.) investigated properties of regular polygons that tessellated in the plane.

Probably the most extensive work with mosaic designs was done by the Moorish artists in Spain from 700–1500. These artists developed tessellating geometric patterns (as well as calligraphy) for ornamentation and are especially known for intricate star designs.

Starting in 1930, the Dutch artist, M.C. Escher, began altering geometric shapes and creating forms such as birds, reptiles, fish, and people. Geometric shapes can be translated, rotated, and reflected to create more unusual tessellations.

SPECIAL CONSIDERATIONS

Many students have difficulty differentiating between designs and tessellations, so it is helpful to introduce concrete examples of tessellations to them, either before or after they build at this station in the GEMS *Build It! Festival* guide.

A beehive (constructed with hexagons) is an example of a tessellation from the natural world. Sample sheets of tiling from a hardware store may spark students to look in their homes for other examples. A walk in the neighborhood can provide examples of both tessellations and designs.

For young students, the concept of a tessellation is sophisticated. However, at the early primary level, an important focus of the math program is patterns. This station will provide experiences with advanced patterns and students will gain an intuitive understanding of tessellations that can be explored in greater depth in later grades.

These pages include excerpts from one of the activities in the GEMS Build It! Festival guide (see page 47). The guide includes eight main activities that can be set up at classroom learning stations. Many math skills are put into play, along with numerous constructive connections to the real world.

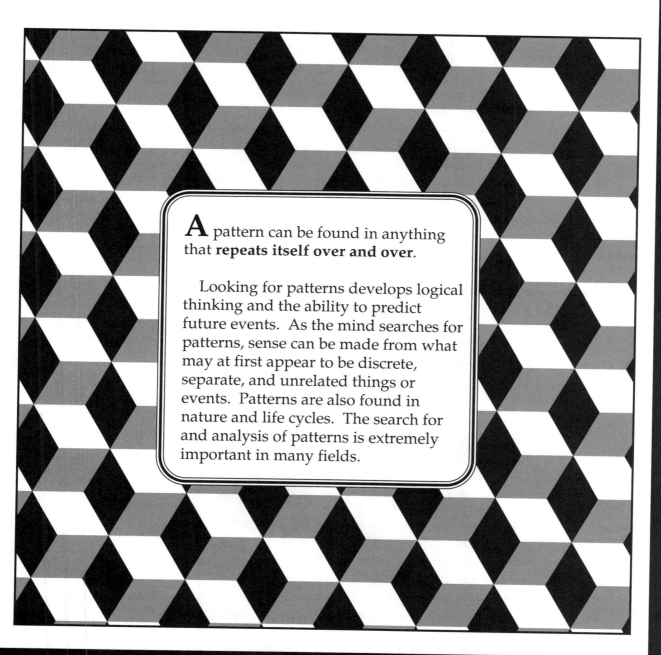

A pattern can be found in anything that **repeats itself over and over**.

Looking for patterns develops logical thinking and the ability to predict future events. As the mind searches for patterns, sense can be made from what may at first appear to be discrete, separate, and unrelated things or events. Patterns are also found in nature and life cycles. The search for and analysis of patterns is extremely important in many fields.

Probability & Statistics

Probability is the part of math that tries to figure out and make predictions about the **chance or likelihood that something will happen**.

Prediction involves **making a guess** about something that is expected to happen. For example, if you are going to toss a coin, you can predict if it will land on the head or the tail side of the coin. Meteorologists try to figure out the probability of whether it will rain or be sunny based on weather patterns. Young students can **predict outcomes** of simple **probability experiments involving coins, dice, and spinners** and through these experiments, explore the **concept of chance**. After these experiences, older students can **assign probabilities** to the possible outcomes of an experiment and go further to investigate **permutations and combinations**. In all of these explorations with probability, students generate data to make observations and predictions so that they can determine the probability of something happening.

Statistics is the collection, classification, analysis, and interpretation of numerical facts and data.

Statistics abound in the world around us, from the average yearly rainfall in the Amazon to the number of children in an average family in the United States, and understanding statistics is a highly important real-life skill. Statistics are generated in many ways including from **surveys, experiments, sports, demographics,** and **observations**, and are usually represented in some type of graph form. From statistics we can glean information such as the favorite ice cream of a class of students or the **mean** (average), **mode** (most frequently occurring number) and **median** (number that falls in the middle of the numbers listed from lowest to highest) from generated data.

Students begin their study of statistics by **collecting, organizing, representing,** and **making interpretations of data** that has meaning to them. For example, students can bring in their favorite stuffed animals to use as data to sort and organize them into a **real-life (or "concrete") graph** using the animals. Students can then interpret the graph from the way it was organized. The same data can be reorganized and interpreted a different ways. In addition, a **pictorial or symbolic graph** can be made to record the concrete graph of the animals. On a more advanced level, students can be given graphs and tables and **extract valid information** from them.

MAKING CONNECTIONS

Most of the books listed here lend themselves to learning more about statistics—especially data collection—and many tie in with concepts in probability as well. Several weather-related books, such as *The Cloud Book* and *Cloudy With A Chance of Meatballs*, encourage further thinking about prediction. Others involve data collection in almanac form, and several books humorously illustrate more advanced probability concepts, such as permutation and combination.

CROSS REFERENCES

> Probable impossibilities are to be preferred to improbable impossibilities.
>
> — *Aristotle*
> *Poetics*

LITERATURE CONNECTIONS

Alligator Shoes

by Arthur Dorros
E.P. Dutton, New York. 1982
Grades: K–2

Being locked in a shoe store by mistake is Alvin Alligator's dream come true. He tries on an endless variety of footwear, but his decision about what footgear he prefers will surprise and delight. Great to read before using students' real shoes to provide the data for a graph.

Anno's Mysterious Multiplying Jar

by Masaichiro and Mitsumasa Anno
Philomel/Putnam & Grosset, New York. 1983
Grades: 3–8

Through an understanding of multiplication, you learn about factorials and the way that numbers can expand. On a second reading, students can follow along using calculators to verify the large number of jars at the end of the story.

The Cloud Book

by Tomie dePaola
Holiday House, New York. 1975
Grades: K–3

The variety of clouds, when they appear, what they mean (includes explanations about the clouds from Native American and Greek legends) are explored. With this introduction, students could make observations about clouds and graph them on a chart once a day for a month, or even over the course of a day. Students can make predictions about which type of cloud is most likely to be seen.

To understand
God's thoughts
we must study
statistics, for these
are the measure of
his purpose.

— *Florence Nightingale*

Cloudy With a Chance of Meatballs

by Judi Barrett; illustrated by Ron Barrett
Atheneum, New York. 1978
Grades: K–3

> A hilarious look at weather conditions in the town of Chewandswallow. This tiny town needs no food stores because daily climactic conditions bring the inhabitants food and beverages—although not always as they would be most appreciated! A humorous look at unusual weather prediction.

Harriet's Halloween Candy

by Nancy Carlson
Puffin/Penguin, New York. 1982
Grades: Preschool–3

> Harriet learns the hard way that sharing her Halloween candy makes her feel much better than eating it all herself. In the process, she sorts, classifies and counts her candy. Tallying candy or other food items always generates high interest.

Jumanji

by Chris Van Allsburg
Houghton Mifflin, Boston. 1981
Scholastic, New York. 1988
Grades: K–5

> A bored brother and sister left on their own find a discarded board game (called Jumanji), which turns their home into an exotic jungle. A final roll of the dice for two sixes helps them escape from an erupting volcano. Students can investigate dice and discover the probability of two sixes being rolled.

Linnea's Windowsill Garden

by Christina Bjork; illustrated by Lena Anderson
R&S Books/ Farrar, Straus & Giroux, New York. 1988
Grades: 3–6

> From seeds to cuttings to potted plants, Linnea describes the care of her indoor garden plants throughout their life cycle. The friendly narration and simple information invite readers to try the activities and games at home. Statistical activities include recording sprouting and growth times and the number of seeds in various fruits and vegetables.

Shoes

by Elizabeth Winthrop; illustrated by William Joyce
Harper & Row, New York. 1986
Grades: Preschool–2

> A survey of the many kinds of shoes in the world concludes that the perfect natural shoes are your feet. Great to read before doing a survey of shoes or sorting and classifying a group of real shoes. Can be used to generate data that can be graphed about size of feet or left- or right-handedness.

Socrates and the Three Little Pigs

by Tuyosi Mori; illustrated by Mitsumasa Anno
G.P. Putnam, New York. 1986
Grades: 4–8

Socrates, a wolf, attempts to catch one of three pigs for his wife's dinner. These three pigs collectively own five cottages. With the help of his frog-friend, the mathematician Pythagoras, Socrates tries to determine the possible cottages the pigs might be in. As the story unfolds, the illustrations show the many possible locations of the pigs and, in doing so, visually and clearly show the difference between permutations and combinations. This type of math, known as combinatorial analysis, forms the basis for computer programming and problem-solving and this connection is explained on a more advanced level in the back of the book. This book is also available under the title *Anno's Three Little Pigs.*

Storytelling

BY LINCOLN BERGMAN

In the introduction to this handbook we note: "In recent years, numerous excellent books celebrating many different cultures have been published. Our selections in no way represent the multitude of these fine books; we chose those among them that also had a strong connection to science and mathematics concepts. We did, however, endeavor to include a diverse spectrum, as one way to represent our commitment to quality *and* equality in education."

This commitment is an important and growing one. As the population of our country changes, and the term "minorities" becomes increasingly inaccurate, as we seek to find ways for peoples of differing cultures to both affirm their own roots, traditions, and values, and work together in socially progressive ways, children's literature can be a powerful and positive illuminator for the road ahead.

Stories and storytelling can open new vistas in student interest and involvement. Learning the treasured stories of many cultures, their own included, can give students a truly healthy respect for different ways of life. Newer books for children feature a wide diversity of stories from many cultures, and we are sure you have your favorites.

Storytelling can, of course, be raised to a high art. In many cultures, storytellers are considered spiritual leaders, bearers and passers-on of sacred beliefs and traditions. In two outstanding books, Michael Caduto and Joseph Bruchac combine Native American stories with environmental and science activities. These books, *Keepers of the Earth* and *Keepers of the Animals*, are highly recommended and can be especially related to GEMS biology and environmental activities and to GEMS guides that integrate myths as explanations for natural phenomena, as in *Investigating Artifacts* and *Earth, Moon, and Stars*. Bruchac, a leading Abenaki storyteller and folklorist, has written many articles on respectful ways for non-Indian storytellers to relate to and tell Native American stories.

There is also a growing national association of storytellers—the National Association for the Preservation and Perpetuation of Storytelling (NAPPS), Box 309, Jonesborough, Tennessee 37659. Their "Storytelling Catalog" lists tapes, records, films, and videos of stories told by storytellers. The organization holds an annual national festival and sponsors regional festivals as well. Intense audience interaction with a dynamic storyteller makes for unique and memorable experiences.

There are also many fine recordings of storytellers. One we recommend highly is by Gayle Ross: *How Rabbit Tricked Otter and Other Cherokee Animal Stories,*

—That Grand Art

Audiotape, Caedmon, an imprint of Harper Audi, HarperCollins Publishers, New York, 1991. This is a wonderful collection of stories told by a nationally known storyteller who is also the great-great-great granddaughter of John Ross, who was the Principal Chief of the Cherokee nation during the infamous "Trail of Tears."

Storytelling in one form or another is a central part of all cultures. When the late Alex Haley, the author of *Roots*, launched his quest for finding his ancestral connections in Africa, it was a present-day *griot*, a highly respected spiritual elder who carried the oral tradition of the tribe, whose memorization of names of tribal members kidnapped many generations ago for the slave trade dovetailed with the research that Haley had already done about his ancestry in the United States. In this way, the African name Kunta Kinte became known to many millions of people.

Among the Iroquois and a number of other peoples around the world, notably in Africa, people who became storytellers have a "storyteller's bag," filled with items, such as a cornhusk doll, that act as mnemonic devices. In their *Keepers of the Earth* teacher's guide, Caduto and Bruchac note: "Making a storyteller's bag is a project that can be adopted by a teacher. You and your students can gather items from the natural world or make things to add to the bag. Feathers, stones, small carvings, animal teeth, anything that can be jostled around in a bag without breaking can be part of your collection." You could make such a bag, for example, when doing the Session 4 story sharing activity in the GEMS guide *Investigating Artifacts*.

On the one hand, we need to recognize that there are individuals who have raised storytelling, and the audience interaction so central to it, to a high and beautiful art. Their accomplishments need to be recognized, and you may want to invite a local storyteller to your school or community center in connection with related class activities. **At the same time, what child is not a storyteller? What teacher cannot integrate some form of storytelling into her presentations to great advantage?** (See the article on the next page for a good example.)

It's worth remembering that many of the greatest stories are not originated by any one person—they grow and change over time from within the rooted burrows of popular culture, from the day-to-day joy, anguish, and humor of the downtrodden and dispossessed—spun against the cold of winter, the pain of disease, and the fear of death, they stand as affirmations of the human spirit.

Lincoln Bergman is the principal editor of the GEMS project.

A Substitute Teacher's

BY CAROLYN WILLARD

As a substitute teacher, I found that there were often periods of from 10 to 20 minutes when most or all of the class had nothing to do—this usually spelled trouble. As everyone remembers and knows, children can and do get "out of hand" with a "sub."

One particularly desperate day, I rushed to the "folklore"section of the school library during recess to find something to read. I happened to choose a book called *The Jack Tales* by Richard Chase. After reading the first tale to the class I knew I had solid gold. Because the tales are written verbatim from the tape recordings of Appalachian storytellers, they are especially good to read aloud. Just a few sentences into the story, the rowdy fifth-graders became silent and were no longer my adversaries. I calmed down too, as the class begged to hear more of the tricks and trials of "Jack in the Giant's Newground," "Jack and Hardy Hardhead," or "Jack and the Doctor's Girl."

Not every tale in the book turned out to be great, but as time went by, I found there were six or seven of these stories that could "tame" any class I encountered, no matter what grade level, no matter how wild! I bought the book,

and kept it in my substitute's "bag of tricks." I grew to depend on *The Jack Tales* to save the day whenever a class started getting negative or rebellious. Since that was a frequent occurrence, I read the stories many times!

So imagine my dismay and panic one crazy morning in a class when I discovered my book was missing. I really needed a "Jack Tale." I decided to try just telling the story. Although I have a memory like a sieve, and can hardly remember a whole phone number long enough to dial, I found I somehow could remember a story that took 10 or 15 minutes to tell. I remembered not only the sequence of events, but the actual words, phrases, and cadences of the tale. The memory came in the telling. (Maybe I was reactivating ancestral memory cells designed in human beings especially to recall and relate an interesting tale. After all, our ancestors depended on oral tradition before the advent of the written word— maybe **listening** to a story was easy for my students for the same reason . . .)

Without the book in front of me, I could make eye contact with the class and of course they found the tale even more engaging this way. It was a new experience for me. What a pleasure! And if I could do it, anybody can. Find a good story, read it aloud a few times, then put it

Storytelling Bag of Tricks

aside and you may just possibly find yourself on your way to becoming a storyteller!

Even though I went on to become a regular teacher, and then a science resource teacher, and have since presented many very engaging activities, such as "Oobleck" and "Bubble-ology," this early storytelling lesson stays with me. In fact, I still run into (now much older) students from that time who look at me and say: "Oh, yeah, I remember you—the Jack Tales!"

Carolyn Willard is a staff development specialist with the GEMS project.

SCIENCE THEMES

For your convenience in locating books listed under a specific thematic category, we have listed the themes in alphabetical order.

In the GEMS handbook *To Build A House: GEMS and the Thematic Approach to Teaching Science* these themes are more fully described, with examples given from many GEMS activities. In the Themes handbook they do not appear alphabetically, but in a more logical pedagogical order: Systems & Interactions; Models & Simulations; Stability; Patterns of Change; Evolution; Scale; Structure; Energy; Matter; Diversity & Unity.

The GEMS program has selected these ten major themes (or major recurring ideas) for the science curriculum. State science frameworks and other educational documents may have a few more or less, and there is no one correct listing. The thematic approach is designed to help teachers present the "big ideas" in science in a dynamic, multidisciplinary way.

Note: Some of the books listed are currently out of print; however, they should be readily available at your local school or public library.

Diversity & Unity → Systems & Interactions → Models & Simulations → Stability → Patterns of Change → Evolution → Scale → Structure → Energy → Matter → Diversity & Unity

Every science begins as philosophy and ends as art.

— *Will Durant*
The Story of Philosophy

Diversity & Unity

Diversity & Unity is a way to comprehend and appreciate both the differences and similarities that are interwoven in the sciences and in our understanding of nature and the physical world. From William Blake's "universe in a grain of sand" to the search for a theory that reflects the unity between molecular structure and the dynamics of the universe itself, diversity and unity runs through many fields of human endeavor and thought.

For example, the variety of living forms is immense, yet all share many basic processes involved in being alive. Many of these diverse species may also be linked and interdependent—in that way they are unified within a larger ecosystem.

MAKING CONNECTIONS

Approximately half of the books selected for this section elaborate **biological diversity and unity** among plants, oviparous animals, mammals, dinosaurs, animals homes, and animal behaviors. Most of the rest of the books illustrate the **diversity and unity among humans**, offering many opportunities to discuss **cultural diversity**, the **uniqueness of individuals**, and the **commonality of humans**. In one book the reader takes a trip through the universe, highlighting **diversity and unity in the realm of astronomy and geography**. Be on the lookout for books for older students with this theme and for books that illustrate diversity and unity in other realms, such as in the **earth and physical sciences**.

CROSS REFERENCES

LITERATURE CONNECTIONS

The Adventures of Connie and Diego

by Maria Garcia; illustrated by Malaquias Montoya
Children's Book Press, San Francisco. 1987
Grades: K–5

The twins Connie and Diego are born different than all the other children because they have "colors all over their little bodies." The other children laugh and laugh so one day the twins decide to run away in search of "a place where no one will make fun of us." They encounter a bear, a whale, a bird (who loves their colors), and a tiger. They learn that although their surface appearance is different from that of other children, they are human beings, no matter what color they may be. This strong anti-racist message, while designed for younger students, becomes a lesson for people of all ages (and colors). This book could touch off a constructive class discussion on the seeing beneath the surface to the common humanity of all people.

Amos and Boris

by William Steig
Farrar, Straus & Giroux, New York. 1971
Penguin Books, New York. 1977
Grades: K–4

Amos, the mouse, and Boris, the whale, discover their common link as fellow mammals as Boris returns Amos to land after he almost drowns at sea. After many years of mousing about, Amos is able to return the favor and rescue Boris when he is stranded by a tidal wave. A tender and comical story of friendship enriched by sophisticated vocabulary.

Ashanti to Zulu: African Traditions

by Margaret Musgrove;
illustrated by Leo and Diane Dillon
Dial Press, New York. 1976
Grades: 3–7

Beautifully illustrated and well-researched alphabet book that describes African ceremonies, celebrations, and day-to-day customs and reflects the richness and diversity of the peoples and cultures. In most of the paintings, a man, woman, child, an artifact, a local animal, and the living quarters are shown so that each page is quite detailed, even though all these elements might not ordinarily be seen together. The border design is based on the Kano Knot, a 17th-century design that symbolizes endless searching.

There is a true yearning to respond to
The singing River and the wide Rock.
So say the Asian, the Hispanic, the Jew
The African, the Native American, the Sioux,
The Catholic, the Muslim, the French, the Greek,
The Irish, the Rabbi, the Priest, the Sheik,
The Gay, the Straight, the Preacher,
The privileged, the homeless, the Teacher.
They hear. They all hear
The speaking of the Tree.

They hear the first and last of every Tree
Speak to humankind today.
Come to me, here beside the River.
Plant yourself beside the River.

— *Maya Angelou*
from On the Pulse of Morning

Chickens Aren't the Only Ones

by Ruth Heller

Grosset & Dunlap, New York. 1981

Grades: Preschool–5

Many animals lay eggs but their eggs and egg-laying behaviors differ. Brightly colored, almost 3-D, illustrations and rhyming text depict the diversity of all egg-laying (oviparous) animals.

Clive Eats Alligators

by Alison Lester

Houghton Mifflin, Boston. 1986

Grades: Preschool–2

Seven children have different styles of eating breakfast, getting dressed, shopping, caring for pets, going to bed, etc. Guess what Clive goes to sleep with? This visual game makes a great read aloud book as you turn the page at the end of each section anticipating the witty punch line.

Dinosaurs are Different

by Aliki

Harper & Row, New York. 1985

Grades: K–5

This book suggests ways that the various orders and suborders of dinosaurs were similar and different in structure and appearance.

A House Is A House For Me

by Mary Ann Hoberman; illustrated by Betty Fraser

Viking Penguin, New York. 1978

Grades: Preschool–3

The dwellings of various animals, peoples, and things such as a shell for a lobster or a glove for a hand are listed in rhyme. The concept of diversity is cleverly illustrated as the notion of a "house" is explored.

The Lamb and the Butterfly

by Arnold Sundgaard; illustrated by Eric Carle

Orchard Books, New York. 1988

Grades: K–3

A protected lamb and an independent butterfly discuss their different ways of living. Spirited introduction to the concept of diversity and acceptance of differences.

Loving

by Ann Morris; photographs by Ken Heyman

Lothrop, Lee & Shepard, New York. 1990

Grades: Preschool–5

Color photographs of a variety of world cultures remind us that the similarities of relationships among people are more important than the differences.

There were never
in the world two
opinions alike, any
more than two hairs
or two grains.
Their most universal
quality is diversity.

— *Montaigne
Essays*

The Mountains of Tibet

by Mordicai Gerstein

Harper & Row, New York. 1987

Grades: K–4

A Tibetan woodcutter lives his life without fulfilling his dream of seeing
other parts of the world. When he dies, he is given the choice of living
another life or going to "heaven." He chooses to live another life, and
given the option to be anyone, anywhere, he chooses his familiar life once
again. Before making his decision he sees the structure of the universe, of
"hundreds of millions of worlds" called galaxies, of stars, planets, and the
thousands of different kinds of people in the world.

On the Day You Were Born

by Debra Frasier

Harcourt, Brace, Jovanovich, San Diego. 1991

Grades: K–6

This beautiful book explores the relativity of time and our interrelationship
with nature and the universe by portraying other phenomena that take
place in nature at the same time as "you" (or any child) are born. The age
range of the book is extended by a special section "More About the World
Around You," which provides concise scientific information about the
natural world.

People

by Peter Spier

Doubleday, New York. 1980

Grades: Preschool–6

The differences between the billions of people that populate the Earth are
celebrated: different noses, different clothes, different customs, different
religions, different pets, and so on. You are left with a feeling of awe and
respect for the differences,
and enhanced appreciation
for the very basic similari-
ties that all humans share.

The Reason for a Flower

by Ruth Heller

Grosset & Dunlap, New York. 1983

Grades: Preschool–5

This beautifully illustrated rhyming book shows the diverse forms flowering
plants take, focusing on the function of seeds and pollen, different methods
of seed dispersal, and some examples of atypical flowers. A wonderful
way to introduce the unity and diversity among plants.

Stellaluna

by Janell Cannon

Harcourt, Brace & Co., New York. 1993

Grades: Preschool–4

A baby fruit bat named Stellaluna is separated from her mother when an owl attacks. Not yet able to fly, Stellaluna falls and clutches a tiny twig. She is adopted and raised by birds. Many of her "bat ways" disappear as she, for example, learns to eat bugs, but, although she tries, she never stops her habit of hanging by her feet. In the end she is reunited with her mother but stays connected to her bird friends. By bringing two quite different species together in a compelling and imaginative way, the book makes a strong connection to the idea of diversity and unity. Two detailed pages of fascinating background material about bats are included.

The Three Astronauts

by Umberto Eco; illustrated by Eugenio Carmi

Harcourt, Brace, Jovanovich, San Diego. 1989

Grades: K–5

An American, a Russian, and a Chinese astronaut take off separately in their own rockets with the goal of being first on Mars. They all land at the same time, immediately distrusting each other. When they encounter a Martian their cultural differences disappear as they unite against him. In a surprise happy ending, they recognize the "humanity" of the Martian after observing his kindness toward a baby bird and extend this understanding to differences between all peoples. Younger children may not get the full benefit of the sophisticated illustrations and humor. The astronauts are all male, with no women characters or references.

Correspondences

Nature is a temple from whose living columns
Commingling voices emerge at times;
We wander here through forests of symbols
That seem to observe us with familiar eyes.

Like long-drawn echoes afar converging
In harmonies darksome and profound,
Vast as the night and vast as light,
Colors, scents, and sounds correspond . . .

— *Charles Baudelaire*
(modified from a translation by Kate Flores)

Whose Footprints?

by Masayuki Yabuuchi

Philomel Books, New York. 1983

Grades: Preschool–4

A good visual guessing game for younger students that depicts the different footprints of a duck, cat, bear, horse, hippopotamus, and goat.

Why Does That Man Have Such a Big Nose?

by Mary Beth Quinsey; photographs by Wilson Chan

Parenting Press, Seattle. 1986

Grades: Preschool–6

Here are the answers to the sometimes awkward and embarrassing questions children often ask in a very matter of fact way. Twelve questions offer a wonderful springboard for discussions about differences among people.

Energy

All physical phenomena and interactions involve **energy**. In physical terms, energy is the capacity to do work or the ability to make things move. In chemical terms, it is the basis for reactions between compounds. In biological terms, energy gives living systems the ability to live, grow, and reproduce.

One of the most important laws in the physical sciences is that energy can never be created nor destroyed, but only transformed from one form to another. One way to transfer **heat energy** from one place to another is through **convection**. Another way to transfer heat is through **radiation**. Energy is very important in the earth sciences, and the uses and availability of some forms of energy have assumed an increasingly significant role in environmental issues and world events.

MAKING CONNECTIONS

A number of the books we list show **how animals use and store energy**. One book focuses on fireflies and the **chemical energy** that causes them to glow; others depict hibernation and how animals both store and preserve energy during the winter; and show how a chameleon reacts to **heat energy**.

Several books focus on the concept of **fuel and the direct relationship between amount of fuel consumed and the amount of work, light, or heat energy obtained**. One of these books is about a girl who must keep a lighthouse light burning with limited fuel, others deal with fueling spaceships with real and imaginary fuel.

Quite a few books deal with **force**, from the simple story of **pulling an enormous turnip from the ground** to calculations of **the number of hot air balloons needed to lift persons of various weight**. A

simple book about a runaway train engine explores the **relationship of mass and energy**, and another wonderful book is about **moving heavy things.**

Heat energy is touched on in a book about a snowball that melts, a book about the tremendous heat within the center of the Earth, a clever book on **refrigeration** and **air conditioning**; and another on the **insulating properties** of a special suit.

Several books on **wind and its effects** go nicely with another about **kites**. There are also books about **thunder** and **lightning, volcanoes, radioactivity,** and **electricity**, each in the context of a compelling story or in one case, a biography.

CROSS REFERENCES

LITERATURE CONNECTIONS

Blueberries for Sal

by Robert McCloskey

Penguin Books, New York. 1976

Grades: Preschool–3

On a summer day in Maine a little girl and a bear cub, wandering away from their blueberry-picking mothers, each mistakes the other's mother for its own. A wonderful first encounter with the need for stored food as energy, both as body fat (to be used during hibernation in the case of the bears), and as canned food (to be used in the winter as food in the case of the humans). Caldecott honor book.

Catch the Wind: All About Kites

by Gail Gibbons

Little, Brown & Co., Boston. 1989

Grades: K–6

When two children visit Ike's Kite Shop they learn about kites and how to fly them. A fun way to make use of the energy produced by wind! Includes instructions for building a kite.

A Chilling Story: How Things Cool Down

by Eve and Albert Stwertka; illustrated by Mena Dolobowsky

Julian Messner/Simon & Schuster, New York. 1991

Grades: 4–8

How refrigeration and air conditioning work is explained in simple language, with sections on heat transfer, evaporation, and expansion. Humorous black and white drawings show a family and its cat testing out these principles in their home.

Choo Choo

by Virginia Lee Burton

Houghton Mifflin, Boston. 1964

Grades: Preschool–2

A little black locomotive pulls trains from the city to the country. One day she runs away. Tired of pulling all those heavy coaches, she thinks she will be able to go much faster and easier by herself. A simple illustration of the interaction between mass (weight) and energy.

Einstein Anderson Sees Through the Invisible Man

by Seymour Simon; illustrated by Fred Winkowski

Viking Press, New York. 1983

Grades: 4–7

In Chapter 7, "A Cold Light," Einstein's ever hopeful friend Stanley is excited about a secret formula for cold light, which would offer an alternative to electric light. Einstein explains that light energy can be chemically produced by using a known formula similar the substance that causes a firefly's luminescence, but production would be too expensive.

> Over all, rocks, wood, and water, brooded the spirit of repose, and the silent energy of nature stirred the soul to its inmost depths.
>
> — *Thomas Cole*
> *Essay of American Scenery*

Einstein Anderson Shocks His Friends

by Seymour Simon; illustrated by Fred Winkowski

Viking Press, New York. 1980

Grades: 4–7

> In the first chapter, "The Electric Spark," Einstein uses static electricity to scare off his nemesis Pat Burns.

Fireflies in the Night

by Judy Hawes; illustrated by Ellen Alexander

Harper & Row, New York. 1963, 1991

Grades: K–5

> A young girl visits her grandfather and they investigate fireflies on summer nights. Describes how and why fireflies make their light, how to catch and handle them, and several uses for firefly light.

Gilberto and the Wind

by Marie Hall Ets

Viking Press, New York. 1963

Grades: Preschool–2

> Gilberto has a conversation with the wind as he investigates what the wind can do. The wind blows his balloon, plays with the clothes on the wash line, and bangs a gate shut. Children see how the wind can be playful and unpredictable at times. A nice introduction to wind as a source of energy.

Have Spacesuit, Will Travel

by Robert A. Heinlein

Charles Scribner's Sons, New York. 1958

Grades: 6–10

> A boy works very hard to win a soap jingle contest—first prize is a trip to the moon! He wins an obsolete model of a spacesuit, described in detail in the early chapters. Because he completely refits his suit and practices wearing it, he becomes caught up in a dramatic adventure. When he is on one of the outer planets (or moons), he has to test his suit against the weather in order to accomplish a task. A good description of insulation and heat energy.

Hill of Fire

by Thomas P. Lewis; illustrated by Joan Sandin

Harper & Row, New York. 1971

Grades: 2–5

> A Mexican villager working in the field opens up a crack in the Earth that erupts and within days births a new volcano, Paricutin. While told in simple language, the book would still be very appropriate for older students who are studying volcanoes. The story strongly conveys the power of the volcano. This is a true story of the 1943 event.

How To Dig a Hole to the Other Side of the World

by Faith McNulty; illustrated by Marc Simont
HarperCollins, New York. 1990
Grades: K–8

> A child takes an imaginary 8,000-mile journey through the earth and discovers what's inside. Depicts the heat energy stored in the center of the earth and makes connections to lava and geysers. (A detailed review of this book is on page 344.)

James and the Giant Peach

by Roald Dahl; illustrated by Nancy E. Burkett
Alfred A. Knopf, New York. 1961
Grades: 3–8

> James escapes from Aunt Sponge and Aunt Spiker via the giant peach. Inside the peach are insects of huge proportions. Many of the book's episodes suggest powerful metaphors for the generation of energy and the dynamics of force and movement, from the incredible growth of the peach itself to its destructive rolling over the countryside. The Cloud Men who create storms and build rainbows in the sky (and who throw hailstones at James and his insect friends) provide yet another energetic metaphor. You could discuss students' ideas of the energies involved in thunderstorms. It is worth noting that some reviewers have strongly criticized this author for negative and stereotypical female characters such as the cruel and greedy aunts in this book, who are crushed to death as the giant peach rolls over them.

Keep Looking!

by Millicent E. Selsam and Joyce Hunt; illustrated by Normand Chartier
Macmillan, New York. 1989
Grades: Preschool–3

> Where do animals spend their time in the winter in order to keep warm, find food, and maintain their energy? As the reader turns each page, a new animal is added to the lively world of an apparently empty country home in the winter.

Keep the Lights Burning, Abbie

by Peter and Connie Roop; illustrated by Peter E. Hanson
Carolrhoda Books, Minneapolis. 1985
Grades: 3–5

> True story of Abbie Burgess, daughter of a lighthouse keeper on the Maine coast in the 1850s, who kept the lights burning in the lighthouse during a four-week storm when her father was gone. Shows how crucial fuel (for both the light and the humans) is as an energy source.

There was a young lady named Bright
Whose speed was much faster than light.
She went out one day,
In a relative way,
And returned the previous night.

Marie Curie

by Mary Montgomery; illustrated by Severino Baraldi

Silver Burdett Press/Simon & Schuster, New Jersey. 1990

Grades: 4–7

Colorful biography of Marie Sklodowska including her childhood in Warsaw, study at the Sorbonne, marriage to scientist Pierre Curie, their joint work discovering radioactivity, and her discovery of radium and polonium. Includes sections describing applications of radioactivity and X-rays.

Mirandy and Brother Wind

by Patricia C. McKissack; illustrated by Jerry Pinkney

Alfred A. Knopf, New York. 1988

Grades: Preschool–5

Mirandy dreams of winning the Junior Cakewalk dance contest and enlists the wind to inspire her partner. Social life, clothing, and African-American dialect of a small town in the early 1900s are portrayed. The dancing, powerful energy of the wind is personified by the whirling figure of "Brother Wind" in a blue top hat.

The Mixed-Up Chameleon

by Eric Carle

Harper & Row, New York. 1975

Grades: Preschool–2

A bored chameleon wishes it could be more like all the other animals it sees, but soon decides it would rather just be itself. Protective coloration, (the chameleon changes color according to the surface on which it rests) and energy, (when the chameleon is warm and full, it turns one color, when it is cold and hungry, it turns another), are woven into the story, as is a discussion of the attributes of various other animals.

Moving Heavy Things

by Jan Adkins

Sandpiper/Houghton Mifflin, Boston. 1980

Grades: 6–Adult

Although this is a how-to guide to moving heavy things, it offers much more. Humorously-written chapters on body mechanics, friction, and techniques and devices (linework, block and tackle, levers, wedges, jacks, and winches) are illustrated with specific drawings and diagrams and include historical examples of problem solving. Winner of the Children's Science Book Award from the New York Academy of Sciences.

The Snowy Day

by Ezra Jack Keats

Viking Press, New York. 1962

Grades: Preschool–2

Peter goes for a walk on a snowy day. He makes different patterns in the snow with his feet, a stick, and then his whole body. He tries to save a snowball in his pocket but is disappointed when it melts. That night Peter dreams that the sun melted all the snow outside, but when he wakes up, it's snowing again! Good example of how heat energy can cause change.

The Space Ship Under the Apple Tree

by Louis Slobodkin

Macmillan, New York. 1952

Out of print

Grades: 5–8

Eddie befriends a stranded junior explorer from another planet. All his gadgets, as well as his spacecraft, are fueled by Zurianomatichrome, an energy source unavailable on Earth. The problem is finally solved when the explorer's people send energy to power his ship.

Thunder Cake

by Patricia Polacco

Philomel Books, New York. 1990

Grades: Preschool–6

Grandma finds a way to dispel her grandchild's fear of thunderstorms. The two race to gather ingredients and bake a "thunder cake" before the storm arrives. By counting the seconds between when they see the lightning and hear the thunder, they are able to measure the distance of the approaching storm. Both counting and cake-making provide an engaging and empowering distraction from the energy, sounds and sights of the thunderstorm. The recipe for thunder cake, with its secret ingredients, is included!

The Turnip: An Old Russian Folktale

by Katherine Milhous and Alice Dalgiesh; illustrated by Pierr Morgan

Philomel Books, New York. 1990

Grades: Preschool–3

One of Dedoushka's turnips grows to such an enormous size that the whole family, including the dog, cat, and mouse, is needed to pull it up. The concept of force is nicely joined with the need for cooperation, the value of persistence, and that even small beings are important. The energy of six different people and animals is needed to create enough force to dislodge the turnip.

Energy is
Eternal Delight.

— *William Blake*
Marriage of Heaven
and Hell

The Twenty-One Balloons
by William P. duBois
Viking Press, New York. 1947
Grades: 5–12

> Professor Sherman leaves San Francisco in 1883 to cross the Pacific by balloon. Three weeks later he is picked up in the Atlantic clinging to the wreckage of a platform that had been flown through the air by 21 balloons. On his trip he briefly passes through Krakatoa just before the historic volcanic eruption. Both the volcanic eruption and numerous descriptions of balloons lifting various weights connect to the theme of energy. Newbery award winner.

The Wonderful Flight to the Mushroom Planet
by Eleanor Cameron; illustrated by Robert Henneberger
Joy Street/Little, Brown & Co., New York. 1954
Grades: 5–8

> Chuck and David respond to an advertisement from the mysterious Mr. Tyco Bass (inventor, astronomer, and mushroom grower): "Wanted: a small space ship about eight feet long, built by a boy, or by two boys." In Chapters 7 and 8, the boys meet Mr. Bass and have their spaceship outfitted and fueled by him. There are details about the rocket motor, invention of a special fuel, and the energy requirements of the space ship.

"GEMSian" Energy

Many GEMS guides examine in depth or are related in some way to the theme of energy. All of the guides about animals and/or nutrition explore how living things obtain the energy to survive and grow. In addition:

- In *Chemical Reactions* the startling results generate a great deal of energy!

- In *Color Analyzers*, students investigate light.

- *Convection: A Current Event* enables students to observe convection currents then extend the concepts to ocean currents, wind, the movement of tectonic plates, and heat transfer within the Sun.

- In *Hot Water and Warm Homes from Sunlight* students learn more about solar energy by conducting experiments and controlling variables.

- A major topic in *Global Warming and the Greenhouse Effect* is the way solar energy is selectively absorbed by carbon dioxide and other greenhouse gases in our atmosphere and students consider how world energy use contributes to global warming. *Acid Rain* helps students understand how acid rain is created from the waste products of burning fossil fuels.

- *The Wizard's Lab* helps students learn more about electrical and solar energy, along with magnets, pendula, light polarizers, and a spinning platform.

The GEMS handbook, *To Build A House: GEMS and the Thematic Approach to Teaching Science*, defines ten major themes and explains how they relate to the entire GEMS series.

Evolution

The main idea of **evolution** is that the present arises, or evolves, from the past; that change takes place over time. Our hands, the fins of whales, the wings of bats, and the paws of cats all appear to have evolved from a set of bones in an ancient reptile through an evolutionary process involving natural selection and numerous other factors. While many people think of evolution as gradual, the process can include sudden leaps, explosions, chemical reactions, mutations, etc., so rate becomes a factor in analyzing the process. Additionally, the process does not take place in isolation. Other organisms are arising as dinosaurs decline; society undergoes an industrial revolution, resulting in changes in the Earth's atmosphere and climate; a theory of political science originates through interaction with the political conditions surrounding it. (The science theme of evolution includes the usual Darwinian biological meaning, but is far from limited to it.)

MAKING CONNECTIONS

The books included show **change over time** in a variety of situations, including cities changing, an oak tree growing, a skunk dying, the life of a tree stump, changes in ecosystems, in airplane designs, and more. A number of books reflect the **biological theory of evolution**, featuring fossils, and imagining what the world might have been like **millions of years ago**. Several books focus on dinosaurs and how they might have evolved.

The **concepts of time and evolution** are shown in *Go Fish*, a wonderfully thought-provoking and tender book about a boy and his grandfather. *A Day in the Tropical Rain Forest* touches on the issue of **endangered species** as a boy and a scientist search for a rare species of butterfly that lives in the rain forest canopy. We've also included two books about **Charles Darwin**.

CROSS REFERENCES

LITERATURE CONNECTIONS

Before Adam

by Jack London

MacMillan, New York. 1962

Grades: 6–Adult

> This riveting short novel takes us back to the dawn of human evolution, as a modern boy uses his genetic memory to recall the time and adventures of Big-Tooth, a child of the Tree People—when people first evolved from ape-like creatures. An afterword by Loren Eiseley points out some inaccuracies based on modern findings, including telescoping three separate stages into one, but highly praises the underlying scientific and humanitarian impulses. This fascinating book is certain to get students thinking about evolution.

The Beagle and Mr. Flycatcher: A Story of Charles Darwin

by Robert Quackenbush

Prentice-Hall, Englewood Cliffs, New Jersey. 1983

Grades: 4–8

> As an unpaid naturalist aboard the brig H.M.S. Beagle on a five-year voyage around South America, Charles Darwin began to formulate his revolutionary theory of evolution. This biography includes brief descriptions of Darwin's specimen collecting , scientific observation, and of a subsequent eight-year study of barnacles. The illustrations are in a cartoon style and the writing uses a humorous approach, but quite a bit of scientific information is conveyed.

Darwin and the Voyage of the Beagle

by Felicia Law; illustrated by Judy Brook

Andre Deutsch, Great Britain. 1985

(Distributed by E.P. Dutton, New York)

Grades: 4–8

> A cabin boy along on Charles Darwin's five-year voyage keeps a diary. Assisting Darwin with his collection of insect, bird, and marine life specimens, the boy learns about their habits and habitats. On one occasion they return with 68 different species of one beetle. They collect fossils in the Andes, straddle Galapagos tortoises, discover the skeleton of the Megatherium, and get to know Fuegan natives. The format is oversized, with many drawings, charts and maps.

Dinosaurs Are Different

by Aliki

Harper & Row, New York. 1985

Grades: K–5

> This book describes how the various orders and suborders of dinosaurs were similar and different in anatomy and appearance. Are birds descended from dinosaurs? Students can compare dinosaurs to identify those that might be related to each other.

Glorious Flight

by Alice and Martin Provensen
Viking Press, New York. 1983
Grades: 2–4

This is the true story of Louis Bleriot, a pioneer of aviation, who developed and flew a plane over the English Channel in 1909. The evolution of his various prototypes of flying machines is described, from Bleriot I, which "flaps like a chicken," Bleriot II, a glider without a motor, Bleriot VII, a "real aeroplane that will fly," to Bleriot XI, which makes the 36-minute flight. The charming watercolor illustrations of a French village, the Bleriot family in period costume, and the very rudimentary aircraft emphasize the audacity of the attempt. Caldecott award winner.

Go Fish

by Mary Stolz; illustrated by Pat Cummings
HarperCollins, New York. 1991
Grades: 4–6

Eight-year-old Thomas and his grandfather go fishing in the Gulf of Mexico, sharing questions and spirited opinions. Grandfather is a collector of shells, petrified wood, and even a sandstone with a fossil fish. On pages 12 and 13, Thomas muses on the fossil fish, trying to imagine how it could have lived 50 million years ago. In Chapter 2, they talk about predators and which animals are most likely to survive. Throughout the book, a child's sense of wonder at imagining a past world without himself in it and grappling with the concepts of time and evolution are well portrayed. (A detailed review of this book is on page 336.)

If You Are a Hunter of Fossils

by Byrd Baylor; illustrated by Peter Parnall
Charles Scribner's Sons, New York. 1980
Grades: 3–6

A fossil hunter looking for signs of an ancient sea in the rocks of a western Texas mountain describes how the area must have looked millions of years ago.

In the Night, Still Dark

by Richard Lewis; illustrated by Ed Young
Atheneum, New York. 1988
Grades: K–6

This poem emphasizes the unity of living creatures and traces the evolution of all life from the depths of darkness to the simplest creatures, to the birth of human life and the day. It is based on the Hawaiian creation song, the Kumulipo, originally chanted at the birth of each royal child. The illustrations are striking.

A Rock, A River, A Tree
Hosts to species long since departed,
Marked the mastodon,
The dinosaur, who left dried tokens
Of their sojourn here
On our planet floor,
Any broad alarm of their hastening doom
Is lost in the gloom of dust and ages.

But today, the Rock cries out to us, clearly, forcefully,
Come, you may stand upon my
Back and face your distant destiny,
But seek no haven in my shadow.
I will give you no hiding place down here.

— *Maya Angelou*
from On the Pulse of Morning

A Quick Stop
at Tinker Creek

I am sitting under a sycamore by Tinker Creek. I am really here, alive on the intricate earth under trees. But under me, directly under the weight of my body on the grass, are other creatures, just as real, for whom also this moment, this tree, is "it." Take just the top inch of soil, the world squirming right under my palms. In the top inch of forest soil, biologists found "an average of 1,356 living creatures present in each square foot, including 865 mites, 265 springtails, 22 millipedes, 19 adult beetles and various members of 12 other forms . . . Had an estimate also been made of the microscopic population, it might have ranged up to two billion bacteria and many millions of fungi, protozoa and algae—in a mere *teaspoonful* of soil." The chrysalids of butterflies linger here too, folded, rigid, and dreamless. I might as well include these creatures in this moment, as best I can . . .

Earthworms in staggering procession lurch through the grit underfoot, gobbling downed leaves and spewing forth castings by the ton. Moles mine intricate tunnels in networks; there are often so many of these mole tunnels here by the creek that when I walk, every step is a letdown. A mole is almost entirely loose inside its skin, and enormously mighty. If you can catch a mole, it will, in addition to biting you memorably, leap from your hand in a single convulsive contraction and be gone as soon as you have it. You are never really able to see it; you only feel its surge and thrust against your palm, as if you hold a beating heart in a paper bag. What could I not do if I had the power and will of a mole! But the mole churns earth . . .

Under my spine, the sycamore roots suck watery salts. Root tips thrust and squirm between particles of soil, probing minutely; from their roving, burgeoning tissues spring infinitesimal root hairs, transparent and hollow, which affix themselves to specks of grit and sip . . . Under the world's conifers—under the creekside cedar behind where I sit—a mantle of fungus wraps the soil in a weft, shooting out blind thread after frail thread of palest dissolved white. From root tip to root tip, root hair to root hair, these filaments loop and wind; the thought of them always reminds me of Rimbaud's "I have stretched cords from steeple to steeple, garlands from window to window, chains of gold from star to star, and I dance." . . . Here the very looped soil is an intricate throng of praise. Make connections; let rip; and dance where you can.

— *Annie Dillard*
Pilgrim at Tinker Creek

Just A Dream

by Chris Van Allsburg

Houghton Mifflin, Boston. 1990

Grades: K–5

When he has a dream about a future Earth devastated by pollution, Walter begins to understand the importance of taking care of the environment. Unique and evocative pictures of what our future may hold provide the powerful backdrop as young Walter becomes enlightened and changes his thinking and actions. He learns that even ecosystems can evolve.

Living With Dinosaurs

by Patricia Lauber; illustrated by Douglas Henderson

Bradbury Press/Macmillan, New York. 1991

Grades: 3–6

In prehistoric Montana 75 million years ago lived the giant reptiles and fishes of the sea; the birds and pterosaurs in the sky; the dinosaurs, tiny mammals, crocodiles, and plants of the lowlands; and the predators of dinosaur nesting grounds in the dry uplands. This book describes their relationships to each other. At the end is a clear and elementary description of how a fossil forms and evolves. The colorful paintings are dynamic.

My Place

by Nadia Wheatley and Donna Rawlins

Kane/Miller Book Publishers, New York. 1987

Grades: 4–8

The history of a neighborhood in an Australian city is chronicled via two-page spreads that each have drawings, a map, and a story written by a child who lives on a certain spot of land—my place. The story goes back decade by decade as you learn who lives on that spot of land, what else is in the vicinity, and what happens in the lives of those who live there. The evolution of lifestyle, vocations, buildings, and the land are beautifully depicted by these accounts and maps. Children's Book Council of Australia Book of the Year award winner.

Nate's Treasure

by David Spohn

Lothrop, Lee & Shepard, New York. 1991

Grades: Preschool–3

After Bruno the family dog tangles with a skunk and kills it, the father buries the skunk near a windbreak offering "a feast for all the creatures." The seasons pass and the skunk's skeleton is retrieved and treasured by the boy Nate. Delicate pointillist paintings portray the cycles of nature, including dying. Shows change over time as the skunk passes from live animal to skeleton.

Evolution is
not a force,
but a process . . .
— *Viscount Morley*
On Compromise

New Providence: A Changing Cityscape
by Renata Von Tscharner and Ronald L. Fleming; illustrated by Denis Orloff
Gulliver Books/Harcourt, Brace, Jovanovich, San Diego. 1987
Grades: 4–7

    The evolution of an American city from 1910 to 1987 is shown in six double-page colored illustrations with a brief text describing the factors involved and the basic architectural styles depicted. Complete down to billboards and graffiti, this sequence of detailed illustrations invites careful observation and comparisons. Students could be asked to list the factors causing the evolution of this city.

An Oak Tree Dies and a Journey Begins
by Louanne Norris and Howard E. Smith, Jr.; illustrated by Allen Davis
Crown, New York. 1979
Out of print
Grades: 3–6

    A storm uproots an old oak tree on the bank of a river, it falls in and its journey to the sea begins. Animals seek shelter in the log, children fish from it, mussels attach to its side. This tree, even after it dies, contributes to the environment. Older students will appreciate the fine pen and ink drawings.

Once There Was a Tree
by Natalia Romanova; illustrated by Gennady Spirin
Dial Books/Penguin, New York. 1985
Grades: Preschool–5

    Rich and detailed color illustrations trace the evolution of a tree which was struck by lightning, cut down, and reduced to a stump. The stump is visited, inhabited, or used by a succession of beetles, birds, ants, bears, frogs, earwigs, and humans who consider it theirs. As a new tree grows from the old stump, the question arises, "Who's tree is it?"

One Day in the Tropical Rain Forest
by Jean C. George; illustrated by Gary Allen
HarperCollins, New York. 1990
Grades: 4–7

    When a section of rain forest in Venezuela is scheduled to be bulldozed, a young boy and a scientist seek a new species of butterfly for a wealthy industrialist who might preserve the forest. As they travel through the ecosystem rich with plant, insect, and animal life, everything they see on this one day is logged beginning with sunrise at 6:29 a.m. They finally arrive at the top of the largest tree in the forest and fortuitously capture a specimen of an unknown butterfly. A glimpse into the ecosystem of the tropical rain forest.

The Roadside

by David Bellamy; illustrated by Jill Dow
Clarkson N. Potter, New York. 1988
Grades: 3–5

Construction of a six-lane highway in the countryside disrupts the balance of nature and forces animals there to change their patterns. The toads lay their eggs in different places, some animals eat the leftovers from the road workers' lunches, a raw mound of earth is soon covered by wildflowers that attracts butterflies. An attempt is made to be evenhanded, "the road builders have done a good job." The toads use the new pipe under the road as a passageway, no hunters with guns come there because walking is not allowed, etc. Students could consider how human interaction with the environment can impact evolutionary adaptation and change.

Sierra

by Diane Siebert; illustrated by Wendell Minor
HarperCollins, New York. 1991
Grades: 4–8

Long narrative poem in the voice of a mountain in the Sierra Nevada, beginning and ending

> *I am the mountain*
> *Tall and grand.*
> *And like a sentinel I stand.*

Dynamic verse and glorious mural-like colored panels depict the forces shaping the earth over time and the plant, animal, and human roles in this ecosystem.

And Still the Turtle Watched

by Sheila MacGill-Callahan;
illustrated by Barry Moser
Dial, New York. 1991
Grades: K–5

A turtle carved by Native Americans in a rock watches with sadness the changes humans bring over the years. After the rock is cleaned of spray paint and installed indoors at a botanical garden, the turtle's vision is restored and he communicates his wisdom to the many children visiting. The watercolor paintings are dramatic.

Gone Fishin'

By Lincoln Bergman

*G*o Fish by Mary Stolz has something very special about it. You will find it annotated under the GEMS guide *Mapping Fish Habitats* and under the theme of "Evolution." Yet like many of the most compelling and compassionate books in this collection, such categorization to some extent "misses the boat." The book, among much else, intertwines a family relationship between an African American boy and his grandfather, a dynamic sense of nature, a recurring awareness of the relativity of time, imaginative fascination with a fossil, intuitive and amusing understanding of and communication with the family cat, and an African folktale about why the east wind and the west wind are always chasing each other around the globe. The contradictions involved in treasuring all life and fishing for your dinner are discussed, and the title of the book becomes a joke, hilarious to Thomas, when he and his grandfather play cards. Throughout there is both an endearing sense of humor and a sacred seriousness about the life cycle and the human role within it:

> "Thomas tried to imagine a world that he, his very own self, wasn't part of—a world that people had been going up and down in, doing the things people do, a world where *Grandfather* had been a boy himself, with a grandfather of his own, and where *that* grandfather had had *his* grandfather, and you could keep going back like that until, as Grandfather often said, you were face to face with a long-long-long ago ancestor in Benin, an ancient town in Africa. And none of those people knowing a thing about Thomas.

> "Well, it was strange.

> "Thomas liked to hear stories about those times. But even as he was listening to and believing Grandfather's tales, he couldn't make himself know that there had once been a world without him in it. It was as hard to believe that North Carolina had been at the bottom of the ocean, or that the tiny fossil fish on Grandfather's desk had lived and swum in and died in that ocean, fifty million years ago."

Perhaps it is the **interweaving** itself—of childhood experience, oral history, fascination with the natural world, awareness of pain and loss, and sheer joy in the rhythm of daily life—that distinguishes this book. We recommend it highly for young and old alike! The author spins another story with the same characters in a highly acclaimed earlier book, *Storm in the Night.*

Lincoln Bergman is the principal editor of the GEMS project.

Matter

atter is the stuff that makes up all of the objects in the universe, including ourselves.

The notion of matter is important in all the sciences. Physicists and chemists conduct experiments and develop mathematical models to explain the behavior of the smallest particles of matter at very high energies. Materials engineers study the interactions of various elements and compounds under a wide range of conditions, and try to develop materials with special properties such as superconductivity or high strength. Biochemists try to unravel the material basis for life processes, and geneticists have begun to manipulate the genetic code within cells to change the properties and products of living organisms.

MAKING CONNECTIONS

Many of the books we list describe the **properties of matter** in various and appealing ways. The composition of the Earth, the nature of hot air, popcorn, quicksand, soap, water, rocks, slime, and even the fictional substance, Oobleck, are investigated in detail through picture books, adventure stories, and books that feature fun and sometimes obscure topics.

The other overlapping category of books listed portray **changing forms of matter**. Phase change and chemical change are featured through adventures involving a hot air balloon, water purification, a quest to make salty coffee drinkable again, a book explaining how refrigeration and air conditioning work, a young genetic engineer's monstrous creation, and a great cookbook. In each situation, matter is changed in many ways, and doing so creates useful products and processes as well as out-of-control results!

An inspirational **biography of Marie Curie**, discoverer of two elements and sleuth of the nature of radioactive matter, is included. Accounts of great scientists, whose work was directed at understanding and utilizing the properties of matter, are well worth sharing.

CROSS REFERENCES

LITERATURE CONNECTIONS

Bartholomew and the Oobleck

by Dr. Seuss
Random House, New York. 1949
Grades: K–9

A king orders his royal magicians to cause something new to rain down from the sky. But when the green gooey material "Oobleck" falls onto the kingdom, its strange properties cause quite a mess until the king learns some humility. (A note about how this book can be used with younger students, and how to modify it for appropriate use with older students, is on page 208.)

A Chilling Story: How Things Cool Down

by Eve and Albert Stwertka; illustrated by Mena Dolobowsky
Julian Messner/Simon & Schuster, New York. 1991
Grades: 4–8

How refrigeration and air conditioning work is simply explained, with sections on heat transfer, evaporation, and expansion. Humorous black and white drawings show a family and its cat testing out these principles in their home. Nice examples of practical applications of changing matter.

Einstein Anderson, Science Sleuth

by Seymour Simon; illustrated by Fred Winkowski
Viking Press, New York. 1980
Grades: 4–7

In the "Universal Solvent," Einstein Anderson's friend Stanley tries to convince him that the cherry soda-looking liquid he has invented will dissolve anything. Einstein's knowledge of the properties of solvents leads him to unravel this mystery.

Elliot's Extraordinary Cookbook

by Christina Bjork; illustrated by Lena Anderson
Farrar, Straus & Giroux, New York. 1990
Grades: 3–6

With the help of his upstairs neighbor, Elliot cooks wonderful foods and investigates what's healthy and what's not so healthy. He finds out about proteins, carbohydrates, and the workings of the small intestine. He learns about the history of chickens and how cows produce milk. His friend shows him how to grow bean sprouts, and he sews an apron. Nice real-life connection to the ways that cooking, by combining many substances at varying temperatures, is an exploration of matter and its properties.

> Life is not easy for any of us. But what of that? We must have perseverance and, above all, confidence in ourselves.
>
> — *Marie Curie*

Everybody Needs a Rock

by Byrd Baylor; illustrated by Peter Parnall
Aladdin Books, New York. 1974
Grades: Preschool–5

> The qualities to consider in selecting the perfect rock for play and pleasure include color, size, shape, texture, and smell. After reading this delightful book, you'll want to rush out and find a rock of your own. Nice introduction or follow-up to a discussion of the properties of solids.

Hot-Air Henry

by Mary Calhoun; illustrated by Erick Ingraham
William Morrow, New York. 1981
Grades: K–3

> Henry, a spunky Siamese cat, stows away on a hot air balloon and accidentally gets a solo flight. He learns that there is more to ballooning than just watching as he deals with air currents, power lines, and manipulating the gas burner. Shows practical use of a gas and how changing the temperature of a gas changes its properties.

How to Dig a Hole to the Other Side of the World

by Faith McNulty; illustrated by Marc Simont
HarperCollins, New York. 1990
Grades: K–8

> A child takes an imaginary 8,000-mile journey through the earth and discovers what's inside. A good opportunity to learn about the composition of the earth as each layer is examined. (A separate feature on this book is on page 344.)

The Lady Who Put Salt in Her Coffee

by Lucretia Hale
Harcourt, Brace, Jovanovich, San Diego. 1989
Grades: Preschool–6

> When Mrs. Peterkin accidentally puts salt in her coffee, the entire family embarks on an elaborate quest to find someone to make it drinkable again. A visit to a chemist, an herbalist, and a wise woman results in a solution, but not without having tried some wild experiments first.

The Magic School Bus at the Waterworks

by Joanna Cole; illustrated by Bruce Degen
Scholastic, New York. 1986
Grades: K–6

> When Ms. Frizzle, the strangest teacher in school, takes her class on a field trip to the waterworks, everyone ends up experiencing the water purification system from the inside. Evaporation, the water cycle, and filtration are just a few of the concepts explored in this whimsical field trip.

The Magic School Bus Inside the Earth

by Joanna Cole; illustrated by Bruce Degen
Scholastic, New York. 1987
Grades: K–6

> On a special field trip in the magic school bus, Ms. Frizzle's class learns first hand about different kinds of rocks and the formation and structure of the earth.

Marie Curie

by Mary Montgomery; illustrated by Severino Baraldi
Silver Burdett Press/Simon & Schuster, New York. 1990
Grades: 4–7

> Colorful biography of Marie Sklodowska including her childhood in Warsaw, study at the Sorbonne, marriage to scientist Pierre Curie, their joint work discovering radioactivity, and her discovery of the elements radium and polonium. Additional sections describe the applications of radioactivity and X-rays.

The Microscope

by Maxine Kumin; illustrated by Arnold Lobel
Harper & Row, New York, 1984
Grades: 4–8

> A beautifully written and illustrated book, which, while poking fun at many things, also portrays Anton van Leeuwenhoek grinding lenses, exploring matter by viewing many common objects under a microscope, and historical information. Although told in rhyme and with few words, its language is fairly sophisticated, and the book can be read with delight by older students and adults.

The Monster Garden

by Vivien Alcock
Delacorte Press, New York. 1988
Grades: 5–8

> Frankie Stein, whose father is a genetic engineer, creates her own special monster, Monnie, from a "bit of goo" her brother steals from the lab. Scientific information is sprinkled throughout the book and Chapter II includes Frankie's experiment log. Very current ideas about how inanimate matter can be given life can be introduced. This combination of fantasy, science fiction, and young adult novel stimulates thinking about the complex issues surrounding biotechnology.

EATS

Under

    this autumn sky   i think of these
ingredients when they were rooted
in the ground
        that pound of flour
                as
      some stalks
         of wheat
this sugar as
      sweet
   sugar canes or beets

   even the chicken eggs
and salt and rising
      yeast

i always use this wooden spoon to stir
      the batter for the bread

it was
   once a tree   or part of a tree
      rooted in the ground

under
   the
   sky

— Arnold Adoff

If you love it
enough, anything
will talk to you.

— *George Washington Carver*

The Popcorn Book

by Tomie dePaola

Holiday House, New York. 1978

Grades: 2–4

> While two brothers pop popcorn, answers are given to questions such as "Why does popcorn pop?" and "Why is it best stored in the refrigerator?" Tidbits of history and folklore include other uses and ways of cooking popcorn, such as in hot sand.

The Quicksand Book

by Tomie dePaola

Holiday House, New York. 1984

Grades: 2–4

> A jungle girl learns about the composition of quicksand, how different animals escape it, and how humans can use precautions to avoid getting stuck. Her teacher, the overly confident jungle boy, turns out not to be so superior. A variety of graphics and a helpful monkey give visual interest. A recipe for making your own quicksand is included.

The Slimy Book

by Babette Cole

Random House, New York. 1986

Out of print

Grades: Preschool–4

> Lighthearted look at slime in all its "sticky, sludgy, slippy, sloppy, ploppy, creepy kind" and where it may be found: around the house, in invertebrate creatures, in foods, and maybe even outer space. Nice way to model good descriptive language of the properties of an intriguing form of matter.

The Snowy Day

by Ezra Jack Keats

Viking Press, New York. 1962

Grades: Preschool–2

> Peter goes for a walk on a snowy day. He makes different patterns in the snow with his feet, a stick, and then his whole body. He tries to save a snowball in his pocket but is disappointed when it melts. That night Peter dreams that the sun melted all the snow outside, but when he wakes up, it's snowing again! Good example of phase change of matter (snow to water).

Splash! All About Baths

by Susan K. Buxbaum and Rita G. Gelman; illustrated by Maryann Cocca-Leffler

Little, Brown & Co., Boston. 1987

Grades: K–6

> Penguin answers his animal friends' questions about baths such as: "What shape is water?" "Why do soap and water make you clean?" "What is a bubble?" "Why does the water go up when you get in?" "Why do some things float and others sink?" Answers to these questions are both clear and simple. Won American Institute of Physics Science Writing Award.

Water's Way
by Lisa W. Peters; illustrated by Ted Rand
Arcade Publishing/Little, Brown & Co., Boston. 1991
Grades: K–3

"Water has a way of changing" inside and outside Tony's house, from clouds to steam to fog and other forms. Innovative illustrations show the changes in the weather outside while highlighting water changes inside the house.

Poem Stew

Poems selected by William Cole
Pictures by Karen Ann Weinhaus
HarperCollins, New York. 1983

This poetic romp through the world of food and related matters contains many poems that could be lightheartedly related to GEMS activities and guides, such as *Vitamin C Testing,* with Ogden Nash's "Tableau at Twilight" where we discover that ice cream

> "Cones are composed of many a vitamin
> My lap is not the place to bitamin."

Issues of density and flow, as in *Discovering Density,* naturally arise from couplets like this anonymous ditty:

> "When you tip the ketchup bottle,
> First will come a little, and then a lot'll."

X.J. Kennedy's hilarious poem "Father Loses Weight" might be used to introduce the theme of Stability, and numerous poems exemplify the visual and tactile observations so basic to the scientific method. The natural tendency for many of these poems to focus on unusual behaviors of certain food substances, including gooey and sticky properties, suggests the GEMS "Oobleck" activities, the theme of Matter, and the GEMS assembly guide *Solids, Liquids, and Gases.*

DIGGING A HOLE

By Jacqueline Barber

In Session 2 of the GEMS unit *Earth, Moon, and Stars*, students grapple with the concepts of the Earth's shape and gravity. They apply their mental models of the Earth to real and imaginary situations as they engage in animated discussions. One of the discussion questions asks students to pretend that a tunnel was dug all the way through the Earth, from pole to pole. Students are asked to imagine that a person holds a rock above the opening at the North Pole, and then drops the rock into the tunnel. What happens to the rock?

While this age-old question is used to stretch students to apply their understanding of the concept of gravity, many a class has enjoyed thinking and talking about the tunnel itself and what the center of the Earth is like.

How to Dig a Hole to the Other Side of the World by Faith McNulty (HarperCollins Publishers, 1979), is a wonderful children's literature book that can capitalize on your students' curiosity about the center of the Earth. This clever book provides instructions for those who want to dig to the other side of the world. Beginning with:

> "Find a soft place. Take a shovel and start to dig a hole. The dirt you dig up is called loam. Loam, or topsoil, is made up of tiny bits of rock mixed with many other things, such as plants and worms that died and rotted long ago."

The digging goes on, through the next layer of clay, gravel, or sand, through rockier soil, to bedrock. But it doesn't stop there! Diggers are instructed on how to get through bedrock, through the water table, through magma, the mantle, the outer core, the inner core, to the center of the Earth. And that's only halfway through the Earth!

The book has practical tips for the diggers, about using a bucket to pull up the soil and rocks as you go, about using a drill when you come to bedrock, an underwater diving suit for when you hit underground rivers and lakes, and yes, a jet-propelled submarine with a super cooling system, a fireproof skin and a drill at the tip of its nose for negotiating through the magma. It warns you to watch out for fossils, diamonds, geysers (that could deliver you suddenly back to the surface), gooey oil deposits, and more. It gives the following advice for when you reach the center of the Earth:

> "At the center of the earth there is nothing under you. Every direction is up. Your feet are pointing up and your head is pointing up, both at the same time. Because there is nothing under you, you will weigh nothing...The weight of the whole world will press down on your ship. Do not stay long."

The final surprise is that you don't end up in China, but at the bottom of the Indian Ocean (if you left from the United States, that is). This delightful 8,000 mile journey through the Earth provides students with information in the most vivid way. It is an ideal use of children's literature to extend the science investigations begun in *Earth, Moon, and Stars*.

Jacqueline Barber is the director of the GEMS project.

Models & Simulations

— Including Physical, Conceptual, and Mathematical Models —

A **model** is a simplified imitation that can help us better understand a phenomenon, system, or process.

The model may be a device, a plan, a drawing, an equation, a computer program, or a mental image. The value of models lies in suggesting how things work or might work. Models, whether physical, conceptual, mathematical, or computer-based, have served invaluable functions in enhancing human scientific understanding. However, models can also be misleading, suggesting characteristics, mechanisms, or analogies that do not exist or are inaccurate. The limitations of models also need to be understood.

Simulations are usually dynamic models. Simulations represent natural processes in the form of, for example, a game, a computer program, or a dramatization.

MAKING CONNECTIONS

Scientists use models and simulations to better understand real phenomena, systems, or processes. Literature can serve the same purpose—to enable the reader to better understand about real relationships, events, topics, etc. In a certain way, most literature can be viewed as models or simulations of life.

Most of the books we chose to list however involve **physical models**: of cars, airplanes, and even a robotic dog. Several books present a **simulation**—in one case exploring what can happen to an ecosystem through overuse and abuse of resources. *Almost the Real Thing: Simulation in Your High-Tech World* is a fascinating behind the scenes explanation of how a wide variety of processes are simulated, from earthquakes and other disasters to flights and computer modeling.

It may also be interesting for you to consider the **models of natural phenomena** that are represented in **myths from other cultures**. Books that include such myths can be found in the listings in this handbook for the GEMS guides *Earth, Moon, and Stars* and *Investigating Artifacts*.

CROSS REFERENCES

LITERATURE CONNECTIONS

Almost the Real Thing:
Simulation in Your High-Tech World
by Gloria Skurzynski
Bradbury Press, New York. 1991
Grades: 4–10

Simulation of flight (airplanes and space ships), accidents and disasters (earthquakes), space flight conditions (astronaut training), computer-image simulation technology (training for pilots and other personnel), computer modeling, and the new virtual reality technology are all explained. With active color photographs and up-to-the-minute examples, this book should interest students in how models and simulations work both inside and outside the classroom.

Einstein Anderson Tells a Comet's Tale
by Seymour Simon; illustrated by Fred Winkowski
Viking Press, New York. 1981
Grades: 4–7

Chapter 10 describes a soapbox derby race in which both teams have to build soapbox racing cars that weigh the same amount and are started in the same way. Our hero identifies the one test variable that allows the team to win the race. Soapbox cars are of course *models* of racing cars.

Galimoto
by Karen L. Williams; illustrated by Catherine Stock
Lothrop, Lee, and Shepard, New York. 1990
Grades: 1–4

In a Malawi village in Africa, "galimoto" means car and also signifies a type of push toy made by children in the shape of cars, trucks, bicycles, trains, etc. A young boy finds materials and recycles them into a galimoto made like a pickup truck, which he might modify to an ambulance, airplane or helicopter tomorrow. Children will appreciate the creation of models and the recycling concept. They will enjoy the watercolor illustrations showing how the whole village becomes involved in the project.

Glorious Flight
by Alice and Martin Provensen
Viking Press, New York. 1983
Grades: 2–4

This is the true story of Louis Bleriot, a pioneer of aviation, who developed and flew a plane over the English Channel in 1909. The evolution of his various prototypes of flying machines is described, from Bleriot I, which "flaps like a chicken," Bleriot II, a glider without a motor, Bleriot VII, a "real aeroplane that will fly," to Bleriot XI, which makes the 36-minute flight. The charming watercolor illustrations of a French village, the Bleriot family in period costume, and the very rudimentary aircraft emphasize the audacity of the attempt. Caldecott award winner.

> Nature has ... some sort of arithmetical-geometrical coordinate system, because nature has all kinds of models. What we experience of nature is in models, and all of nature's models are so beautiful.
>
> — *Buckminster Fuller*

The Pet of Frankenstein

by Mel Gilden; illustrated by John Pierard
Avon Books, New York. 1988
Grades: 5–7

> This book is one in an entertaining series of adventures involving Danny Keegan and his monstrous classmates. Frankie's particular expertise is electronics, so there is a lot of it and some nicely interwoven technical language. From video games and computers to Tesla coils and robotic animals, the bizarre narrative jumps amusingly along. In the end, Frankie succeeds in creating a robotic dog from an old design of his namesake, Baron Frankenstein. Definitely not heavy reading, the book will tickle some students' imagination and funny bones.

The Wump World

by Bill Peet
Houghton Mifflin, Boston. 1970
Grades: Preschool–5

> The Wump World is invaded by Pollutians from the planet Pollutus, who had left their worn-out planet to start a life in a new world. They move in and take over, building cities, creating pollution, and using up resources. Eventually, the Pollutians move to another planet and leave the Wumps to try and revive their ecosystem. This at first silly, then sobering, scenario presents a simulation of what can happen to an ecosystem through overuse and abuse of resources.

Patterns of Change

— Trends, Cycles, Random and Unpredictable Changes —

Analyzing **patterns of change** is of great importance in the sciences. Patterns of change can be classified into three general categories: (1) Changes that are steady, predictable trends, such as the increasing speed of a falling rock or the rate of radioactive decay; (2) Changes that are cyclical, occurring in a sequence that happens over and over, such as the seasonal cycles of weather or the vibration of a guitar string; and (3) Changes that appear to be random and unpredictable, or so complicated as to have apparently chaotic results.

A system may have all three kinds of change occurring at once. In general, describing change helps us make predictions about future changes; analyzing change helps us understand what is going on as well as predict what may happen, and control of change is important in the design of technological systems.

MAKING CONNECTIONS

All literature deals with change, whether it be focused on life changes to which a character must adapt, the changing nature of a town, or a series of events and how they cause people and places to change. Thus, this "literature connections" section could be infinitely large! Indeed, you could ask your students to list examples of change in any book you read with them or they read on their own.

The books listed here include very explicit examples of change. Because scientists and mathematicians analyze change, and there is great value

in having students do this as well, most of the books clearly exemplify many of the **patterns of change** described above.

For the youngest students, several books deal with the most **simple notion of change**—that some things change and some stay the same. There are also quite a few examples of cyclical change or **cycles**. This most elegant form of change appears everywhere we look, from the life cycles of plants, animals and humans, to such processes as the water cycle and seasonal change. There are two books which deal with non-natural cycles, a cycle of requests. Students of all ages can relate more easily to the concept of a cycle when they have experienced or read about a wide variety of cycles.

Many of the books focus on the environment and how humans have caused **reversible or irreversible change**. Death is an irreversible change—this common literary theme appears in several of the books. A crazy concoction that can't be "undone," and a hatched egg are two other instances of irreversible change.

Trends can be identified in many change situations. If you look at a portion of a cycle, let's say from seed to seedling to plant, there's often a good example of a trend. Students get good at identifying trends in stories: "The days are getting colder and colder" "There is less and less oil," "The chipmunk gets fatter and fatter as autumn approaches."

In one book a woman attempts to change things by making them more beautiful, in another the changing uses of a rock are depicted. These are examples of **random change**. Books dealing with **evolutionary change** (change over time) are listed under the separate theme of Evolution in this handbook, starting on page 329.

Many of the books listed below provide great opportunities for class discussions of change in general—what it is, how it happens, and its implications. Change in and of itself is interesting, even when it is hard to classify into a specific category. Through these and your other favorites, enjoy your explorations of the theme of change!

CROSS REFERENCES

Saul and Metamorphosis

In the GEMS handbook *A Parent's Guide to GEMS*, GEMS Director Jacquey Barber describes her then 1-year-old son Saul's fascination with butterflies and the process of metamorphosis. They find three caterpillars in the fennel bush in their yard, place them with lots of leaves inside a flower vase on the kitchen table and begin "extended observations of this scientific centerpiece, leading to many discoveries!" As they observe the changes that take place, some of Saul's favorite books come into play. She writes:

"After watching a while, we discovered that the caterpillars seemed to eat the stems of the plant, rather than the foliage. It's no wonder they poop so much, given the amount they ate! We made daily trips to the fennel bush to replenish our bouquet in the vase. Watching the caterpillars eat so much gave new meaning to one of Saul's favorite books, *The Very Hungry Caterpillar* by Eric Carle.

"I don't know what Saul understands about all this. I don't know if he ever knew the connection between a caterpillar and a chrysalis, or if he understands that creatures metamorphose, though we now have several books that picture this process, including Ruth Heller's *Chickens Aren't The Only Ones*. He seems to linger on the pages that picture an egg changing to a caterpillar changing to a chrysalis changing to a butterfly."

LITERATURE CONNECTIONS

Badger's Parting Gifts

by Susan Varley

Lothrop, Lee & Shepard, New York. 1984

Grades: K–3

Badger's friends are sad when he dies, but they treasure the legacies he left them. Mole, in particular, has difficulty adjusting to the loss of his dear friend and finds solace in the memories of badger. A sensitive look at death as part of the life cycle.

Chipmunk Song

by Joanne Ryder

E.P. Dutton, New York. 1987

Grades: K–5

A chipmunk goes about its cycle of activities in late summer, prepares for winter, and settles in until spring. You are put in the place of a chipmunk through food gathering, hiding from predators, hibernation, and more. Detailed illustrations show roots invading the chipmunk's hole and underground stashes of acorns.

An Egg is an Egg

by Nicki Weiss

G.P. Putnam's Sons, New York. 1990

Grades: Preschool–3

A poetic explanation of how many things change: eggs to chicks, branches to sticks, and day to night; but some things stay the same.

If You Give a Moose a Muffin

by Laura J. Numeroff; illustrated by Felicia Bond

HarperCollins, New York. 1991

Grades: Preschool–3

Chaos can ensue if you give a moose a muffin and start him on a cycle of urgent requests. A great way to introduce young children to the concept of a cycle.

If You Give a Mouse a Cookie

by Laura J. Numeroff; illustrated by Felicia Bond

HarperCollins, New York. 1985

Scholastic, New York. 1989

Grades: Preschool–3

The requests a mouse is likely to make after you give him a cookie soon gets out of hand. "He's going to ask for a glass of milk … And that's only the beginning!" A great way to introduce young children to the concept of a cycle. The Big Book edition includes a teacher's guide.

What man that sees the ever-whirling wheel
Of Change, the which all mortal things doth sway,
But that thereby doth find, and plainly feel,
How Mutability in them doth play
Her cruel sports, to many men's decay?

— Edmund Spenser
The Faerie Queen

The world's a scene of changes, and to be Constant, in Nature were inconstancy.

— Abraham Cowley
Inconstancy

The Lady Who Put Salt in Her Coffee

by Lucretia Hale
Harcourt, Brace, Jovanovich, San Diego. 1989
Grades: Preschool–6

When Mrs. Peterkin accidentally puts salt in her coffee, the entire family embarks on an elaborate quest to find someone to make it drinkable again. Visits to a chemist, an herbalist, and a wise woman result in a solution, but not without having tried some wild experiments first. Good example of irreversible change.

Leese Webster

by Ursula K. LeGuin; illustrated by James Brunsman
Atheneum, New York. 1979
Out of print
Grades: 3–5

A spider in a deserted palace spins extraordinary webs, copying paintings, carvings, and tapestries. The authorities cover the webs with glass to save them, which is good for tourists, but prevents the spider from catching any insects for food! She eventually resettles in the palace garden. Kids could be challenged to compare the patterns of various spiders' webs and to create some designs of their own.

Linnea's Almanac

by Christina Bjork; illustrated by Lena Anderson
R&S Books/Farrar, Straus & Giroux, New York. 1989
Grades: 3–6

Linnea writes an almanac to keep track of her indoor and outdoor investigations of nature over a year's time. She opens a bird restaurant in January and goes beach combing in July. The almanac is written in journal form with simple monthly activities for young readers to do at home.

Linnea's Windowsill Garden

by Christina Bjork; illustrated by Lena Anderson
R&S Books/Farrar, Straus & Giroux, New York. 1988
Grades: 3–6

From seeds to cuttings to potted plants, Linnea describes the care of her plants through their life cycle in her indoor garden. The friendly narration and simple information invite readers to try the activities and games at home.

Listen to the Rain

by Bill Martin Jr. and John Archambault; illustrated by James Endicott
Henry Holt and Co., New York. 1988
Grades: Preschool–1

An inviting poem-like story of a rainstorm cycle in which young readers and listeners will enjoy repeating the words that describe the rain, as the "tiptoe pitter-patter" storm builds, stops, and finally becomes the silent after-time of rain.

The Magic School Bus at the Waterworks

by Joanna Cole; illustrated by Bruce Degen
Scholastic, New York. 1986
Grades: K–6

> When Ms. Frizzle, the strangest teacher in school, takes her class on a field trip to the waterworks, everyone ends up experiencing the water purification system from the inside. Evaporation, the water cycle, and filtration are just a few of the concepts explored in this whimsical field trip.

Miss Rumphius

by Barbara Cooney
Puffin/ Viking Penguin, New York. 1982
Grades: K–4

> Great-aunt Alice was once a little girl who loved the sea, traveled to far-away places, and longed to do something wonderful to make the world more beautiful. She decided to sow lupine seeds as she walked and became known as the "Lupine Lady," though some might have called her crazy. Illustrates the relationship between life and art, and how older generations can change the world of the future in positive ways.

Never Cry Wolf

by Farley Mowat
Atlantic Monthly/Little, Brown & Co., Boston. 1963
Bantam, New York. 1984
Grades: 6–Adult

> Wolves are killing too many of the Arctic caribou, so the Wildlife Service assigns a naturalist to investigate these changes. Farley Mowat is dropped alone onto the frozen tundra of Canada's Keewatin Barrens to live among the wolf packs and study their ways. His interactions with the packs, and his growing respect for and understanding of the wild wolf will captivate all readers.

New Providence: A Changing Cityscape

by Renata Von Tscharner and Ronald L. Fleming;
illustrated by Denis Orloff
Harcourt, Brace, Jovanovich, San Diego. 1987
Grades: 4–7

> The evolution of an American city from 1910 to 1987 is shown in six double-page colored illustrations with a brief text describing the factors involved and the basic architectural styles depicted. Complete down to billboards and graffiti, this sequence of detailed illustrations invites careful observation and comparisons. The cycle of thriving urban growth, flight to the suburbs, and return to a renovated city is well summarized by the Carl Sandburg quote cited, "Put the city up; tear the city down; put it up again."

Nicky The Nature Detective
by Ulf Svedberg; illustrated by Lena Anderson
R&S Books/Farrar, Straus & Giroux, New York. 1988
Grades: 3–8

Nicky loves to explore the changes in nature. She watches a red maple tree and all the creatures and plants that live on or near the tree through the seasons of the year. Her discoveries lead her to look carefully at the structure of a nesting place, why birds migrate, who left tracks in the snow, where butterflies go in the winter, and many more things. This book is packed with information.

Ox-Cart Man
by Donald Hall; illustrated by Barbara Cooney
Viking Press, New York. 1979
Grades: Preschool–3

The day-to-day life of an early-nineteenth century rural New England family throughout the changing seasons is described as they grow and make food, clothing, and other items for sale and for their own consumption. Caldecott award winner.

Pumpkin Pumpkin
by Jeanne Titherington
Greenwillow Books, New York. 1986
Grades: Preschool–3

Jamie plants a pumpkin seed, watches the pumpkin grow, carves it, and saves some seeds to plant in the spring. A very simple rendition of the cycle from seed to sprout to plant to flower to fruit to seed, with lots of action in the detailed and humorous illustrations.

Rain Forest
by Helen Cowcher
Farrar, Straus & Giroux, New York. 1988
Grades: K–3

A cry of alarm sounds through the rain forest and the animals know something more powerful than the jaguar is threatening to change their world. The earth-moving machines brought in by man are in turn affected by the power of the environment, and must leave at the threat of flooding. Eventually the animals are able to return to their homes in a somewhat simplistic happy ending.

The Rock
by Peter Parnall
Macmillan, New York. 1991
Grades: 2–6

Over the years a rock provides shelter, a hiding place for animal and human hunters, and protects food and water sources until it is struck by lightning. Then a white birch tree sprouts from the blackened rock, growing into a stand of trees. Affirmative portrait of nature's multiplicity and endurance.

The Sky was Blue

by Charlotte Zolotow; illustrated by Garth Williams

Harper & Row, New York. 1963

Grades: Preschool–3

> A young child and her mother look at family photo albums going back several generations. They discover that while the clothes, houses, vehicles and toys looked different, many things were the same: "the sky was blue, grass was green, snow was white and cold, the sun was warm and yellow, just as they all are now."

The Tiny Seed

by Eric Carle

Picture Book Studio, Saxonville, Massachusetts. 1990

Grades: Preschool–3

> A flowering plant's life cycle through the seasons is simply described. The various methods of seed dispersal and how some seeds are lost or exist in conditions that prevent them from growing are shown.

The Very Hungry Caterpillar

by Eric Carle

Philomel Books/Putnam & Grosset, New York. 1969

Grades: Preschool–3

> A hungry little caterpillar eats its way through a varied and very large quantity of food (and through the pages of the book). Full at last, it forms a chrysalis around itself and goes to sleep. A good opportunity to correct the common misuse of the word "cocoon" (moths emerge from cocoons), with the correct term of "chrysalis," in the case of butterflies.

The Watching Game

by Louise Borden; illustrated by Teri Weidner

Scholastic, New York. 1991

Grades: Preschool–2

> Nana's four grandchildren visit her in the country during the different seasons and play a watching game with a sneaky fox. This book approaches the cycle of the seasons with warmth and a sensitivity to the natural environment.

Water's Way

by Lisa W. Peters; illustrated by Ted Rand

Arcade Publishing/Little, Brown & Co., Boston. 1991

Grades: K–3

> "Water has a way of changing" inside and outside Tony's house, from clouds to steam to fog and other forms. Innovative illustrations show the changes in the weather outside while highlighting water changes inside the house. Could begin a discussion of the phase changes of matter.

O, swear not
by the moon,
the wandering moon,
that monthly changes
in her circled orb.

— *William Shakespeare*

Where Butterflies Grow

by Joanne Ryder; illustrated by Lynne Cherry
Lodestar Books/E.P. Dutton, New York. 1989
Grades: Preschool–5

What might it feel like to change from a caterpillar into a butterfly? Structure, metamorphosis, locomotion, camouflage, and feeding behaviors are all described from the perspective of the emerging swallowtail butterfly. A beautiful book, unusual in its detailed drawings of metamorphosis.

Why the Tides Ebb and Flow

by Joan C. Bowden; illustrated by Marc Brown
Houghton Mifflin, Boston. 1979
Grades: K–4

A feisty old woman bargains with the Sky Spirit, finally gaining a hut, a daughter and son-in-law, and the loan of a very special rock to beautify her yard. The result of her action explains why the tides ebb and flow, and incidentally, why dogs have cold noses!

The Wump World

by Bill Peet
Houghton Mifflin, Boston. 1970
Grades: Preschool–5

The Wump World is invaded by Pollutians from the planet Pollutus, who had left their worn-out planet to start a life in a new world. They move in and take over, building cities, creating pollution, and using up resources. Eventually, the Pollutians move to another planet and leave the Wumps to try and restore their ecosystem.

The Year at Maple Hill Farm

by Alice and Martin Provensen
Macmillan, New York. 1978
Grades: K–3

Each month in the cycle of seasons on a farm throughout the year is depicted on a two-page spread filled with detailed pictures and observations about pigs sleeping in cool mud puddles, geese feet never freezing in February's icy pond water, and animals getting haircuts in May.

In winter, when the fields are white,
I sing this song for your delight—

In spring, when woods are getting green,
I'll try and tell you what I mean:

In summer, when the days are long,
Perhaps you'll understand this song:

In autumn, when the leaves are brown,
Take pen and ink, and write it down.

— *Lewis Carroll*
Through the Looking Glass

Scale

$\mathcal{S}$cale concerns the range, comparison, and measurement of various magnitudes, such as sizes, speeds, distances, or lengths of time.

While some distances or magnitudes (such as the speed of light) are so immense (or in other cases, so minute) that it is difficult for our brains to truly grasp them, we can represent such amounts symbolically and manipulate them mathematically, thus further exploring them and their interaction. **Scale is relative.** An ant finds the distance around a tree to be much greater than a squirrel does; a mound of earth to us may be a small mountain to the ant. In some thematic listings, scale is linked with structure.

MAKING CONNECTIONS

Most of the books in this section pertain to the concept of **size**, while some focus on **distance, time,** and **amount**.

In several books, tiny beings interact in a large world and one case it's the reverse! From Gulliver to the Borrowers there are many size-related lessons demonstrating the concepts of **length, volume, weight, area,** and in a couple of cases, the magnitude of **noise**. A fun book about dinosaurs compares the length, volume, and weight of a dinosaur with common objects, as well as speculating on the amount of food dinosaurs might have eaten. Another book compares the strength of various animals. One imaginative story takes place in a town where enormous vegetables fall from the sky, providing a silly situation in which to speculate on the results.

Some books feature microscopes and a view at a magnified level, such as ants carrying enormous sugar crystals. These books focus on the **structure of small things**. Other books deal with the concept of **orders of magnitude**, showing the **relationship of small to big, of near to far, and of a few to a lot**. On a simple level, one book begins with a rain drop and reaches the sea. *The Yellow Button* presents an elementary version of the classic film and book *Powers of Ten*, as the pages go from a button to a dress to a child, a country, a planet, the Universe! *The King's Chessboard* leads to the sobering discovery of **exponential growth**. One book deals with **time** and time zones, and another examines **the scale of the Earth**, via an imaginary journey through its center. Be on the lookout for opportunities to explore the magnitude/scale of speed, noise, size, amount, acidity, weight, time, and lots else in your own favorite books.

CROSS REFERENCES

LITERATURE CONNECTIONS

Alice's Adventures in Wonderland

by Lewis Carroll

Viking Penguin, New York. 1984

Grades: 4–Adult

> In Chapter 4, Alice encounters a strange drink that changes her size as well as her perception of everything around her. It isn't until Chapter 5, and a discussion with a curmudgeonly caterpillar, that she finds a mushroom that also is size altering.

As the Crow Flies: A First Book of Maps

by Gail Hartman; illustrated by Harvey Stevenson

Bradbury Press, New York. 1991

Grades: Preschool–2

> The different worlds and favorite places of an eagle, rabbit, crow, horse, and seagull, each from its own perspective, are charted on maps. The last two pages form a large map incorporating all the geographical areas: the mountains, meadow, lighthouse and harbor, skyscrapers, hot dog stand, etc. A wonderful book with which to introduce or reinforce an appreciation of scale.

The Borrowers

by Mary Norton; illustrated by Beth and Joe Krush

Harcourt, Brace, Jovanovich, San Diego. 1953

Grades: 5–12

> The first book in a classic series about the Clock family, also known as the "Borrowers," and their miniature world. They are the original recyclers, their portraits on the wall are postage stamps, the chest of drawers is a matchbox and the washbasin is an aspirin tin lid. Children are intrigued by the detail and by the totally convincing creation of another world.

The Borrowers Aloft

by Mary Norton; illustrated by Beth and Joe Krush

Harcourt, Brace, Jovanovich, San Diego. 1961

Grades: 5–12

> The Borrowers are kidnapped by a mercenary couple and make their escape via a balloon. They design it using a diagram in a London newspaper and there are wonderful details about their calculations and adaptations.

Cloudy With a Chance of Meatballs

by Judi Barrett; illustrated by Ron Barrett
Atheneum, New York. 1978
Grades: K–3

> A hilarious look at weather conditions in the town of Chewandswallow. This tiny town needs no food stores because daily climactic conditions bring the inhabitants food and beverages—although not always as they would be most appreciated! One day the wind blows in a storm of gigantic pancakes and a downpour of maple syrup that nearly floods the town. This story provides a comical introduction to the concept of scale for young students. As a follow-up activity, they can make huge paper maché foods and compare them to life-size items.

The Dinosaur Who Lived in My Backyard

by B.G. Hennessy; illustrated by Susan Davis
Penguin Books, New York. 1988
Grades: Preschool–4

> A young boy imagines what it was like long ago when a dinosaur lived in his backyard. Nearly every page has a scale comparison dealing with length, volume, weight, area, etc. "An egg as big as a basketball," "By the time he was five, he was as big as our car," "Just one of his dinosaur feet was so big it wouldn't even have fit in my sandbox," and so on.

George Shrinks

by William Joyce
HarperCollins, New York. 1985
Grades: Preschool–3

> Taking care of a cat and a baby brother turns into a series of comic adventures when George wakes up to find himself shrunk to the size of a mouse. Full-page illustrations depicting George making his bed, brushing his teeth, and washing the dishes, make the concept of relative scale meaningful.

The Girl on the Hat

by Jane Jacobs;
illustrated by Karen Reczuch
Oxford University Press, Toronto. 1989
Grades: 2–6

> Tiny Tina, who can fit in a peanut shell, has adventures in a dangerous cave, falls into a pond, fights off a hawk who thinks she is a mouse, takes photographs inside bureau drawers and animal burrows. The black and white detailed drawings portray Tina's size in relation to many backgrounds.

Goldilocks and the Three Bears

retold and illustrated by Jan Brett
Dodd, Mead & Co., New York. 1987
Grades: Preschool–3

This classic tale introduces a consistent and predictable scale comparison, as Goldilocks encounters the three bowls of porridge, the three chairs, the three beds, and finally, the three bears. The gorgeous illustrations in this version (caterpillars changing to butterflies, many varieties of birds' eggs, seeds and leaves of various trees, and forest scenes that show a system of interacting plants and animals) are captivating.

Greg's Microscope

by Millicent E. Selsam; illustrated by Arnold Lobel
Harper & Row, New York. 1963
Grades: 2–4

Greg's father buys him a microscope and he finds an unlimited array of items around the house to observe, even the hair of Mrs. Broom's poodle. The illustrations show the salt and sugar crystals, threads, hair, and other materials as they appear magnified. This is not a high tech, state-of-the-art representation, but the fun and empowering experience of playing with scale are well portrayed.

The Grouchy Ladybug

by Eric Carle
Harper & Row, New York. 1986
Grades: Preschool–3

A grouchy ladybug who is looking for a fight from sunrise to sunset challenges everyone she meets, regardless of their size or strength. This book would be a great way to lead off classroom measurement activities involving size and time. There is a clock on each page to chronicle the day in hours.

Gulliver's Travels

by Jonathan Swift
Penguin Books USA, New York. 1983
Grades: 4–Adult

Classic tale of an Englishman who becomes shipwrecked in a land where people are six inches high and then stranded in a land of giants. His first two voyages provide a wonderful springboard into activities related to scale and measurement. This book is available from a variety of publishers.

Thematic investigation . . . cannot be reduced to a mechanical act. As a process of search, of knowledge, and thus of creation, it requires the investigators to discover the interpenetration of problems, in the linking of meaningful themes. The investigation will be most educational when it is most critical, and most critical when it avoids the narrow outlines of partial or "focalized" views of reality, and sticks to the comprehension of *total* reality. Thus, the process of searching for meaningful thematics should include a concern for the links between themes, a concern to pose these themes as problems, and a concern for their historical-cultural context.

— *Paulo Freire*
Pedagogy of the Oppressed

How To Dig A Hole to the Other Side of the World

by Faith McNulty; illustrated by Marc Simont
HarperCollins, New York. 1990
Grades: K–8

>   A child takes an imaginary 8,000-mile journey through the earth and discovers what is inside. As he passes through the earth's crust, then the mantle, and into the core, the book describes how far he's come and how far he has to go. You are left with an awesome appreciation of the size of the earth. See the article on page 344 for a more detailed look at this fun and instructive book.

The King's Chessboard

by David Birch; illustrated by Devis Grebu
Dial Books, New York. 1988
Grades: K–6

>   A proud king, too vain to admit what he does not know, learns a valuable lesson when he readily grants his wise man a special request. One grain of rice on the first square of a chessboard on the first day, two grains on the second square on the second day, four grains on the third square on the third day and so on. After several days the counting of rice grains gives way to weighing, then the weighing gives way to counting sackfuls, then to wagonfuls. The king soon realizes that there is not enough rice in the entire world to fulfill the wise man's request.

Many Moons

by James Thurber; illustrated by Louis Slobodkin
Harcourt, Brace & World, New York. 1943
Grades: K–5

>   A little princess wants the moon. Neither the King, the Lord High Chamberlain, the Royal Wizard, the Royal Mathematician, nor the Court Jester are able to solve the problem. She figures out how to get it herself. A debate about the distance between the Earth and the moon is included.

The Microscope

by Maxine Kumin; illustrated by Arnold Lobel
Harper & Row, New York. 1984
Grades: 4–Adult

>   A beautifully written and illustrated book, which, while poking fun at many things, also portrays Anton van Leeuwenhoek grinding lenses, the appearance of many common objects under a microscope, and historical information. Although told in rhyme and with few words, its language is fairly sophisticated, and could be read with delight by older students.

The Mitten

by Jan Brett

G.P. Putnam's Sons, New York. 1989

Grades: Preschool–5

In this Ukranian folk tale, animals of increasing size, from a mouse no bigger than an acorn to a great bear, squeeze their way into a boy's mitten lost in the snow. The illustrations are distinctive.

My Place

by Nadia Wheatley and Donna Rawlins

Kane/Miller Book Publishers, New York. 1987

Grades: 4–8

The history of a neighborhood in an Australian city is chronicled via two-page spreads that each have drawings, a map, and a story written by a child who lives on a certain spot of land—my place. The story goes back decade by decade as you learn who lives on that spot of land, what else is in the vicinity, and what happens in the lives of those who live there. The evolution of lifestyle, vocations, buildings, and the land are beautifully depicted by these accounts and maps. Comparing the different maps to see the changes over time would be a wonderful lesson in scale and structure. Children's Book Council of Australia Book of the Year award winner.

Nine O'Clock Lullaby

by Marilyn Singer; illustrated by Frane Lessac

HarperCollins, New York. 1991

Grades: Preschool–6

Bedtime story transports children through many lands showing what people might be doing on different parts of the globe at the same time. The pictures of the various cultures are fresh and lively from cooking on a "barbie" in Australia to conga drumming and coconut candy in Puerto Rico. If it's 9:00 p.m. in New York, what time is it in Japan? Understanding time zones and why they exist relates to the theme of scale. Try having your students label the time of day on the class globe to gain a more concrete working sense of the international time scale.

The Old Man and the Astronauts

by Ruth Tabrah; illustrated by George Suyeoka

Island Heritage Limited, Honolulu. 1975

Grades: All

An old Bongu man in New Guinea, upon hearing the news reports of Americans walking on the moon, becomes very concerned that moon rocks are being taken since he relies on the light from the moon to fish at night. The village headman helps the old man and his grandson understand that the astronauts can no more take all the rocks from the moon than he could scoop water away from the sea. This comparison communicates vast orders of magnitude in a down-to-earth way. This lesson of scale and stability is accompanied by the wise message of "taking no more than your share." A true story.

Rain, Drop, Splash

by Alvin Tresselt; illustrated by Leonard Weisgard
Lothrop, Lee & Shepard, New York. 1946
Grades: Preschool–3

Raindrops fall to make puddles. Puddles become larger and larger to form ponds. Ponds overflow to brooks which lead to lakes. The rainstorm continues, falling on plants and animals, making mud, flooding a road. The last scene leads to the ocean, when at last the rain stops and the sun emerges.

Two Bad Ants

by Chris Van Allsburg
Houghton Mifflin, Boston. 1988
Grades: Preschool–4

When two curious ants set off in search of the beautiful sparkling crystals (sugar), they have a dangerous adventure that convinces them to return to the former safety of their ant colony. Drawn from an ant's perspective, the illustrations show the ants lugging individual sugar crystals and other views from "the small."

The Yellow Button

by Anne Mazer;
illustrated by Judy Pedersen
Alfred A. Knopf,
New York. 1990
Grades: K–8

Begins with a button in a pocket, worn by a girl, lying on a couch, in a living room, in a house, at the edge of a field. Continues to the edge of a mountain range, in a country, on a planet, in a universe. This book communicates the relationship of small to big in a very simple fashion.

One Inch Tall

If you were only one inch tall, you'd ride a worm to school.
The teardrop of a crying ant would be your swimming pool.
A crumb of cake would be a feast
And last you seven days at least,
A flea would be a frightening beast
If you were one inch tall.

If you were only one inch tall, you'd walk beneath the door,
And it would take about a month to get down to the store.
A bit of fluff would be your bed,
You'd swing upon a spider's thread,
And wear a thimble on your head
If you were one inch tall.

You'd surf across the kitchen sink upon a stick of gum.
You couldn't hug your mama, you'd just have to hug her thumb.
You'd run from people's feet in fright,
To move a pen would take all night,
(This poem took fourteen years to write—
'Cause I'm just one inch tall).

— *Shel Silverstein*
Where the Sidewalk Ends

— Scale —

Stability

— Equilibrium, Conservation, Symmetry —

Most physical systems settle into a state of **equilibrium** at a particular point, in which the forces within it are balanced until something new is done to the system. Many systems include feedback subsystems that serve to keep some aspect of the system constant. Whatever changes may happen inside a system, there are some total quantities that remain the same, or constant. **Stability** (or constancy) is a study of the ways in which systems do **not** change.

Stability is related to the repeatability of results in science and the ability to derive "laws" of physical behavior. Given exactly the same experimental conditions, with no uncontrolled variables, results are expected to be replicable. Similarly, the same law of gravity applies not only to bringing a pop fly back to the ball field, but also to keeping the moon in its orbit, and the mutual attraction between galaxies billions of light years away. The law of gravity is only one of many such principles that seem to be constant and unchanging. Without stability of this sort, causality and predictability would have no meaning, and science as we know it would be an impossible task.

MAKING CONNECTIONS

Quite a few of the books we list here are for younger children, with the notion that **some things don't change**: snow is cold, grass is green, the family quilt remains in the attic through generations of change. While some of the examples represent more sentimental life-related themes, others center around the concept of an **unchanging property or quality**. One book shows how **even a cycle can be viewed as unchanging**—the cycle of birth, life, and death is predictable in its constancy.

Other books paint a more complex picture of stability through depiction of **equilibrium** or **balance**, in which the parts of a system are often changing but the total picture remains constant. Several books show the need for **balance in nature**, including an ecological folktale based on a story by Charles Darwin, and an outstanding "ecological mystery" based in the Everglades. A science fiction book for older children presents a simulation of what can happen to an ecosystem subjected to abuse and overuse.

CROSS REFERENCES

LITERATURE CONNECTIONS

An Egg Is An Egg
by Nicki Weiss
G.P. Putnam's Sons, New York. 1990
Grades: Preschool–3

A poetic explanation of how many things change: eggs to chicks, branches to sticks, day to night; but some things stay the same.

Lifetimes: A Beautiful Way to Explain Death to Children
by Robert Ingpen and Bryan Mellonie
Bantam Books, New York. 1983
Grades: K–3

The birth, life, and death of many living things are discussed with the idea that each has "its own special lifetime." Illustrations are detailed, realistic depictions of plants and animals, including some more exotic examples such as a kookaburra and an emu.

The Missing 'Gator of Gumbo Limbo: An Ecological Mystery
by Jean C. George
HarperCollins, New York. 1992
Grades: 4–7

Sixth-grader Liza K and her mother live in a tent in the Florida Everglades. She becomes a nature detective while searching for Dajun, a giant alligator who plays a part in a waterhole's oxygen-algae cycle, yet is marked for extinction by local officials. The book is full of detail about the region's flora and fauna and habitats. Liza and her neighbor are especially interested in the many changes taking place in their environment due to human forces. The theme of balance and equilibrium is developed with many examples from her world.

The Old Ladies Who Liked Cats
by Carol Greene; illustrated by Loretta Krupinski
HarperCollins, New York. 1991
Grades: K–6

When the old ladies are no longer allowed to let their cats out at night, the delicate balance of their island ecology is disturbed, with disastrous results. Based on Charles Darwin's story about clover and cats, this ecological folk tale demonstrates the interrelationships of plants and animals.

> Every body perseveres in its state of rest or of uniform motion in a straight line except in so far as it is compelled to change that state by impressed forces.
>
> — *Isaac Newton*
> *First Law of Motion*
> *Principia*

The more things change, the more they remain the same.

— *Alphonse Karr*
Les Guepes

The Old Man and the Astronauts
by Ruth Tabrah; illustrated by George Suyeoka
Island Heritage Limited, Honolulu. 1975
Grades: All

An old Bongu man in New Guinea, upon hearing the news reports of Americans walking on the moon, becomes very concerned that moon rocks are being taken since he relies on the light from the moon to fish at night. The village headman helps the old man and his grandson understand that the astronauts can no more take all the rocks from the moon than he could scoop water away from the sea. This lesson of scale and stability is accompanied by the wise message of "taking no more than your share." A true story.

The Quilt Story
by Tony Johnston and Tomie dePaola
G.P. Putnam's Sons, New York. 1985
Grades: Preschool–3

A pioneer mother stitches a beautiful quilt to warm and comfort her daughter. After many years of use, the quilt is stored in the attic where it is used by a raccoon, a cat, and a mouse family. Years later, the quilt is rediscovered by a granddaughter, and provides the same loving security and continuity. When the family moves to a new house, "everything smells of white paint and boxes. Everything but the quilt."

The Roadside
by David Bellamy; illustrated by Jill Dow
Clarkson N. Potter, New York. 1988
Grades: Preschool–5

The construction of a six-lane highway in the countryside disrupts the balance of nature and forces animals there to change their patterns. The toads lay their eggs in different places, some animals eat the leftovers from the road workers' lunches, a raw mound of earth is soon covered by wildflowers which attract butterflies. An attempt is made to be even-handed, "the road workers have done a good job." The toads use the new pipe under the road as a passageway, no hunters with guns come there because walking is not allowed, etc. The book depicts the need for ecological balance.

The Sky Was Blue
by Charlotte Zolotow; illustrated by Garth Williams
Harper & Row, New York. 1963
Grades: Preschool–3

A young child and her mother look at family photo albums going back several generations. They discover that while the clothes, houses, vehicles, and toys looked different, many things were the same: "the sky was blue, grass was green, snow was white and cold, the sun was warm and yellow, just as they all are now."

The Song in the Walnut Grove

by David Kherdian; illustrated by Paul O. Zelinsky

Alfred A. Knopf, New York. 1982

Grades: 4–6

A curious cricket meets a grasshopper. Together they learn of each other's daytime and nighttime habits while living in an herb garden. The friendship between them grows when the cricket saves the grasshopper's life and they learn to appreciate each other's differences. This story weaves very accurate accounts of insect behavior and their contributions to the ecology of Walnut Grove and shows the need for balance in nature.

When I'm Sleepy

by Jane R. Howard; illustrated by Lynne Cherry

E.P. Dutton, New York. 1985

Grades: Preschool–2

A young girl speculates about sleeping in places other than her bed and is shown sleeping with 12 different animals in a nest, a swamp, standing up, hanging upside down, etc. The witty, glowing illustrations present a variety of possibilities, but end with the comforting security of settling into one's own bed each night.

Who Really Killed Cock Robin?

by Jean C. George

HarperCollins, New York. 1991

Grades: 3–7

This compelling ecological mystery examines the importance of keeping nature in balance, and provides an inspiring account of a young environmental hero who becomes a scientific detective.

Whose Footprints?

by Masayuki Yabuuchi

Philomel Books, New York. 1983

Grades: Preschool–4

The footprints of a duck, cat, bear, horse, hippopotamus, and goat are examined. Good guessing game format for younger students as they learn some things don't change.

A Wizard of Earthsea

by Ursula K. LeGuin

Houghton Mifflin, Boston. 1968

Bantam Books, New York. 1984

Grades: 6–Adult

The terms "balance," and "equilibrium" are used throughout the book to express the fundamental principles that guide the compelling world of the Wizard. The book illustrates how opposite forces operate within us, and that a balance of those forces is necessary for harmony.

The Wump World
by Bill Peet
Houghton Mifflin, Boston. 1970
Grades: Preschool–5

The Wump World is invaded by Pollutians from the planet Pollutus, who had left their worn-out planet to start a life in a new world. They move in and take over, building cities, creating pollution, and using up resources. Eventually, the Pollutians move to another planet and leave the Wumps to try and restore the balance in their ecosystem.

STABILITY

Within this wondrous world of change
As atoms clash and re-arrange
There yet remains some constancy
The anchor of stability.

Experiments one lab refines
Can be performed time after time
If variables are well-controlled
The same results are sure to hold.

A chemical reaction's change
Is also only fair exchange
Though liquid may turn into gas
There's no change in the total mass.

Rock at the bottom of a hill
Earth's gravity keeps it there still
In every system balance comes
When reaching equilibriums.

In all these ways and many more
Stability's like sandy shore
A centered place, a balanced range,
Beneath the ocean roar of change.

— *Lincoln Bergman*

Structure

Structure is the way in which things are put together, whether it be the hexagonal shapes used in a beehive or an airplane's wing, the internal configuration of an atom, or the geologic structure of the Earth. There is an immense diversity of structure in the natural world. Structure is directly related to both function and to scale. A 50-story skyscraper cannot be constructed in exactly the same way as a two-story house. A beam that is twice as big is four times as strong. Structure, like scale, can have a central mathematical component.

MAKING CONNECTIONS

A rich array of books exists elaborating on the **structure of things both animate and inanimate**. Quite a few of the books listed here focus on the structure of buildings including the building of a simple house by some hippopotamuses, to a fantasy about "unconstructing" the Empire State Building, and a voyage through a typical American city with a focus on different architectural styles. One book describes the structure of a tipi, another the structure of a pueblo. Several books feature the construction of bridges and kites.

Many books elaborate on the **structure and anatomy of living things**, including plants, snails, butterflies, feet, dinosaurs, caribou, and more. Several books focus on the structure of things made by living things, such as footprints and spider webs.

A couple of books depict the **composition of the Earth** and **quicksand**. Another book considers the **structure of the Universe**.

The relationship between **structure and function**—that the structure of an object or creature is closely related to it's function—is the subject of a number of books. One book humorously examines why animals shouldn't wear clothes. In another, a pair of trousers is used by many animals in a wide variety of ways. One book compares and contrasts the different homes of animals, showing how the structure of each animal's home is related to how it is used. Finally, there is a story of a boy who was flattened, and all the new things he can do thanks to his new shape.

The relationship between **scale and structure** is important. One book follows the fate of a piece of land in Australia, decade by decade for 150 years, complete with kid-drawn maps of the area. A comparison of the maps makes an interesting lesson in how scale is related to structure; from the size of the buildings to the sprawl of the neighborhood. Be on the lookout for more books which illustrate this important relationship.

CROSS REFERENCES

LITERATURE CONNECTIONS

Animals Should Definitely Not Wear Clothing
by Judi Barrett; illustrations by Ron Barrett
Macmillan, New York. 1974
Grades: Preschool–3

Pictures of animals wearing clothes show why this would be a ridiculous custom for them to adopt. Much of the text focuses on the structure of particular animals and how this makes them unsuited to being covered by clothing. Hens laying eggs in their pants, possums wearing clothing upside down, and porcupines poking through their pajamas make for hilarious images.

Bones, Bones, Dinosaur Bones
by Byron Barton
Thomas Y. Crowell, Minneapolis. 1990
Grades: Preschool–2

In bold, block illustrations the process of paleontology is shown—from field work to the steps involved in collecting and mounting skeletons in a museum exhibit. The last page names and shows eight dinosaurs.

The Bridge
by Emily C. Neville; illustrated by Ronald Himler
Harper & Row, New York. 1988
Grades: Preschool–3

When the old wooden bridge breaks, a young boy is delighted to be able to watch, from his front yard, the many different machines at work building the new bridge across the brook. The new bridge is a dirt road over a culvert. This new structure is more sturdy, but not as unique and memorable as the old rattling wooden bridge.

A Caribou Alphabet
by Mary Beth Owens
Dog Ear Press, Brunswick, Maine. 1988
Farrar, Straus & Giroux, New York. 1990
Grades: K–5

An alphabet book depicting the characteristics and ways of caribou. At first glance this book may seems a primary-level "A, B, C" book, but it includes a compendium of information about caribou, including intricacies of their behavior, habitat requirements, and physical features.

Catch the Wind: All About Kites
by Gail Gibbons
Little, Brown & Co., Boston. 1989
Grades: K–6

When two children visit Ike's Kite Shop, they learn about the structure of kites and how to fly them. Includes instructions for building a kite.

Dinosaurs are Different

by Aliki
Harper & Row, New York. 1985
Grades: K–5

This book explains ways in which the various orders and suborders of dinosaurs were similar and different in structure and appearance.

Flat Stanley

by Jeff Brown; illustrated by Tomi Ungerer
HarperCollins, New York. 1964
Grades: 2–5

One night a giant bulletin board falls on Stanley Lambchop while he is sleeping and he becomes Flat Stanley, four-feet-tall and half-an-inch thick. Advantages to his new shape include being able to slip under doors and through gratings, to be mailed in an envelope for a trip to California or to fly like a kite, and finally, to apprehend art thieves while disguised as a painting in a museum. He doesn't like being ridiculed by the other kids for being different, so his brother blows him up to his former size and shape with a bicycle pump. A good example of the relationship of structure to function.

Harry Builds A House

by Derek Radford
Macmillan, New York. 1990
Grades: Preschool–6

Harry Hippo and his friends build a house, step by step, from digging a ditch for the plumbing, to putting in the last joints and plugs. While the illustrations and characters are cartoon style, much of the information about the process of building a house may be new to many adults.

> Total grandeur of a total edifice,
> Chosen by an inquisitor of structures
> For himself. He stops upon this threshold
> As if the design of all his words takes form
> And frame from thinking and is realized.
> — *Wallace Stevens*
> *To an Old Philosopher in Rome*

A House Is A House For Me

by Mary Ann Hoberman; illustrated by Betty Fraser
Viking/Penguin, New York. 1978
Grades: Preschool–3

Lists in rhyme the dwellings of various animals, peoples, and things such as a shell for a lobster or a glove for a hand. The concept of structure and function is illustrated as the notion of a "house" is explored.

How to Dig A Hole to the Other Side of the World

by Faith McNulty; illustrated by Marc Simont
HarperCollins, New York. 1990
Grades: K–8

A child takes an imaginary 8,000-mile journey through the earth and discovers what's inside. The structure of the earth becomes apparent as he passes through the crust, the mantle, the core, etc. Details about the composition of each layer are included. See page 310 for a more detailed review of this book.

Leese Webster

by Ursula K. LeGuin; illustrated by James Brunsman
Atheneum, New York. 1979
Out of print
Grades: 3–5

A spider spins extraordinary webs, copying paintings, carvings, and tapestries. The authorities cover the webs with glass to save them, which is good for tourists, but prevents the spider from catching any insects for food! The spider eventually resettles in a garden. Kids could be challenged to compare the patterns and structure of various spiders' webs and to create some designs of their own.

Linnea's Windowsill Garden

by Christina Bjork; illustrated by Lena Anderson
R&S Books/Farrar, Straus & Giroux, New York. 1988
Grades: 3–6

Linnea has an indoor garden. From seeds to cuttings to potted plants, Linnea investigates the structure and describes the care of her plants through their life cycle. The friendly narration and simple information invite readers to try the activities and games at home.

Little Bear's Trousers

by Jane Hissey
Philomel Books/Putnam & Grosset, New York. 1987
Grades: K–3

While looking for his beloved red trousers, Little Bear discovers that other animals have found many uses for them: sails for a boat, a container for dog bones, a hump-warmer or a hat, a flag, even a cake frosting bag! A very creative application of the concept of structure and function.

Lizard in the Sun

by Joanne Ryder; illustrated by Michael Rothman
William Morrow, New York. 1990
Grades: Preschool–2

Imagine spending a day as a lizard: you are camouflaged from hungry birds and hidden from insects that become your next meal. Children enjoy seeing the natural world from a lizard's viewpoint, and learn interesting facts about the lizard's anatomy and lifestyle.

The Mountains of Tibet

by Mordicai Gerstein
Harper & Row, New York. 1987
Grades: K–4

A Tibetan woodcutter lives his life without fulfilling his dream of seeing other parts of the world. When he dies, he is given the choice of living another life or going to "heaven." He chooses to live again, and is given the option to be anyone, anywhere. He chooses his own life once more. Before deciding, he sees the structure of the universe, of "hundreds of millions of worlds," and the thousands of kinds of people in the world.

The most beautiful thing we can experience is the mysterious. It is the source of all true art and science.

— *Albert Einstein*

My Feet

by Aliki

Thomas Y. Crowell, Minneapolis. 1990

Grades: Preschool–1

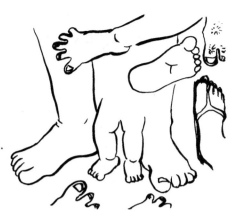

A fun look at the various parts of the foot and all the things feet enable us to do. We see big feet, small feet, foot prints, the arch, the ball, socks, etc.

My Place

by Nadia Wheatley and Donna Rawlins

Kane/Miller Book Publishers, New York. 1987

Grades: 4–8

The history of a neighborhood in an Australian city is chronicled via two-page spreads that each have drawings, a map, and a story written by a child who lives on a certain spot of land—my place. The story goes back decade by decade as you learn who lives on that spot of land, what else is in the vicinity, and what happens in the lives of those who live there. The evolution of lifestyle, vocations, buildings, and the land are beautifully depicted by these accounts and maps. Comparing the maps to see how the location and its uses changed over time would be a wonderful lesson in scale and structure.

New Providence: A Changing Cityscape

by Renata Von Tscharner and Ronald L. Fleming; *illustrated by* Denis Orloff

Gulliver Books/Harcourt, Brace, Jovanovich, San Diego. 1987

Grades: 4–7

Six double-page colored illustrations show a small imaginary American city evolving from 1910 to 1987. A brief text describes the basic architectural styles depicted. Complete down to billboards and graffiti, this sequence of detailed illustrations invites careful observation and comparisons. The attention to specific architectural styles could lead to a focus on structure and buildings.

The Pueblo

by Charlotte and David Yue

Houghton Mifflin, Boston. 1986

Grades: 5–8

This book describes the Pueblo people and their environment, their dwellings and daily life, and other structures such as kivas, corrals, cooking pits and ovens. Chapter 2, on the structure of the dwellings, goes into detail about materials and tools, masonry and adobe construction, and how the structure is designed to perform. The next chapter focuses on the interiors, furnishings, and details, including hatchways, doors and windows, ladders, fireplaces, and chimneys. Includes an excellent bibliography. The same authors have written about the igloo and the tipi (see *The Tipi* on the next page).

The Quicksand Book

by Tomie dePaola

Holiday House, New York. 1977

Grades: 2–5

> A jungle girl learns about the composition of quicksand, how different animals escape it, and how humans can use precautions to avoid getting stuck. Her teacher, an overly confident jungle boy, turns out not to be so superior. A variety of graphics and a helpful monkey give visual interest. A recipe for making your own quicksand is included.

Rubber Bands, Baseballs and Doughnuts: A Book about Topology

by Robert Froman; illustrated by Harvey Weiss

Thomas Y. Crowell, Minneapolis. 1972

Out of print

Grades: 4–10

> An introduction into the world of topology through active reader participation. The activities provide concrete examples and insights into abstract concepts relating to the structure of various shapes.

The Snail's Spell

by Joanne Ryder; illustrated by Lynne Cherry

Puffin Books, New York. 1988

Grades: Preschool–6

> The reader imagines how it feels to be a snail and in the process learns something about the anatomy and locomotion of a snail. Although the picture-book format has a primary-level feel, the exercise of imagining you are a snail is exciting for older students as well. Winner of the Outstanding Science Book for Young Children Award from the New York Academy of Sciences.

The Tipi: A Center of Native American Life

by David and Charlotte Yue

Alfred A. Knopf, New York. 1984

Grades: 5–8

> This excellent book discusses not only the structure and uses of tipis, but the Plains Indians social and cultural context as well. Some of the cultural language and oversimplification are less vital than they might be, but it is written in an accessible style. Includes excellent material on structure, cone shape and its advantages. The central role played by women in constructing the tipi and in owning it is discussed. There are good charts, exact measurements, and descriptions of tipis. While this book includes some mention of the negative consequences of European conquest, noting that in some places tipis were outlawed, it is much weaker in this important area, and should be supplemented with other books.

Unbuilding

by David Macaulay

Houghton Mifflin, Boston. 1980

Grades: 6–12

> Fictional account of the dismantling and removal of the 86-floor Empire State building. After the long, intricate process is over, it is loaded onto a ship, which sinks during a storm. At the end of the book, an Arab prince who bought it is eyeing the Chrysler building. Incredible draftsmanship highlights the step-by-step dismantling. An excellent portrayal of architectural structure.

Where Butterflies Grow

by Joanne Ryder; illustrated by Lynne Cherry

Lodestar Books/ E.P. Dutton, New York. 1989

Grades: Preschool–5

> What might it feel like to change from a caterpillar into a butterfly? Structure, metamorphosis, locomotion, camouflage, and feeding behaviors are all described from the point of view of the butterfly. This beautiful book is unusual in its detailed drawings of metamorphosis. It also includes gardening tips to attract butterflies.

Whose Footprints?

by Masayuki Yabuuchi

Philomel Books, New York. 1983

Grades: Preschool–4

> This nice guessing game format for younger students depicts the footprints of a duck, cat, bear, horse, hippopotamus, and goat. How do the footprints of animals relate to their structures?

> **W**hen from a long distant past nothing subsists, after the people are dead, after the things are broken and scattered, still, alone, more fragile, but with more vitality, more insubstantial, more persistent, more faithful, the smell and taste of things remain poised for a long time, like souls, ready to remind us, waiting and hoping for their moment amid the ruins of all the rest, and bear unfaltering, in the tiny and almost impalpable drop of their essence, the vast structure of recollection.
>
> — *Marcel Proust*
> *Remembrance of Things Past*

Systems & Interactions

A **system** is any collection of things that have some influence on one another and appear to constitute a unified whole.

In defining a system, enough parts must be included so the relationship of the parts to each other makes sense. Where we draw the boundaries of any system makes a difference in terms of understanding what is going on. Any part of a system can also be considered as a system or subsystem, with its own internal parts and interactions. Systems often overlap and are not mutually exclusive. Change in the operation or structure of any part of a system has an impact on the rest. Analyzing the input and output of a system and investigating instances of feedback are also important to understanding systems.

MAKING CONNECTIONS

All literature depicts interactions, so a careful look at almost any book can uncover a collection of interacting parts, or a system. We have included books that all have clear examples of interactions and systems. Some of the books focus on **simple interactions**—between the wind and the objects it blows, garden flowers and animals, the seasons and farm animals, or even in a cookbook where ingredients and heat interact to create new products. Experiences with simple interactions are an important first step to understanding a system of interacting parts.

The most common **system** appearing in many of the books is an **ecosystem**: how plants and animals interact with each other and their environment. Some books show **small-scale ecosystems**: a close-up view of a saguaro cactus and all the animals that live in and around it, a walnut grove, a forest in India, or an old tree stump. Other books paint a broader picture through their depiction of **large ecosystems**, focusing on the fate of a planet, the entire food chain in Borneo, or the rain forests. Many of the books feature **human disruption of ecosystems**: what happens to a meadow when a road is built, to a stream ecosystem when toxic waste water is leaked, etc.

There are also a number of systems we often forget are "systems" such as the **digestive system**, the **circulatory system**, and the **planetary system**. Too often these are presented as static entities whose names children must memorize. Understanding that these are moving systems of interacting parts offers far greater depth of understanding. There are books pertaining to the **solar system**, the **water purification system,** a **mechanical system,** and a **system of urban development.**

CROSS REFERENCES

LITERATURE CONNECTIONS

Bringing the Rain To Kapiti Plain

by Verna Aardema; illustrated by Beatriz Vidal
Dial Books, New York. 1981
Grades: K–2

> "This is the great Kapiti Plain, all fresh and green from the African rains."
> But one year the rains are very belated and a terrible drought descends,
> driving out all of the big wild creatures. The beautiful plain grows barren
> and dry until Ki-pat, watching his herd of hungry and thirsty cows, spies a
> cloud hovering above and comes up with an ingenious way to "green-up
> the grass, all brown and dead, that needed the rain from the cloud over-
> head." Written with cumulative rhyming.

Cactus Hotel

by Brenda Z. Guiberson; illustrated by Megan Lloyd
Henry Holt & Co., New York. 1991
Grades: 3–5

> The life cycle of the saguaro cactus and especially its role as a "hotel" for
> birds and animals in the desert is described. The illustrations and text
> dramatize the beneficial interaction between the plants and animals and
> give very specific instances: the paloverde plant protects the cactus
> seedling from the sun; the pack rat drinks the water that drips off the tree;
> the jack-rabbit gnaws on the pulp; birds, bees and bats drink the nectar
> from the cactus flower.

Chitty Chitty Bang Bang: The Magical Car

by Ian Fleming; illustrated by John Burningham
Alfred A. Knopf, New York. 1964
Grades: 6–Adult

> Wonderful series of adventures featuring a
> magical transforming car, an eccentric
> explorer and inventor, and his 8-year-old
> twins. Nice combination of technical and
> scientific information, much of it accurate,
> with a magical sense of how some machines
> seem to have a mind of their own. This one
> definitely does, as it flies when it encounters
> traffic jams, becomes a boat when the tide
> comes in, senses a trap, and helps catch some
> gangsters. The car is a mechanical system to
> figure out. What can this car do? A humor-
> ous and fantastic literature accompaniment to
> activities involving technology and inven-
> tions.

The Day They Parachuted Cats on Borneo: A Drama of Ecology
by Charlotte Pomerantz; illustrated by Jose Aruego
Young Scott Books/Addison-Wesley, Reading, Massachusetts. 1971
Out of print
Grades: 4–7

> This cautionary verse, based on a true story, explores how spraying for mosquitoes in Borneo eventually affected the entire food chain, from cockroaches, rats, cats, geckoes, the river and the farmer. Although DDT may not be an up-to-date example, opportunities for tracing the ecological consequences of actions which have not been fully thought out abound (for example, rabbits have become major pests in Australia and New Zealand since they were introduced). The strong, humorous text makes the book a success whether read out loud or performed as a play.

The Desert Is Theirs
by Byrd Baylor; illustrated by Peter Parnall
Macmillan, New York. 1975
Grades: K–5

> The interaction in the desert of plant and animal life, terrain, weather, and other systems is poetically described. Plants that can do without water, birds that nest in cactus, hawks and lizards that like lonely canyons and hot sand, pack rats hiding their treasure are examples. The rain is a blessing and plants "don't waste it on floppy green leaves." The Papago Indians, characterized as "Desert People," also share the land and partake of this wisdom. "This is no place for anyone who wants everything green." Caldecott Honor Book.

Elliot's Extraordinary Cookbook
by Christina Bjork; illustrated by Lena Anderson
Farrar, Straus & Giroux, New York. 1990
Grades: 3–6

> With the help of his upstairs neighbor, Elliot cooks wonderful foods and investigates what's healthy and what's not so healthy. He finds out about proteins, carbohydrates, and the workings of the small intestine. He learns about the history of chickens and how cows produce milk. His friend shows him how to grow bean sprouts, and he sews an apron. Cooking is really all about the interactions of ingredients!

Gilberto and the Wind
by Marie Hall Ets
Viking Press, New York. 1963
Grades: Preschool–2

> Gilberto has a conversation with the wind as he investigates what the wind can do. The wind blows his balloon, plays with the clothes on the wash line, and bangs a gate shut. Children see how the wind can be playful and unpredictable at times. Shows simple interactions for the youngest students.

The House That Crack Built

by Clark Taylor; illustrated by Jan Thompson Dicks

Chronicle Books, San Francisco. 1992

Grades 3–Adult

> This book makes use of the familiar "House that Jack Built" rhyme to expose, with strong words and powerful illustrations, the interacting system that brings cocaine to the streets of our cities. The emotional impact is profound. A cogent "Afterword" emphasizes the responsibility of all people to make choices, and in this case to choose NOT to ignore this problem, but to do something about it. The book is seen as "a tool that can be used to open up discussion and to help children learn to make the right choices."

Just A Dream

by Chris Van Allsburg

Houghton Mifflin, New York. 1990

Grades: 1–6

> When he has a dream about a future Earth devastated by pollution, Walter begins to understand the importance of taking care of the environment.

The Magic School Bus at the Waterworks

by Joanna Cole; illustrated by Bruce Degen

Scholastic, New York. 1986

Grades: K–6

> When Mrs. Frizzle, the strangest teacher in the school, takes her class on a field trip to the waterworks, everyone ends up experiencing the water purification system from the inside. Evaporation, the water cycle, and filtration are just a few of the concepts explored in this whimsical field trip.

The Magic School Bus Lost in the Solar System

by Joanna Cole; illustrated by Bruce Degen

Scholastic, New York. 1990

Grades: K–6

> On a special field trip in the magic school bus, Ms. Frizzle's class goes into outer space and visits each planet in the solar system. They make a chart with facts on planet size, rotation time, distance from the sun, and the number of moons or rings.

Michael Bird-Boy

by Tomie dePaola

Prentice-Hall, Englewood Cliffs, New Jersey. 1975

Grades: Preschool–3

> A young boy who loves the countryside determines to find the source of the black cloud that hovers above it. When he discovers the source of this pollution, a sugar and artificial honey factory, he aids the "boss-lady" in setting up beehives, so she can make natural honey without creating pollution. Simple depiction of a system involving a product, its production, and the environment.

Finally we shall place the Sun himself at the center of the Universe. All this is suggested by the systematic procession of events and the harmony of the whole Universe, if only we face the facts, as they say, "with both eyes open."

— *Nicholas Copernicus*

Minn of the Mississippi
by Holling C. Holling
Houghton Mifflin, Boston. 1951
Grades: 5–9

> The journey of Minn, a snapping turtle, is followed from northern Minnesota to the bayous of Louisiana. Her adventures with people, animals, and the changing seasons are vividly described. Wonderful drawings and maps of her travels accompany the engaging true-life story on the Mississippi River. Presents a detailed picture of a river system.

The Missing 'Gator of Gumbo Limbo: An Ecological Mystery
by Jean C. George
HarperCollins, New York. 1992
Grades: 4–7

> Sixth-grader Liza K and her mother live in a tent in the Florida Everglades. She becomes a nature detective while searching for Dajun, a giant alligator who plays a part in a waterhole's oxygen-algae cycle, and is marked for extinction by local officials. The book is full of detail about the region's flora and fauna and its interaction with humans. In her foreword to the book, the author states that human beings did not weave the web of life, but are a strand in it. "Now that we know what we have done to the web, we see that our role as an intelligent animal is to mend it. There are millions of Gumbo Limbo Holes on this earth, from a city window box or vacant lot to streams and lakes to the wilderness areas of Alaska."

My Place
by Nadia Wheatley and Donna Rawlins
Kane/Miller Book Publishers, Brooklyn, New York. 1987
Grades: 3–8

> The history of a neighborhood in an Australian city is chronicled via two-page spreads that each have drawings, a map, and a story written by a child who lives on a certain spot of land—my place. The story goes back decade by decade as you learn who lives on that spot of land, what else is in the vicinity, and what happens in the lives of those who live there. The story progresses through who built the house, how it has changed over time, how the land was gradually settled, that different ethnic groups lived in the neighborhood at different times, how wars affected the people who lived there, the settlement of Australia, the displacement of Aboriginal peoples, and more.

Observe how system into system runs,
What other planets circle other suns.

— *Alexander Pope*
An Essay on Man

Never Cry Wolf

by Farley Mowat
Atlantic Monthly/Little, Brown & Co., Boston. 1963
Bantam, New York. 1984
Grades: 8–Adult

> Wolves are killing too many of the Arctic
> caribou, so the Wildlife Service assigns a
> naturalist to investigate. Farley Mowat is
> dropped alone onto the frozen tundra to live
> among the wolf packs to study their ways. His
> interactions and growing respect for and under-
> standing of how wolf society works will captivate
> all readers.

An Oak Tree Dies and a Journey Begins

by Louanne Norris and Howard E. Smith, Jr.;
illustrated by Allen Davis
Crown, New York. 1979
Out of print
Grades: 3–5

> A storm uproots an old oak tree on the bank of a river, it falls in and its
> journey to the sea begins. Animals seek shelter in the log, children fish
> from it, mussels attach to its side; all of which shows how a tree, even after
> it dies, contributes to the environment. A good, accessible example of
> environmental interactions. The book includes fine pen and ink drawings.

The Old Ladies Who Liked Cats

by Carol Greene; illustrated by Loretta Krupinski
HarperCollins, New York. 1991
Grades: K–6

> When the old ladies are no longer allowed to let their cats out at night, the
> delicate balance of their island ecology is disturbed, with disastrous results.
> Based on Charles Darwin's story about clover and cats, this ecological folk
> tale demonstrates the interrelationships of plants and animals.

Once There Was a Tree

by Natalia Romanova; illustrations by Gennady Spirin
Dial Books, Penguin, New York. 1983
Grades: K–6

> A tree is struck by lightning, cut down, and survives as a stump though the
> seasons. A bark beetle lays her eggs under its bark; its larvae gnaw
> tunnels. Ants make their home there. A bear uses the stump to sharpen
> her claws. The stump is visited and used by birds, frogs, earwigs, time,
> weather, humans. As a new tree grows from the stump, the question arises
> "Whose tree is it?"

The People Who Hugged the Trees

This lovely book by Deborah Lee Rose is based on a classic folktale about a small village in Rajasthan in western India.

Amrita was a young girl who loved the forest that kept the nearby sands of the desert from her village. She especially loved one favorite tree, and to show her love for the tree she would often hug it. As she grew older and had children of her own, she taught them to love the forest as she did.

One day, the Maharajah's axemen came to chop down the forest. Amrita rallied the villagers, and they all went to the forest and hugged the trees to stop the axemen. When the Maharajah heard what the villagers had done, he was furious, so he sent an army to collect his wood. Here is an excerpt from The People Who Hugged the Trees.

The villagers raced to the forest as the soldiers flashed their swords. Step by step the soldiers drew closer, as the sand swirled around their feet and the leaves shivered on the trees. Just when the soldiers reached the trees the wind roared in from the desert, driving the sand so hard they could barely see.

The soldiers ran from the storm, shielding themselves behind the trees. Amrita clutched her special tree and the villagers hid their faces as thunder shook the forest. The storm was worse than any the people had ever known. Finally, when the wind was silent, they came slowly out of the forest.

Amrita brushed the sand from her clothes and looked around. Broken tree limbs were scattered everywhere. Grain from the crops in the field littered the ground.

Around the village well drifts of sand were piled high, and Amrita saw that only the trees had stopped the desert from destroying the well and the rest of the village.

After seeing how much the people loved the trees, and how well the forest protected the village, the Maharajah took back his order to cut down the trees.

The annotation for this book with grade levels and publishing information is on the next page.

One Day in the Tropical Rain Forest

by Jean C. George; illustrated by Gary Allen
HarperCollins, New York. 1990
Grades: 4–7

When a section of rain forest in Venezuela is scheduled to be bulldozed, a young boy and a scientist seek a new species of butterfly for a wealthy industrialist who might preserve the forest. As they travel through the ecosystem rich with plant, insect, and animal life, everything they see on this one day is logged beginning with sunrise at 6:29 a.m. They finally arrive at the top of the largest tree in the forest and fortuitously capture a specimen of an unknown butterfly. A glimpse into the ecosystem of the tropical rain forest.

The People Who Hugged the Trees

adapted by Deborah Lee Rose; illustrated by Birgitta Säflund
Roberts Rinehart, Niwot, Colorado. 1990
Grades: K–5

This Rajasthani (India) folktale tells of a girl who so loves the trees outside her desert village that she thanks them daily for shade from the sun, protection from the sandstorms, and guidance in finding water. When the Maharajah's axemen try to cut down the trees, the people hug them to save the forest. The Chipko (Hug the Tree) Movement is still strong in India today. Beautiful paintings adorn the text and make clear the contrast between desert and forest systems, as well as the trees' role in the life of the village and ecosystem. An excerpt from this book is on the preceding page.

The Pied Piper of Hamelin

by Mercer Mayer
Macmillan, New York. 1987
Grades: Preschool–4

In a village infested with too many rats, a Pied Piper pipes the village free of the rats. When the villagers refuse to pay him for the service he pipes away their children as well. A classic illustration of an ecosystem gone out of whack, as well as a challenge requiring creative problem solving.

The River

by David Bellamy; illustrated by Jill Dow
Clarkson N. Potter/Crown, New York. 1988
Grades: 3–5

Plants and animals coexist in a river and have to struggle for survival when a man-made catastrophe strikes. Details about stream ecology include a description of the effects of waste water discharged from a factory and how the bacteria, algae, and oxygen interact in the dam area and beyond. The ending seems overly optimistic with the river "back to normal" a month after the waste was discharged. "Everyone hopes the factory owners will be more careful in the future."

The Roadside

by David Bellamy; illustrated by Jill Dow
Clarkson N. Potter/Crown, New York. 1988
Grades: 3–5

The construction of a six-lane highway in the countryside disrupts the balance of nature and forces animals there to change their patterns. The toads lay their eggs in different places, some animals eat the leftovers from the road workers' lunches, a raw mound of earth is soon covered by wildflowers that attract butterflies. An attempt is made to be evenhanded: "the road builders have done a good job." The toads use the new pipe under the road as a passageway, no hunters with guns come there because walking is not allowed, and so on.

The Rose in My Garden

by Arnold Lobel; illustrated by Anita Lobel
Greenwillow Books, New York. 1984
Grades: Preschool–2

Each page adds a new rhyming line as a beautiful garden of flowers and animals grows. A surprising interaction among the garden residents takes place at the end of the book. Young readers will enjoy the repeated patterns in the story.

The Salamander Room

by Anne Mazer; illustrated by Steve Johnson
Alfred A. Knopf, New York. 1991
Grades: K–3

A little boy finds an orange salamander in the woods and thinks of the many things he can do to turn his room into a perfect salamander home. In the process, the habitat requirements of a forest floor dweller are nicely described, providing a simple example of a system of interacting parts.

Seven Blind Mice

by Ed Young
Philomel Books, New York. 1992
Grades: K–4

This strikingly illustrated book retells the classic fable of the blind men and the elephant, only this time with brightly colored mice. There is a Mouse Moral at the end: "Knowing in part may make a fine tale, but wisdom comes from seeing the whole."

No man is an island, entire of itself, every man is a piece of the continent, a part of the main; if a clod be washed away by the sea, Europe is the less, as well as if a promontory were, as well as if a manor of thy friends or of thy own were; any man's death diminishes me, because I am involved in mankind; and therefore never send to know for whom the bell tolls; it tolls for thee.

— *John Donne*
Devotions upon Emergent Occasions

Sierra
by Diane Siebert; illustrated by Wendell Minor
HarperCollins, New York. 1991
Grades: 4–8

> Long narrative poem in the voice of a mountain in the Sierra Nevada, beginning and ending with the lines: *I am the mountain/Tall and grand./ And like a sentinel I stand.* Dynamic verse and glorious mural-like colored panels depict the forces shaping the earth and the plant, animal, and human roles in this ecosystem.

The Song in the Walnut Grove
by David Kherdian; illustrated by Paul O. Zelinsky
Alfred A. Knopf, New York. 1982
Grades: 4–6

> A curious cricket meets a grasshopper. They learn of each other's daytime and nighttime habits while living in an herb garden. The friendship between them grows when the cricket saves the grasshopper's life. They learn to appreciate each other's differences. The story weaves accurate accounts of insect behavior and their contributions to the ecology.

The Wump World
by Bill Peet
Houghton Mifflin, Boston. 1970
Grades: Preschool–5

> The Wump World is invaded by Pollutians from the planet Pollutus, who had left their worn-out planet to start a life in a new world. They move in and take over, building cities, creating pollution, and using up resources. Eventually, the Pollutians move to another planet and leave the Wumps to try and restore the balance in their ecosystem.

The Year at Maple Hill Farm
by Alice and Martin Provensen
Aladdin Books/Macmillan, New York. 1978
Grades: K–3

> Each month of the changing seasons on a farm and surrounding country-side is described on a two-page spread filled with pictures and details such as: pigs sleeping in cool mud puddles; what it must be like to pick your way out of an egg; and that geese feet never freeze when standing on February's frozen ponds.

Robert Taylor's sixth grade class is investigating systems and interactions. They used *Terrarium Habitats* and *River Cutters.* Robert gave his students a list of the age-appropriate books listed in the Systems & Interactions section of this handbook. His students each chose one book to read, and then drew a poster-sized picture, depicting the elements that interacted as part of a system.

Going Further with Library Resources

By Valerie Wheat

If you would like to go further in finding literature connections to your own special themes or topics, the following resources should be of interest. Should you be lucky enough to still have a funded school library, the librarian can help you and may already have many of these tools in the library.

Children's Books in Print, published by R.R. Bowker, prepares a subject guide volume each year. The last edition included a staggering 60,000+ entries. The computer-generated index is prepared from advance information supplied by book publishers, thus giving an overview of publishing without any selection or evaluation. Forthcoming Books, published six times a year by R.R. Bowker, previews this information.

To try a more targeted, personal approach, check to see if your local public or university library offers online access to its catalog. The database usually allows you to search by subject, and sometimes even by key word. You can have great fun and sometimes surprising results searching for fiction on a particular topic such as "Vitamin C" or finding a listing for "Earthworms—Juvenile Fiction." These systems are user-friendly so you don't have to be an expert on Library of Congress subject headings or even have

Valerie Wheat is a consultant to GEMS.

complete author and title information when looking for a specific book (although that always helps!)

If you want to augment your reading of book reviews beyond journals in your subject or grade area or get a wider exposure to new fiction, following are several standard book selection resources used by children's and school librarians and some additional resources.

LIBRARY JOURNALS

Book Links and Booklist
American Library Association
50 East Huron Street
Chicago, IL 60611
(312) 944-6780

The American Library Association's bimonthly "book-based" magazine *Book Links* is aimed at teachers and librarians particularly concerned with integrating literature into the curriculum. *Book Links* includes features such as "Book Strategies" and "Classroom Connections," "Reading the World" about geographic awareness, and "The Inside Story," an in-depth look at how a particular book came to be. Also available is the journal *Booklist*, published by the ALA since 1905 for small and medium-sized public libraries and school library media centers, which offers a broad-based review of current print and nonprint materials.

Bulletin of the Center for Children's Books

University of Illinois Press
54 East Gregory Drive
Champaign, IL 61820

About 70 titles are reviewed in each monthly issue and are listed alphabetically by author. The annotations are particularly attentive to the style of each book, from both literary and artistic standpoints.

Treating children's literature as literature, the adjective-laden reviews avoid a bland rehashing of jacket copy and give a good feel for each book and its potential audience. Entries include age range, subject terms, a code for recommended or marginal, and indicators for "curricular use" or "developmental values."

Horn Book

14 Beacon Street
Boston, MA 02108

Published six times a year, each issue reviews about 65 titles, some read in galley or proof stage. This journal is very cozy with the publishing world, featuring articles about artists and authors at work or a typical nine-page article comparing two editions of *The Adventures of Huckleberry Finn*. Both fiction and nonfiction are reviewed from picture books through grade 12 and up, with recommended listings of titles published in paperback. A one-volume *Horn Book Index 1924–1989* by Serenna F. Day was published by Oryx Press, Phoenix, Arizona in 1990.

School Library Journal

249 West 17th Street
New York, NY 10011

Publishes a supplement twice a year called "Star Track" which reprints reviews of the best of the "starred" books out of the 4,000 evaluated annually by School Library Journal. The listing is separated into pre-primary, grades 3–6, and junior high and up, and further divided into fiction and nonfiction. Free copies are available from the Circulation Department, School Library Journal.

ADDITIONAL RESOURCES

Best Books for Children: Preschool through Grade 6

by John T. Gillespie and Corinne J. Naden
R.R. Bowker, New Providence, New Jersey. 1990

Lists over 12,000 titles in print with very brief annotations and a fairly detailed subject index. Each annotation can lead you further by referencing book reviews in *School Library Journal, Horn Book, Bulletin of the Center for Children's Books*, and *Booklist*. The book is organized by eight main categories (literature; biography; arts and language; history and geography; social institutions and issues; personal development; physical and applied sciences; and recreation). Within each category, books are divided by grade (preschool–3; 2–4; 4–6; and 5–8), and then by genre. The one or two-line annotations mainly give a plot summary. Because of the many genre subdivisions, this book is easier to approach by table of contents than through the subject index which includes lots of omnibus terms such as "personal problems" or more specific terms with only one book listed. It is not particularly noteworthy for coverage of specific animals or multicultural aspects. Picture books are very strong, and adventure and mystery fiction are well represented. Other titles in the series are *Best Books for Junior High Readers* and *Best Books for Senior High Readers*, by the same author and published in 1991.

Books We Love Best:
A Unique Guide to Children's Books
by Bay Area Kids
Foghorn Press, San Francisco. 1990

Written, edited, and published by students from the Bay Area with reviews by kids from ages 5–17. Co-published with the San Francisco Bay Area Book Festival, the next volume's theme is environment and world geography. The book is divided into three main sections by grade (K–3, 4–6, and 7–12); then by categories such as fantasy/science fiction, animal, people, mystery, poetry, and miscellaneous. Annotations range from one to three paragraphs and are illustrated by children's drawings interpreting some of the books listed. Positive comments about books range from "I like stories with little towns in them" to "it made me think that I have so much and I hardly need any of it." Negative reactions include "the evil man got killed so fast and there's not enough fighting" to "I know there is no such thing as a special touch" and "when they play all these weird tricks they might hurt someone." Includes author and title indexes.

Children's Book Council
568 Broadway, Suite 404
New York, NY 10012

Publishes a series of current awareness bulletins called CBC Features including interviews with authors of fiction and nonfiction, vignettes by teachers, book listings, and more. Of special interest is the January/February 1992 theme issue on "The Environment." They charge a one-time handling fee to be on the mailing list, instead of an annual subscription fee.

Annual reading lists produced by the Children's Book Council in conjunction with other organizations include:

Outstanding Science Trade Books for Children is a joint project with the National Science Teachers Association, and reprinted from the annual March issue of *Science and Children*.

Notable Children's Trade Books in Social Studies is a joint project with the National Council for the Social Studies.

"Children's Choices" reprinted from the October issue of *Reading Teacher* is a joint project with the International Reading Association.

Children's Book Supplement

Hungry Mind Review
648 Grand Avenue
St. Paul, MN 55105

Published quarterly, $12/year. Distributed free at many independent children's bookstores. In a tabloid newspaper format, it includes book reviews and feature articles about children's books. Reviews are very current, but there are also retrospective roundups on topics such as storytelling or representation of landscape in picture books. Offers some of the best coverage of multicultural topics as well as thought provoking words from some of the finest writers of books for young people.

Kids' Favorite Books:
Children's Choices 1989–1991

International Reading Association
800 Barksdale Road
Box 8139, Newark, DE 19714-8139

A total of 300 titles selected from the 1989, 1990, and 1991 annual lists published in *Reading Teacher*. These annual lists usually include about 500 books which have been evaluated by 2,000 children. This guide is divided into sections by age group (all ages, beginning reading, younger readers, middle grades, and older readers) and then lists titles alphabetically by the year published. The short annotations are written by adults, with an occasional quote from a child's review.

Math and Literature

by Marilyn Burns
Cuisenaire Company of America
P.O. Box 5026
White Plains, New York 10602-5026
(800) 237-3142

This Math Solutions Publication includes ten sample lessons showing ways to combine mathematics with literature, with more than twenty creative "additional idea" sections. For grades K–3.

Recommended Readings in Literature: Kindergarten Through Grade Eight

California State Department of Education
P.O. Box 271
Sacramento, CA 95802-0271
(Annotated edition, 1988)

Listings of over 1,000 core and extended materials are divided into major sections such as picture books, folklore, fantasy and science fiction, realistic or historical fiction, and nonfiction; with a grade span and ethnic culture (if appropriate) indicated. The annotated edition includes one or two-line summaries.

Science Through Children's Literature: An Integrated Approach

by Carol M. and John W. Butzow
Teacher's Ideas Press/Libraries Unlimited
Englewood, Colorado. 1989

This collection includes instructional units for more than thirty children's fiction books, emphasis on the whole language approach, and many good ideas for integrating science and literature.

STORYTELLING AND PICTURE BOOKS

A to Zoo: Subject Access to Children's Picture Books
by Carolyn W. Lima and John A. Lima
R.R. Bowker, New Providence,
New Jersey. 1989

This guide has 12,000 titles indexed by 700 subjects, including some out-of-print works. Some of the broad headings are not particularly useful ("folk and fairy tales," "family life," "nature") but if you can go to a specific heading such as "space and spaceships," results can be quicker. The index is more specific than most on type of animal, insect, bird, or reptile; and includes a subject heading for problem solving. The peculiar world of picture books is reflected in the many listings for royalty, sibling rivalry, witches, trains and trucks, and the behavioral sins (boasting, bullying, carelessness) and virtues (seeking better things, collecting things). The second half of the book is a bibliographic guide arranged by author. Title and illustrator indexes.

Books Kids Will Sit Still For: The Complete Read-Aloud Guide
by Judy Freeman
R.R. Bowker, New Providence,
New Jersey. 1990

Written by a librarian/storyteller for parents, teachers, and librarians, this guide covers 2,000 of the author's favorite read-aloud books for preschool through 6th grade and includes picture books, fiction, poetry, folklore, and nonfiction. The annotated listings are organized by grade category (preschool–K, K–1, fiction by grade level in increments of two years)

with sections on folktales and fairy tales, poetry, nonfiction, and biography. In addition to a plot summary, each entry suggests teaching extensions and refers to other subject headings and related titles (especially helpful for folklore). The author is scrupulous about specifying re-tellings of folklore and recommends a liberal approach to grade level, merely suggesting the two most suited grades for each title. Supplementary material includes hints on reading aloud and storytelling, giving book talks, using individual chapters of books, creative dramatics, and "101 Ways to Celebrate Books."

Just Enough To Make A Story: A Sourcebook for Storytelling
by Nancy Schimmel
Sisters' Choice Books and Recordings
1450 Sixth Street, Berkeley, CA 94710

Storyteller Nancy Schimmel shares tips on finding, learning, and telling a story; how to tell tandem and participation stories; and includes five stories. Lists at the end of each chapter offer sources of stories, professional literature about storytelling and folktales, songs and song books, sources of stories to tell to adults, etc.

This new edition has unique resource lists of active heroines in folktales; ecology and peace stories; and author's favorites.

**Storyteller's Sourcebook:
A Subject, Title, and Motif Index to
Folklore Collections for Children**
by Margaret R. MacDonald
Neal-Schuman, New York. 1982

This first edition, still in print and
available from Gale Research, Chicago,
should be found in large public library
children's collections.   A second edition is scheduled for publication by
Gale Research in 1995.

This classic reference work indexes
556 folktale collections and 389 picture
books, including all folktale titles that
appeared in the *Children's Catalog* from
1961–1981 and does not include epic,
romance, and tall tale hero materials.
The Motif Index is an adaptation of the
Stith Thompson motif index. For example, under animals you would find
subheadings such as animal languages,
mythical animals, animals advise men,
helpful animals, devastating animals,
marriage to a person in a particular
animal's form, etc. Other common
motifs are tests, the wise and the fool-
ish, deceptions, magic, and marvels.

The more accessible Subject Index
could help you find many tales about
bees, bears, spiders, and various types
of birds, with a few on earthworms.
Examples of the many subject terms of
potential interest are reflection, clouds,
snow, blood, stone, egg, smell, thread,
and of course sun, moon, stars, and sea.
Two additional indexes are Tale Title
and Ethnic and Geographic Index.

SCIENCE FICTION AND FANTASY

**Anatomy of Wonder:
A Critical Guide to Science Fiction**
edited by Neil Barron
R.R. Bowker, New Providence,
New Jersey. 1987

For science fiction fans, this book is a
gold mine. Unfortunately, because of the
date of publication, many of the titles
listed are no longer in print and many of
them were published in Great Britain.
Still, it makes a nice shopping list for a trip
to a used book store, flea market, or library book sale.

VOYA (Voice of Youth Advocates)
Scarecrow Press
52 Liberty Street, Metuchen, New Jersey 08840

This bimonthly newsletter, published
for young adult librarians, runs an annual
listing of "Best Science Fiction, Fantasy,
and Horror" in their April issue including
about 50 annotated titles and an additional
list of honor books. It's heavy on
cyberpunks and sorcery, darkness and
divergence, druids, and dragons. Other
issues include book reviews with irreverent ratings for quality (lowest is "hard to
understand how it got published"); popularity; and age level, from grades 6–12.
Articles have included "Science Fiction
with Rivets" about the portrayal of hard
science, an overview of Ursula LeGuin's
Earthsea trilogy, an article about popular
sequels, and more.

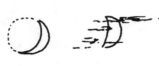

MULTICULTURAL

The Multicultural Catalog

Baker & Taylor Books
A Grace Distribution Company
Five Lakepointe Plaza, Suite 500
2709 Water Ridge Parkway
Charlotte, NC 28217
Customer Service: (800) 233-9811

This catalog (subtitled "Many Voices, Many Books ... Strength Through Diversity") contains an excellent and quite extensive listing of multicultural books and related resources. Each listing includes a brief annotation. Major sections, with separate listings for adults and children in each, include books by and about African-Americans, American Indians, Asian Americans and Asians, Hispanic/Chicano/Latino, and other Multicultural. Many major publishers also advertise in this catalog, promoting their new multicultural releases. The 1991–1992 catalog includes an essay from the Before Columbus Foundation on cultural diversity and the reasons why an understanding of different cultures is so important, with comments on key topics that remain relatively unexplored in both adult and children's literature. There is also an introduction by Charles Johnson, the recipient of the 1990 National Book Award in fiction (author of *Middle Passage* and other novels) who describes his childhood love for literature of all kinds, and the special role Richard Wright's *Black Boy* played in his intellectual and literary development.

Multicultural Literature for Children and Young Adults: A Selected Listing of Books 1980–1990 by and about People of Color

by Ginny M. Kruse and Kathleen T. Horning
Cooperative Children's Book Center
Wisconsin Department of Public Instruction
P.O. Box 7841, Madison, WI 53707-7841
(refer to Bulletin #1923)

This annotated bibliography of about 400 titles is divided into topic sections such as "Seasons and Celebrations," "Issues in Today's World," and "Understanding Oneself and Others." There are sections about fiction (divided into new readers, young readers, and teenagers); books for babies and for toddlers; picture books; and concept books. The section on "Folklore, Mythology and Traditional Literature" is especially good. Appendices note authors and illustrators of color; ethnic/cultural groups represented; and other resources. A combined index includes authors, illustrators, translators, compilers, editors, and book titles. The selection of books is nicely eclectic, including some small presses and nonprofit groups, and the annotations are balanced and comprehensive.

**Our Family, Our Friends, Our World:
An Annotated Guide to Significant
Multicultural Books
for Children and Teenagers**
by Lyn Miller-Lachmann
R.R. Bowker, New Providence,
New Jersey. 1992

Book selection guide has 1,000 listings for books published between 1970 and 1990. The book is arranged geographically with entries in each section arranged by age from preschool–12. All regions of the world are included so the book casts a wider net than those usually defined as minority cultures. Here you will find those, but also Ireland or Iceland. The annotations are informative, usually three paragraphs long, and they are critical. Appendices include professional sources, series titles, and directory of publishers.

**Through Indian Eyes: The Native
Experience in Books for Children**
edited by Beverly Slapin and Doris Seale
New Society Publishers, Philadelphia. 1990

The first half of this compilation is made up of essays and poetry about Native American peoples and how they have been represented (and misrepresented) through the written word. The second half includes about 100 book reviews, "How To Tell the Difference," a checklist for recognizing stereotypes of Native Americans, and resource lists. The authors' expectations are high. They don't mince words when criticizing authors or materials (including some widely used old favorites). Often the reviewer may still recommend a title for use, but with reservations. "While it is disappointing that dePaola did not manage to avoid some of the more common pitfalls, I would not hesitate to use this book with small children. I do not think that white children can pick up any unfortunate attitudes from it; there is nothing here to shame or hurt an Indian child. Whatever else, the author does not condescend to his material." (review of *The Legend of the Bluebonnet).* One of the nicest results of reading this is to find out about materials available from less-publicized sources and Native American publishers.

SCIENCE

**Appraisal:
Science Books for Young People**
Children's Science Book Review Committee
605 Commonwealth Avenue
Boston, MA 02215

This quarterly newsletter reviews about 100 titles in each issue, mostly nonfiction and biography, and includes a review article such as one on "The Juvenile Dinosaur Book." Each book is reviewed by both a librarian and a subject specialist. Reviewers are frank and entertaining, sometimes crotchety. In a review of *The Salamander Room,* one subject specialist stated that initially he found "the story silly and the illustrations dark." He read it again with his six-year-old daughter who loved it and decided it was appropriate for the intended age group. An additional handy feature is a section evaluating series books.

Astronomical Society of the Pacific

390 Ashton Avenue
San Francisco, CA 94112

The society publishes a quarterly news-letter, "The Universe in the Classroom," which is free to teachers who request it on school stationery. They also publish occasional listings or bibliographies such as "Interdisciplinary Approaches to Astronomy," which include science fiction, art, music, and other connections. Write to request an information packet or for information on your specific interest.

Best Science Books and A-V Materials for Children

edited by Susan M. O'Connell,
Valerie J. Montenegro, and Kathryn Wolff
American Association for the
Advancement of Science
Washington, D.C. 1988

Selected listings of materials reviewed in *Science Books and Films from 1982–1987* includes over 800 recommended science books (almost all nonfiction) and over 400 audiovisual materials. Arranged by Dewey decimal classification, subjects include philosophy; mathematics and physical sciences; life sciences; zoological sciences (the largest section); technology; medicine; engineering; and more. Materials focus on kindergarten through junior high and annotations are relatively brief. The title and subject indexes are combined.

E For Environment: An Annotated Bibliography of Children's Books with Environmental Themes

by Patti K. Sinclair
R.R. Bowker, New Providence,
New Jersey. 1992

Includes 517 titles with coverage of preschool through age 14. Most of the books listed are recent, but some classics are also included. Divided into five sections: environmental awareness; specific environments; issues (such as endangered species, pollution, recycling, and energy); people and nature; and activities. Indexed by author, title, and comprehensively by subject. Annotations are long, thoughtful, and usually positive in tone.

Bay Views

The Association of Children's
Librarians of Northern California
P.O. Box 12471
Berkeley, CA 94701

This "journal of book reviews and opinions with a western perspective" is published 10 times a year. Annually *Bay Views* contains about 2500 book reviews covering the entire range of books for children and young adults. The materials are previewed by a panel of experts in the field and then reviewed by librarian subject specialists. Although there is some regional focus, there are subscribers across the country and much of general interest in these detailed reviews. A rating system is used, and an index is included.

Science and Children

National Science Teachers Association
1742 Connecticut Avenue N.W.
Washington, D.C. 20009-1171

The March issue includes an annual listing of "Outstanding Science Trade Books for Children," selected by NSTA in cooperation with the Children's Book Council. The resulting selection is less specialized than the American Association for the Advancement of Science annual list.

Science and Technology in Fact and Fiction: A Guide to Children's Books

by DayAnn M. Kennedy, Stella S. Spangler, and Mary Ann Vanderwerf
R.R. Bowker, New Providence, New Jersey. 1990

The book has two main sections, "Science" and "Technology," and each is divided into fiction and nonfiction. The 350 titles, some out of print, are indexed by author, title, illustrator, subject, and readability. Most annotations are over half a page and include a plot summary and an evaluation. Citations include both an age and a grade level rating.

Science and Technology in Fact and Fiction: A Guide to Young Adult Books

by DayAnn M. Kennedy, Stella S. Spangler, and Mary Ann Vanderwerf
R.R. Bowker, New Providence, New Jersey. 1990

Because of the older age level, many of the entries are much longer than those in the companion children's guide volume. As might be expected, the availability of fiction relating to sci-tech topics is somewhat limited.

Science Books and Films

American Association for the Advancement of Science
1333 H Street N.W.
Washington, D.C. 20005

Divides books by broad age category (adult, junior high/young adult, children's) and then by Dewey decimal number. The coverage by field is quite broad, including not only mathematics, technology, and the sciences but social science, medical science, architecture, agriculture, education, and energy and the environment. Reviews are detailed. A few selected titles are evaluated by "Young Reviewers." The December issue includes an annual "Best Children's Science Booklist," compiled from these reviews. Almost all of the books reviewed are nonfiction. The June/July 1992 issue had a cover story on "Biodiversity." Appears nine times a year.

Science Books and Films' Best Books for Children 1988–91

edited by Maria Sosa and Shirley M. Malcom
American Association for the Advancement of Science
1333 H Street N.W.
Washington, D.C. 20005

This publication's sections include archaeology and paleontology; astronomy and space science; biological sciences; computers and information science; earth sciences; general science; mathematics; physical sciences; technology and engineering.

> The hearts of little children are pure,
> and therefore, the Great Spirit may show
> to them many things which older people miss.
>
> — *Black Elk*
> *Black Elk Speaks*

Our Mother's Keeper

By Lincoln Bergman

Two outstanding books provide clearly-written, nature-based science and environmental activities for children with beautiful, sensitively told, and stirring Native American stories, identified as to tribal origin. These books are very useful in connection with the GEMS guide *Investigating Artifacts*, but their excellence and the vast spectrum of topics and issues they cover make them great teaching resources in much more fundamental ways.

Written by Michael J. Caduto and Joseph Bruchac, *Keepers of the Earth: Native American Stories and Environmental Activities for Children* and *Keepers of the Animals: Native American Stories and Wildlife Activities for Children* are wonderful resources—for helping all of us understand much more about the diverse indigenous cultures of North America—for containing much information and insight into the natural world—and, as related to this GEMS literature handbook, for gifting us with truly exemplary and well-spelled out ways to interweave literature and storytelling with science, mathematics, and environmental concerns. Both books have accompanying and excellent teacher's guides.

The first story in *Keepers of the Earth* is nothing less than a story that explains where stories come from, and it establishes a pattern that is followed throughout both books. Each chapter begins with a strong and evocative story with full-page illustration that can be read out loud and contains key lessons as well as serving as an introduction to the well-organized instructions that follow, with clear diagrams for the activities, discussions, and imaginative extensions.

Every chapter combines a love for children and learning and a sense of the interconnectedness of all who inhabit Mother Earth with creative lessons and solid educational content. Many new ideas leap out. Take, for example, a section that includes two stories: one from the Hopi people about Grandmother Spider naming the animal clans and the other an Osage story "How the Spider Symbol Came to the People." After discussing the stories, children create anatomically correct spider models; look for spiders in their natural habitat; and make a yarn web using a hula hoop as a frame. "Insects" are made by putting velcro strips on ping pong balls, and children try to throw the balls through the "web" to see where in the web most "insects" are caught.

Just two of many extension ideas for the same section include: "Rewrite the story of Little Miss Muffet from the spider's point of view, or rewrite it describing what Little Miss Muffet would do and say if she were not afraid of spiders" and "Create a giant web as a room divider to better understand how intricate and challenging web building really is and to see and appreciate the beauty and symmetry of a web."

Indeed, the metaphor of an interconnected web of life fits these two fine resource books very well. The famous lines attributed to Chief Sealth (Seattle) are referred to by the authors and, most importantly, made manifest by the experiences children have as they do these activities and learn the stories.

> **"This we know. The earth does not belong to people; people belong to the earth. We did not weave the web of life, we are merely a strand in it. Whatever we do to the web, we do to ourselves. All things are connected like the blood that unites one family. All things are connected."**

The strong Native American (and modern environmental) concept of reciprocity between people and the rest of the natural world is emphasized. As the authors put it, "The cyle of giving and receiving—maintaining the cyle of life—is fundamental to Native North American culture."

In one of the introductory comments, the authors summarize their approach in this way: "As the stories unfold and you help the children bring the activities to life, a holistic, interdisciplinary approach to teaching about the animals and Native North American cultures begins. With their close ties to the animals, Native North American cultures are a crucial link between human society and animals. The story characters are voices through which the wisdom of Native North Americans can speak in today's language, fostering listening and reading skills and enhancing understanding of how the native people traditionally live close to the animals. Each story is a natural teaching tool, which becomes a springboard as you dive into the activities designed to provoke curiosity among children and facilitate discovery of the animals and their environments and the influence people have on those surroundings. Pedagogically sound, these activities have been extensively field-tested. They involve the children in creative arts, theater, reading, writing, science, social studies, mathematics and sensory awareness, among other subjects. The activities engage a child's whole self: emotions, senses, thoughts and actions. They emphasize creative thinking and synthesis of knowledge and experiences. Because of the active and involving nature of the experiences found in this book, children who have special needs physically, mentally, and emotionally respond well with proper care and skilled instruction."

With words like these, the authors promise a lot—to their great credit, and thanks as well to the richness and depth of the traditions from which they so respectfully draw, these books not only provide all of the above, but a great deal more. Read a few of the stories to your family around a campfire and see for yourself.

Lincoln Bergman is the principal editor of the GEMS project.